6
Topics in Organometallic Chemistry

Topics in Organometallic Chemistry

Recently Published and Forthcoming Volumes

Springer-Verlag Berlin Heidelberg GmbH

Organometallics in Process Chemistry

Volume Editor: R.D. Larsen

With contributions by
A. F. Abdel-Magid • J. Bien • A. J. Delmonte • E. D. Dowdy
M. Huang • D. L. Hughes • E. N. Jacobsen • J. Kant • A. O. King
H. Kumobayashi • G. C. Lane • J. F. Larrow • C. A. Maryanoff
B. P. Medaer • S. J. Mehrman • M. R. Oberholzer • M. Prashad
K. Sumi • D. J. Watson • G. G. Wu • N. Yasuda

Springer

The series *Topics in Organometallic Chemistry* presents critical overviews of research results in organometallic chemistry, where new developments are having a significant influence on such diverse areas as organic synthesis, pharmaceutical research, biology, polymer research and materials science. Thus the scope of coverage includes a broad range of topics of pure and applied organometallic chemistry. Coverage is designed for a broad academic and industrial scientific readership starting at the graduate level, who want to be informed about new developments of progress and trends in this increasingly interdisciplinary field. Where appropriate, theoretical and mechanistic aspects are included in order to help the reader understand the underlying principles involved.
The individual volumes are thematic and the contributions are invited by the volumes editors.
In references Topics in Organometallic Chemistry is abbreviated
Top. Organomet. Chem. and is cited as a journal

Springer WWW home page: springeronline.com
Visit the TOMC contents at springerlink.com

ISSN 1436-6002
DOI 10.1007/978-3-540-36966-0

Bibliographic information published by Die Deutsche Bibliothek
Die Deutsche Bibliothek lists this publication in the Deutsche Nationalbibliographie; detailed bibliographic data is available in the Internet at <http:// dnb.ddb.de>.

springeronline.com

Originally published by Springer-Verlag Berlin Heidelberg New York in 2004
MyCopy version of the original edition 2004

Typesetting: Medio Technologies AG, Berlin
Production editor: Christiane Messerschmidt, Rheinau
Cover: design & production GmbH, Heidelberg

Printed on acid-free paper 02/3020 – 5 4 3 2 1 0
www.springer.com/mycopy

Volume Editor

Dr. Robert D. Larsen
Department of Process Research
Merck and Co., Inc.
P.O. Box 2000
Rahway, NJ 07065
USA
E-mail: rob_larsen@merck.com

Editorial Board

Topics in Organometallic Chemistry is also Available Electronically

For all customers who have a standing order to Topics in Organometallic Chemistry, we offer the electronic version via SpringerLink free of charge. Please contact your librarian who can receive a password for free access to the full articles by registration at:

springerlink.com

If you do not have a subscription, you can still view the tables of contents of the volumes and the abstract of each article by going to the SpringerLink Homepage, clicking on "Browse by Online Libraries", then "Chemical Sciences", and finally choose Topics in Organometallic Chemistry

You will find information about the

- Editorial Board
- Aims and Scope
- Instructions for Authors
- Sample Contribution

at springeronline.com using the search funktion.

Preface

The impact of organometallics on the synthesis of medicinal agents during the last decade cannot be overstated. Although metals or organometallic species have been used for sometime in the pharmaceutical industry, the advances in the variety of reactions and the tremendous selectivity that is often achieved has expanded the toolbox of reagents available to the medicinal and process chemist. These reagents have helped to streamline the synthetic approaches developed for medicinal agents as can be seen by the increasingly sophisticated strategies used to synthesize complex drug products. In the areas of asymmetric synthesis, pharmaceuticals can be prepared using methodology only dreamed of a few decades ago.

This volume will highlight some of the more active areas where organometallics are playing an important role in process chemistry. With the chelation effects of many chiral ligands, organolithium reagents have become key intermediates in asymmetric synthesis. Similarly, because of the great propensity of titanium to chelate and bind to heteroatoms, organotitanium reagents are extremely useful and versatile in carrying out a number of selective organic transformations. Although the asymmetric hydrogenation of α-amidoacrylates to prepare amino acids has been known for sometime, the discovery and application of new ligands and reaction classes have expanded the number of substrates that can be converted to chiral products with Rh and Ru-mediated reactions. The cyclopropyl group is a common moiety found in pharmaceutical agents. Metal-mediated methods for preparing this ring are reviewed herein. Non-reductive means to prepare the enantiomers of oxy-systems remained elusive until the reports of chiral salen ligands and osmium-mediated reactions over the past two decades. Organopalladium reactions are now so commonly used in the synthesis of complex molecules that one can forget that palladium was once used only in hydrogenation reactions. The use of metals does come with a price when preparing pharmaceuticals for human consumption. Because of the potential toxicities of the metals, only low ppm levels can remain in the active pharmaceutical ingredient. Some of the methods that have been utilized in process chemistry to remove residual metals are presented. Organometallics will certainly continue to be used extensively in the pharmaceutical industry as newer methods and applications are discovered.

Rahway, USA, January 2004 — Robert D. Larsen

Contents

Topics Organomet Chem (2004) 6: 1–35
DOI 10.1007/978-3-540-36966-0

Organolithium in Asymmetric Processes

George G. Wu · Mingsheng Huang
Chemical Process Research and Development, Schering-Plough Research Institute, Union, New Jersey 07083, USA
E-mail: george.wu@spcorp.com

Abstract The development of asymmetric processes has been the focus of industrial research as most of the molecules of pharmaceutical interest contain chiral center(s). Many of the reported processes employ organometallic reagents in their key transformations. This review surveys chemical processes involving organolithium species in their enantioselective steps published in the past decade.

Keywords Organolithium · Asymmetric process · Chiral alkylation · Chiral imine addition · Chiral aldol/Michael addition

1 Introduction

The development of asymmetric processes for drug candidates has become increasingly important as most of the structures of interest contain one or more chiral centers. There is a wealth of literature on asymmetric processes published in the past decade. Many of the enantioselective processes reported involve organolithium species in the asymmetric step. This review will focus on industrial applications involving the use of organolithium as a nucleophile.

The most commonly used lithium reagents are lithium diisopropyl amide (LDA), butyllithium, phenylithium, and lithium hexamethyldisilylamide (LiHMDS). These lithium reagents are commercially available in bulk quantities and are easy to handle in the plant. When appropriate, reactions with organolithium will be contrasted with related cations, such as organosodium, organopotassium, and organozinc reagents. Certain additives, such as LiCl, CuX, and water are often introduced to organolithium reactions to enhance either the reactivity or the enantioselectivity.

While this review focuses on the industrial applications of organolithium in asymmetric syntheses, there are some good reviews of general applications of organolithium reagents. Advances in chiral lithium amides and enantioselective protonation as well as asymmetric synthesis via lithium intermediates have been reviewed previously [1, 2].

2 Organolithium in Enantioselective Alkylations

A majority of the reported enantioselective alkylations can be divided into two distinct groups – those using a covalently bonded chiral auxiliary and those using a chiral additive. The auxiliary is generally removed at the end of induction. However, in some examples, the auxiliary is incorporated as part of the molecule, the so-called chiral pool-based approach. Alkylation using a chiral additive is more straightforward as it does not require the attachment and removal of the auxiliary. While primary electrophiles are used in the majority of chiral alkylations, secondary electrophiles also work well under certain conditions. Chiral auxiliary-based chemistry has been reviewed previously [3]. New approaches such as 1,3-asymmetric induction, the use of sterically rigid templates, and intramolecular *trans*-alkylation have also been developed.

2.1 Chiral Auxiliary-Mediated Alkylation

Diastereoselective processes using covalently bonded chiral auxiliaries have been widely used in enantioselective alkylation of lithium enolates for asymmetric C-C bond formation. Reactions involving organolithium species often afford chelation-controlled products. Lithium can be replaced with potassium or sodium as the countercation when chelation is not preferred.

Oxazolidinones are frequently used as chiral auxiliaries for enantioselective alkylation. For example, Holla et al. of Hoechst have described a short and stereoselective synthesis of (2*S*)-3-(1′,1′-dimethylethylsulfonyl)-2-(1-naphthylmethyl)-propionic acid **1**, a very potent *N*-terminal component in aspartyl protease inhibitors, using an oxazolidinone **2** as an auxiliary [4]. The enantioselectivity was achieved by a stereoselective alkylation of a lithium enolate of the thioether oxazolidinone carboximide **3**. Enolization with LDA followed by treatment with 1-(bromomethyl)-naphthalene and removal of the auxiliary with concurrent oxidation of the thio group produced the alkylated product **1**. An overall yield of 37% and 99% optical purity was obtained on a 100-g scale (Scheme 1).

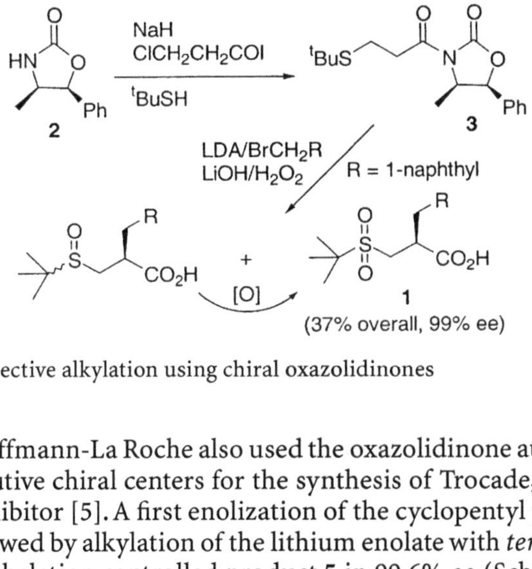

Scheme 1 Stereoselective alkylation using chiral oxazolidinones

Hilpert of Hoffmann-La Roche also used the oxazolidinone auxiliary to establish two consecutive chiral centers for the synthesis of Trocade, a matrix metalloproteinase inhibitor [5]. A first enolization of the cyclopentyl propionic amide **4** with LDA followed by alkylation of the lithium enolate with *tert*-butyl bromoacetate gave the chelation controlled product **5** in 99.6% ee (Scheme 2). After removal of the auxiliary, the free acid **6** was converted to its corresponding piperidine amide **7**. A chemoselective enolization of the amide carbonyl in **7** with a base followed by alkylation with bromomethyl hydantoin **8** gave either *syn*- or *anti*-succinate **9**, depending on the cation of the base used. A lithium base such as LDA furnished preferentially *syn*-**9** (*anti*-/*syn*-=15:85); whereas, potassium bases like KHMDS afforded predominantly *anti*-**9** (*anti*-/*syn*- ratio up to 99:1). In the second alkylation, it was assumed that the chelation of the lithium enolate with the adjacent carbonyl group induces the alkylation from the sterically less hindered face, leading to the *syn*-**9** isomer. In contrast, the potassium enolate

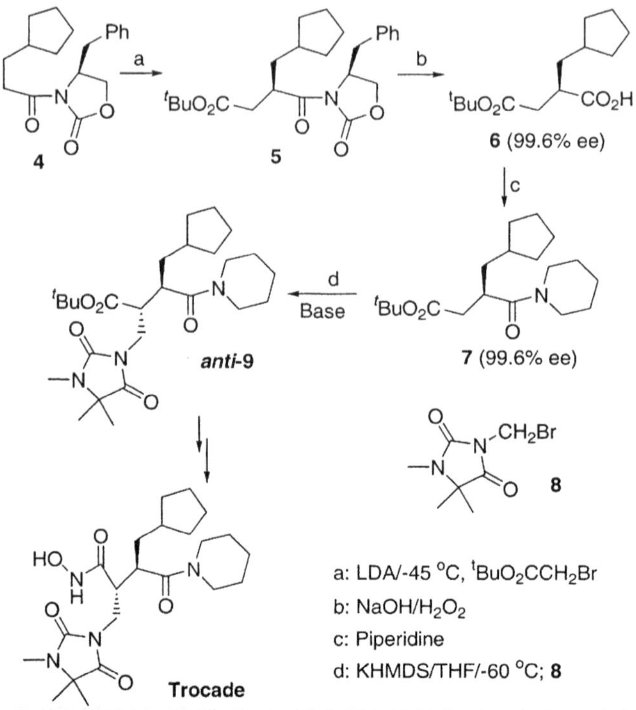

4 5 6 (99.6% ee) 7 (99.6% ee) 8 *anti*-9 Trocade

a: LDA/-45 °C, ${}^{t}BuO_2CCH_2Br$
b: $NaOH/H_2O_2$
c: Piperidine
d: KHMDS/THF/-60 °C; **8**

Scheme 2 Diastereoselective alkylation process for trocade

was assumed to favor the non-chelated, thermodynamically more stable conformation, consequently affording *anti*- alkylation.

In a related investigation, Hilpert has shown that alkylation of the lithium enolate of succinic acid **10** with cinnamyl bromide gave a 93:7 mixture favoring the *syn*-isomer **11** [5]. As shown in Scheme 3, the *syn*-isomer **11** was converted to its corresponding *anti*-isomer via a second enolization with LDA. These selectivities were rationalized by a chelation effect of the lithium enolate which is alkylated on the sterically less hindered side leading to *syn*-isomer **11**. A second deprotonation of **11** to the chelated enolate **12** and protonation again from the sterically less hindered side affords the *anti*-**13**.

Hilper's methods have provided efficient and practical approaches to the matrix metalloproteinase inhibitor Trocade and to TNF-α converting enzyme inhibitor TACE. Both of the processes have been operated in the plant on multi-ton and multi-kilogram scale [5].

cis-Aminoindanols are another important class of chiral auxiliaries that have been extensively investigated by various Merck groups. They were originally investigated because the aminoindanol structure was part of the target molecule, but have since become important auxiliaries in their own right. The first example was reported by Askin et al. for the synthesis of hydroxyethylene dipeptide isostere (HDI) inhibitors of HIV-1 protease **14** [6]. As shown in Scheme 4, enolization

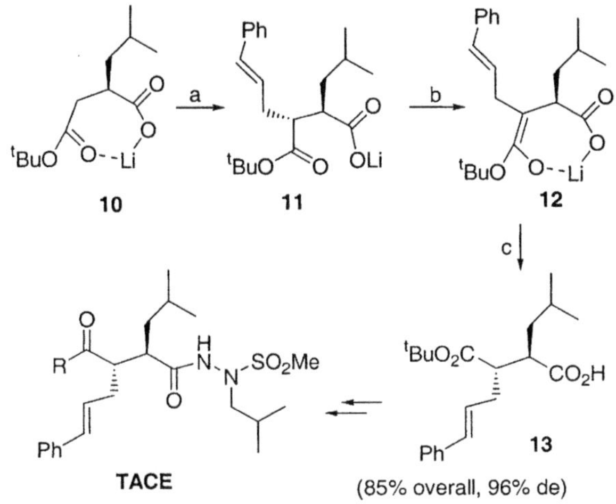

10 11 12 13 TACE

(85% overall, 96% de)

a: LDA/-20 °C, $PhCH{=}CHCH_2Br$; b: LDA/THF/-20 then 22 °C

c: CF_3CONH_2/THF/-90 °C

Scheme 3 Asymmetric synthesis of TACE

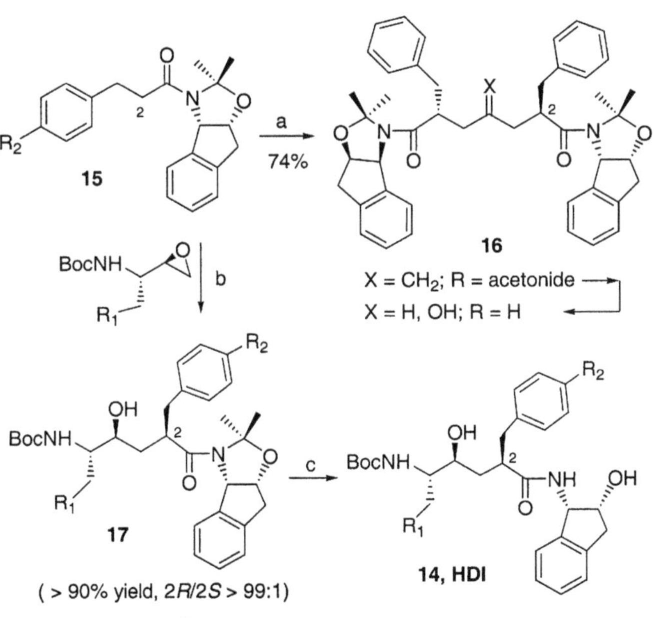

Scheme 4 Diastereoselective synthesis of HIV-1 protease inhibitor

of 3-phenylpropionic amide of aminoindanol **15** with *n*-BuLi followed by alkylation with $H_2C{=}C(CH_2I)_2$ gave a C_2 symmetry dimer **16**. The lithium enolate could also be trapped with an epoxide to give γ-hydroxyamide **17** in 90% yield and 98% ee. The outcome of the *R*-stereochemistry at the C-2 positions of products **16** and **17** suggested that both the alkyliodide and the epoxide electrophiles approached from the least hindered face of the lithium enolate. The facial selectivity observed here is contrasted by those reactions with prolinol amide enolates reported earlier [6e]. In the epoxide coupling reactions, Grignard reagents such as isopropyl magnesium chloride, gave lower yields (~60%) of the desired product.

Another successful application of the rigid tricyclic aminoindanol lithium acetonide was the asymmetric synthesis of the orally active HIV protease inhibitor Crixivan, one of the leading drugs for the treatment of AIDS [6]. Two alternative approaches were developed. Both approaches started with enolization of indanol amide **15** with LiHMDS. In the first approach, the lithium enolate was reacted with epoxy tosylate **18** to give the epoxy derivative **20**. Chemoselectivity was obtained between the displacement of the tosylate and the opening of the epoxide. In the second approach, the lithium enolate was alkylated with an allyl bromide followed by epoxidation to afford **20**. In these syntheses, the *cis*-aminoindanol unit remains as a part of the molecule (Scheme 5).

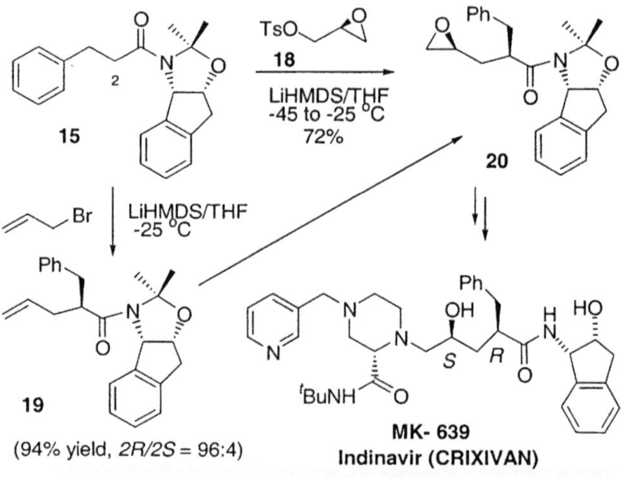

TsO 18 LiHMDS/THF -45 to -25 °C 72%

15 20

Br LiHMDS/THF -25 °C

19 (94% yield, 2R/2S = 96:4)

MK- 639 Indinavir (CRIXIVAN)

Scheme 5 Diastereoselective alkylation process for crixivan

Alkylation of the *cis*-aminoindanol-modified glycine enolate **21** with a number of alkyl halides in the presence of LiCl gave the corresponding alkylated product **22** in 90~99% diastereoselectivity [7]. Benzylic or naphthyl bromide gave 99% de and 85% yield. Allylic or primary alkyl halides typically gave 95 to 98% de. The diastereoselectivity was slightly lower (91%) with a secondary acyclic iodide. Both the ee and the yield suffered in the absence of LiCl. The auxiliary could be effectively removed under epimerization-free conditions. This provided a practical synthesis of α-amino acids (Scheme 6).

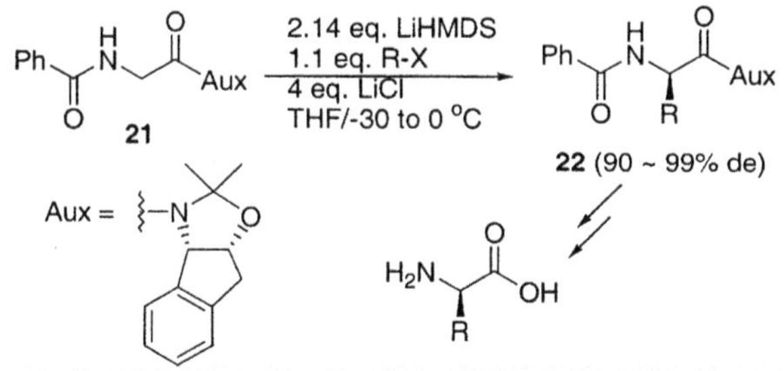

Scheme 6 Diastereoselective alkylation for chiral α-amino acids

In the synthesis of the sidechain of an endothelin receptor antagonist, Song et al. used (1*R*,2*S*)-*cis*-aminoindanol for a chiral alkylation [8]. As shown in Scheme 7, enolization of a propionyl amide **23** with LiHMDS followed by alkylation with benzylchloride gave 2-methyl-3-phenylpropionic amide **24** in 96% de. Removal of the auxiliary by hydrolysis gave the free acid in 60% overall yield.

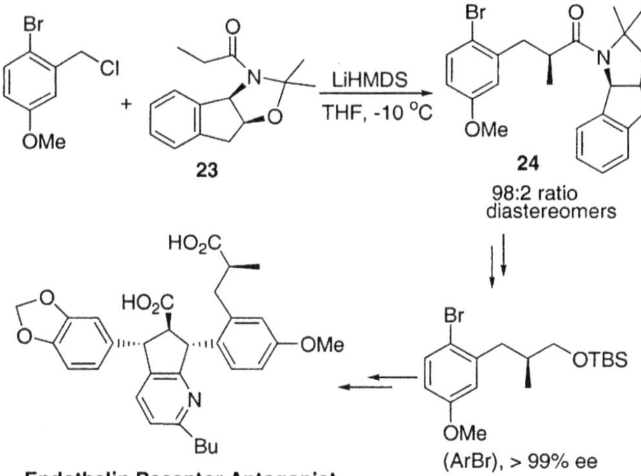

Scheme 7 Asymmetric synthesis of endothelin receptor antagonist

(1*S*,2*R*)-1-Aminoindanol was also used by Kress et al. as a chiral auxiliary for an efficient, diastereoselective [2, 3]-Wittig rearrangement of the α-allyloxyamide lithium enolate **25** [9]. LiHMDS was found to be a much better base than NaHMDS, KHMDS, and *n*-BuLi. Addition of additives, such as HMPA and DMPU, was also necessary. After removal of the auxiliary, the resulting optically active α-hydroxyl acids were transformed to their corresponding functionalized amino acid derivatives (Scheme 8).

Pseudoephedrine is another class of chiral auxiliary used in alkylations. For example, Dragovich et al. of Agouron have employed pseudoephedrine as a chi-

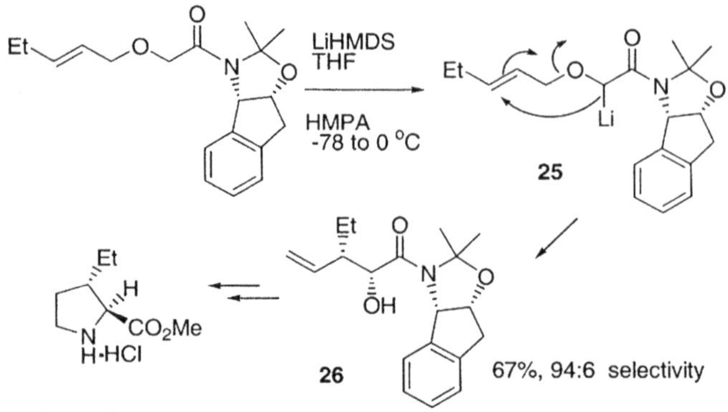

LiHMDS THF HMPA -78 to 0 °C

25

26 67%, 94:6 selectivity

Scheme 8 [2,3]-Wittig rearrangement of aminoindanol-derived lithium amide enolates

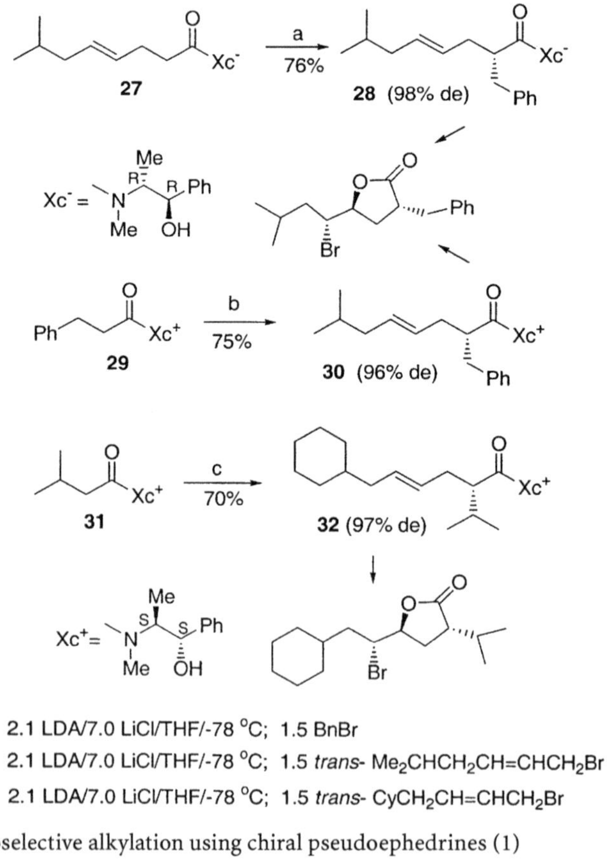

27 a 76% 28 (98% de)

29 b 75% 30 (96% de)

31 c 70% 32 (97% de)

a: 2.1 LDA/7.0 LiCl/THF/-78 °C; 1.5 BnBr

b: 2.1 LDA/7.0 LiCl/THF/-78 °C; 1.5 *trans*- $Me_2CHCH_2CH{=}CHCH_2Br$

c: 2.1 LDA/7.0 LiCl/THF/-78 °C; 1.5 *trans*- $CyCH_2CH{=}CHCH_2Br$

Scheme 9 Stereoselective alkylation using chiral pseudoephedrines (1)

ral auxiliary for the synthesis of nonpeptidic enzyme inhibitors via alkylation of benzyl and allylic halides [10a]. Two equivalents of LDA were required for the formation of a dianion of 7-methyloct-4-enoic amide **27** (Scheme 9). The first equivalent of LDA deprotonates the alcohol and the second enolizes the amide. Alkylation of the lithium dianion derivative with benzyl bromide gave the alkylated product **28** in 98% de. Both enantiomers of pseudoephedrine can be used as effective chiral auxiliaries. Alkylation of the 3-phenylpropionyl amide **29** derived from 1*S*,2*S*-pseudoephedrine produced the enantiomer **30** in 96% de [10a].

The high diastereoselectivities in the alkylation of pseudoephedrine amide enolates can be rationalized by Myers' reactive conformer [10c]. As shown in Fig. 1, the lithium alkoxide and, perhaps more importantly, the solvent molecules associated with the lithium cation were proposed to block the β-face of the (*Z*)-enolate, forcing the alkylation to occur from the α-face.

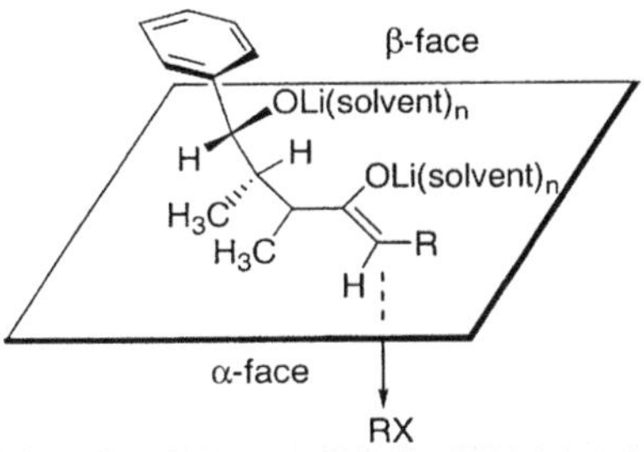

Fig. 1 Myers' reactive conformer of pseudoephedrine amide enolates

Similarly, pseudoephedrine amide alkylation was used by Sandham et al. of Novartis for the synthesis of a novel orally active renin inhibitor **CGP6053B** (Scheme 10) [10b]. As shown in Schemes 11 and 12, (+)-pseudoephedrine was used to control alkylations of the two fragments **33** and **34** in high ee. It is worth noting that LiCl was used as an additive in conjunction with LDA presumably due to the more difficult alkylation with isopropyl iodide. In addition, refluxing

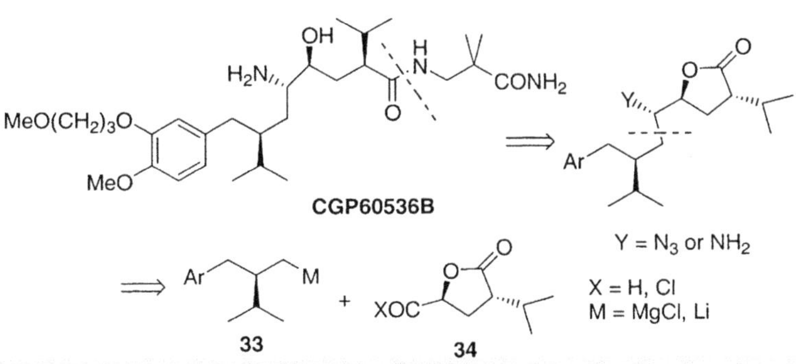

Scheme 10 Asymmetric synthesis of renin inhibitor CGP6053B

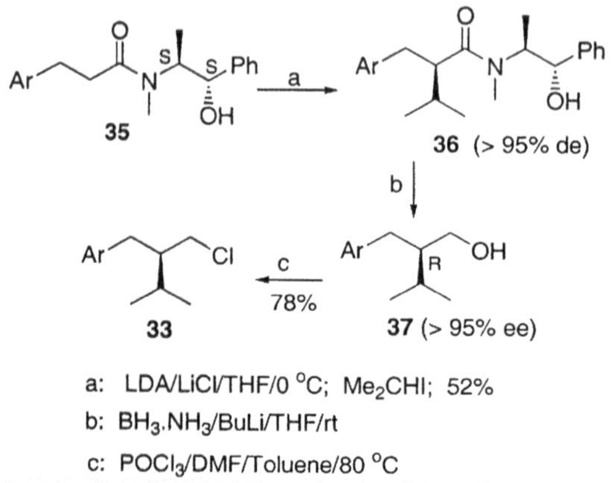
Ar
S
S
Ph
N
OH
a
35
Ar
N
Ph
OH
36 (> 95% de)
b
Ar
Cl
c
78%
Ar
R
OH
33
37 (> 95% ee)
a: LDA/LiCl/THF/0 °C; Me2CHI; 52%
b: BH3.NH3/BuLi/THF/rt
c: POCl3/DMF/Toluene/80 °C

Scheme 11 Stereoselective alkylation using chiral pseudoephedrines (2)

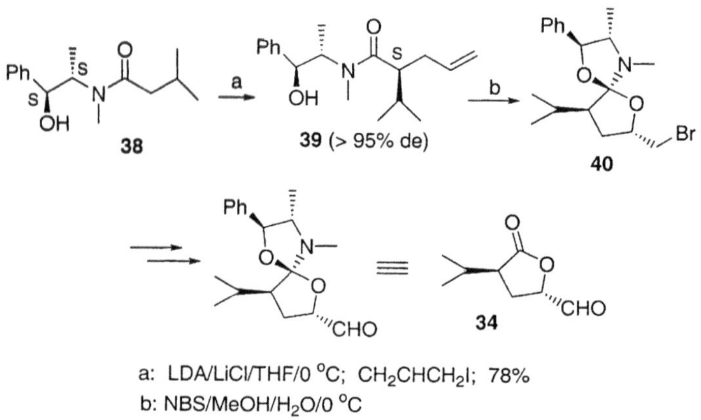
Ph
S
S
OH
N
38
a
Ph
OH
N
S
39 (> 95% de)
b
Ph
N—
O
O
Br
40
Ph
N—
O
O
CHO
O
O
CHO
34
a: LDA/LiCl/THF/0 °C; CH2CHCH2I; 78%
b: NBS/MeOH/H2O/0 °C

Scheme 12 Stereoselective alkylation using chiral pseudoephedrines (3)

temperatures in THF were necessary. The formation of the amide spiroacetal **40** was exploited as a novel lactone-protecting group for a stereoselective Grignard addition reaction. The chiral auxiliary could be efficiently recovered. The process was reported to be amenable to large-scale operation, underscoring the utility of pseudoephedrine as an inexpensive and versatile chiral auxiliary.

In a tandem asymmetric transformation process, Armstrong et al. of Merck used various chiral auxiliaries to prepare a zincate homoenolate, a useful intermediate in the preparation of HIV protease inhibitors and renin inhibitors [11]. In a series of sequential transmetalations, an amide derivative was converted to a β-hydroxyl amide **41** with two new chiral centers. Enolization of the amide with an alkyllithium followed by transmetalation with *bis*(iodomethyl)zinc resulted in a chiral carbon-bonded enolate of zincate. Subsequent treatment of the resulting zincate with an alkoxy lithium reagent followed by a 1,2-migration

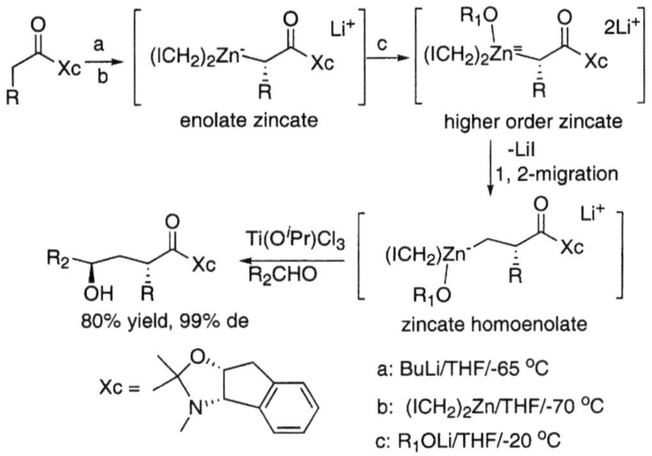

Scheme 13 Tandem asymmetric transformation

gave a chiral zincate homoenolate. The homologation step can be coupled with a homoaldol reaction, as well as with other transmetalation reactions, in a one-pot process. Chiral auxiliaries, such as *cis*-aminoindanol derivatives, oxazolidinones, Meyer's auxiliaries, camphor derivatives, diamines, amino alcohols and their derivatives, were suited for this process. When a *cis*-aminoindanol derivative was used, an 80% yield with 99% de of the homoaldol product **41** was obtained (Scheme 13). This type of sequential use of transmetalation is rare but potentially useful in other industrial processes.

2.2 Chiral Pool-Based Chemistry

Chiral pool-based chemistry is more atom efficient as the starting material not only acts as a chiral template but also ends up as part of the product. Both 1,2- and 1,3-induction have been reported. For example, Tian et al. of Pfizer have extended Hanessian's method [12] and reported a 1,3-asymmetric induction in dianionic alkylation of amino acid esters (Scheme 14) [13]. Two equivalents of LiHMDS were used to generate a dianion of *N*-Boc-*L*-(+)-glutamate **42**. Alkylation of the lithium dianion with cyanomethylbromide proceeds with remarkable stereoselectivity in the absence of additives to give almost exclusively the corresponding *anti* isomers **43** in 90% yield. This asymmetric dianionic cyanomethylation has been applied to an efficient synthesis of the chiral γ-lactam derivative **44**, a key intermediate for the preparation of a rhinovirus protease inhibitor **AG7088**. This process was scaled up to produce multi-kilogram quantities of **AG7088**.

The enantioselectivity of this type of 1,3-induction was rationalized by a general transition state that may involve *Z*(O) or *E*(O) enolate geometry

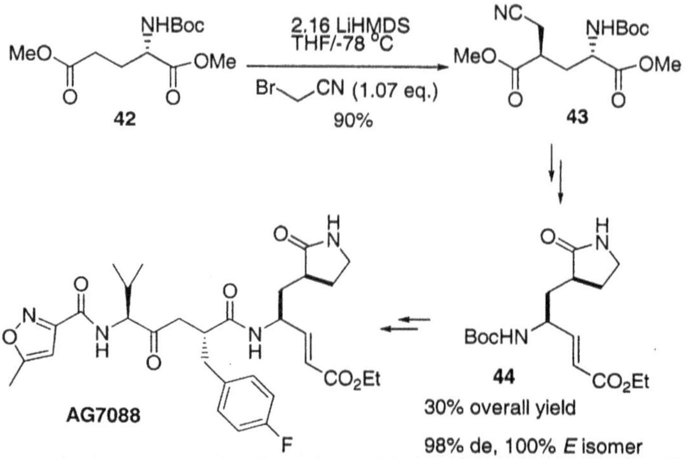

Scheme 14 1,3-Asymmetric induction in alkylation of amino acid esters

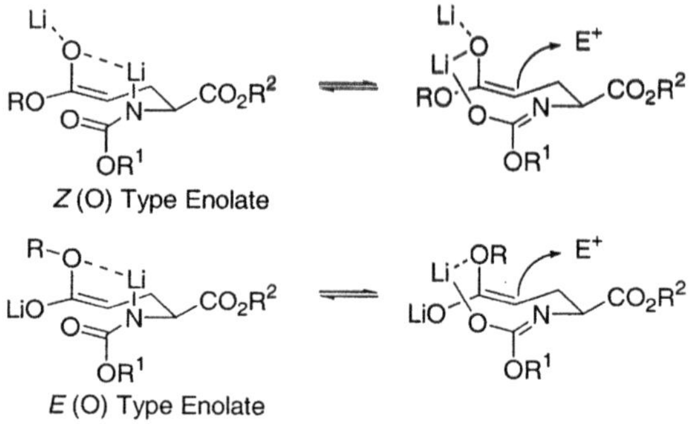

Scheme 15 Possible transition states for 1,3-asymmetric induction in alkylation of amino acid esters

(Scheme 15). In both enolates, the more preferred equatorial attack takes place to generate the desired enantiomer [12].

Yee et al. of Boehringer Ingelheim have extended Seebach's principle of self-regeneration of stereocenters (SROSC) to a practical enantiospecific synthesis of *N*-aryl-hydantoin LFA-1 antagonists **BIRT-377** [14]. The process relied on the stereospecific alkylation of lithium enolate of the 2-*tert*-butyl- or 2-isopropyl-imidazolidinone template **45** as shown in Scheme 16. LiHMDS was used for the enolization. The two templates gave the same yield and the same enantiospecificity, but the latter one was more cost effective. This was the first successful use of a 2-isopropyl-imidazolidinone template for the asymmetric stereospecific synthesis of a α-disubstituted amino acid derivative.

Yee's colleagues, Napolitano et al. further extended the SROSC approach to a *cis*-oxazolidinone system [15]. Benzylation of the lithium enolate of *N*-alkoxy-

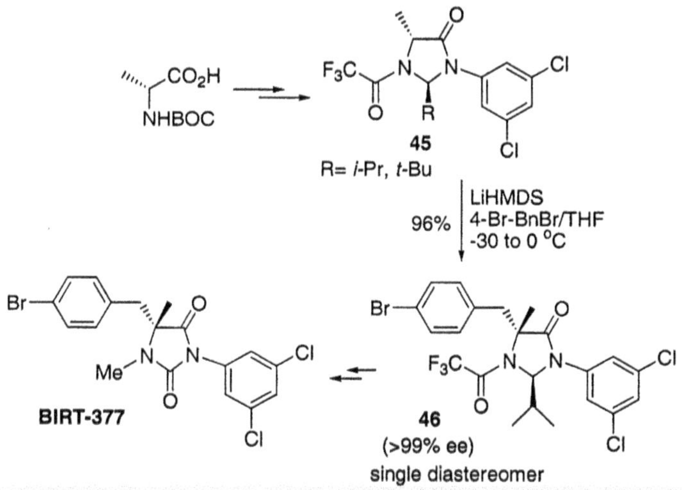

Scheme 16 Stereospecific alkylation on an isopropyl-imidazolidinone template

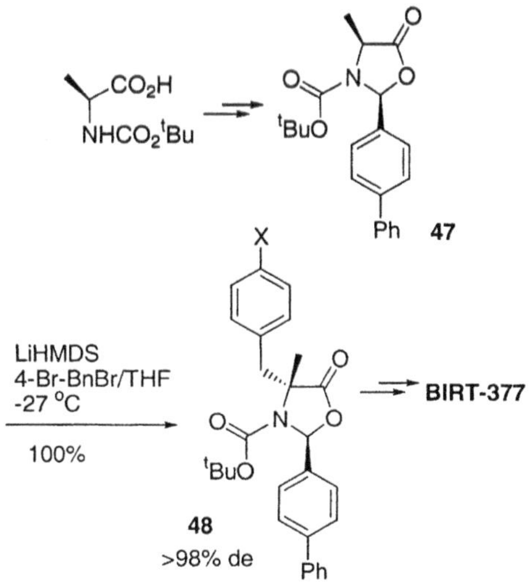

Scheme 17 Stereospecific alkylation on a *cis*-oxazolidinone template

carbonyl-2-aryloxazolidinone **47** at -27 °C gave the alkylation product **48** in excellent yield and high (>98%) diastereoselectivity. Subsequent treatment produced **BIRT-377** in 40% overall yield and >99.9% ee (Scheme 17).

2.3 Intramolecular Alkylation

Schering-Plough scientists have been interested in a series of azetidinone structures as novel cholesterol absorption inhibitors. Chen et al. have reported the first example of an intramolecular *trans*-alkylation of a 2-azetidinone **49** [16]. The diastereomeric control of substituents on the cyclohexyl ring of 2-azaspiro [3,5]-nonan-1-one was directed by the preformed stereochemistry of the β-lactam ring. A *trans*-attack by the lithium enolate on the electrophile would account for the remarkably high diastereoselectivity. This approach has led to an efficient and highly enantioselective synthesis of (+)-**SCH 54016**, a potent cholesterol absorption inhibitor (Scheme 18).

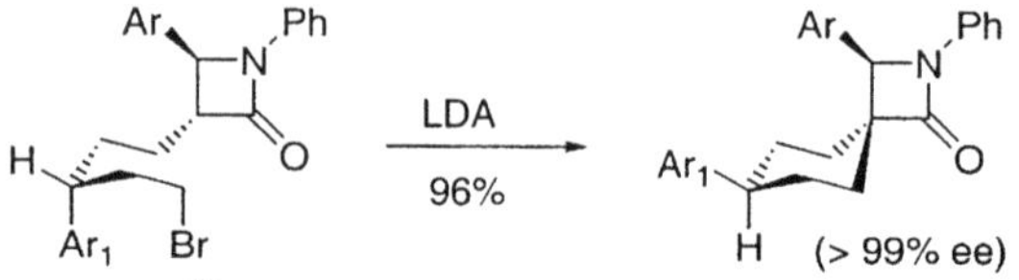

Scheme 18 Intramolecular *trans*-alkylation

2.4 Chiral Additive-Mediated Alkylation

Chemical processes with external additive-mediated chiral alkylation are more efficient than that with auxiliary-based chemistry because the former type does not require the additional attachment and removal of the auxiliary. The challenge is to obtain high enantioselectivity with external additive. In one of the first examples of this approach, Dolling et al. of Merck reported in 1984 an efficient asymmetric alkylation for the synthesis of (+)-indacrinone via chiral phase transfer catalysis. The desired alkylation product was obtained in 95% yield and 92% ee [17a]. In the last decade, chiral amines, such as sparteine-mediated chiral alkylations have been extensively studied in academic research [17b–17m].

Recently, a novel enantioselective alkylation of double benzylic substrates with secondary electrophiles and its application to the synthesis of Lonafarnib, an anticancer agent, was reported by Schering-Plough's Process Development scientists [18]. The key to this synthesis was the development of a chiral additive-mediated chiral alkylation of benzylic lithium species (Scheme 19). Initial studies of the racemic version of the reaction indicated that LDA is superior to PhMgBr and toluene is a better solvent than THF to prevent the elimination of secondary electrophile.

Over 40 commercially available chiral ligands were examined and some selected results are summarized in Fig. 2. For those ligands with a hydroxy group, two equivalents of LDA are required. The first one neutralizes the alcohol and

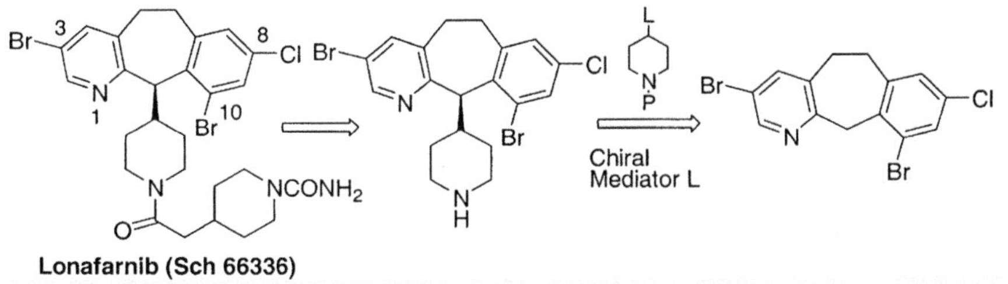

Lonafarnib (Sch 66336)

Chiral Mediator L

Scheme 19 Asymmetric synthesis of lonafarnib

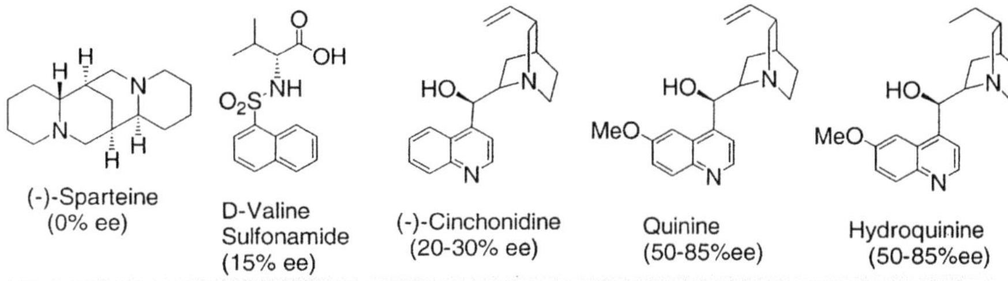

(-)-Sparteine (0% ee)

D-Valine Sulfonamide (15% ee)

(-)-Cinchonidine (20-30% ee)

Quinine (50-85%ee)

Hydroquinine (50-85%ee)

Fig. 2 Commercially available chiral ligands

the second one deprotonates the benzylic substrates. As shown in Fig. 2, the most impressive result has been obtained with the cinchonidine family. Whereas cinchonidine itself only induced 20–30% ee, its methoxy derivatives, quinine and hydroquinine, gave up to 85% ee (Fig. 2). Apparently, the extra methoxy group on the quinoline moiety exerts great influence on the enantioselectivity.

Based on the results in Fig. 2, a three-point interaction model for the asymmetric induction with quinine and hydroquinine was postulated (Scheme 20). The alkoxy and the bridging nitrogen are the first two sites to chelate with lithi-

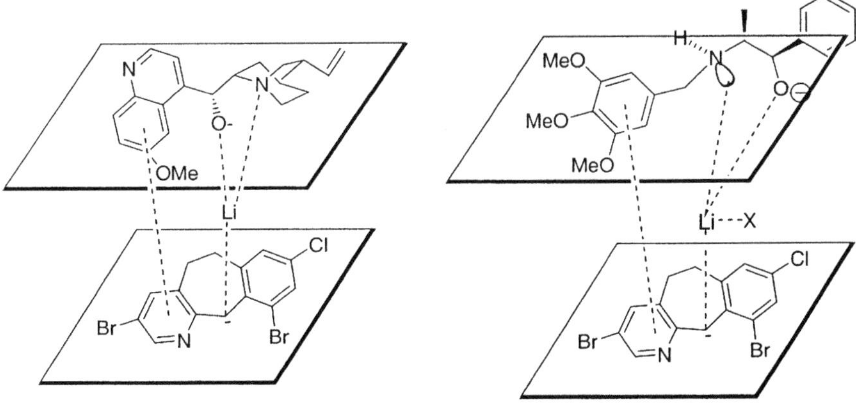

Scheme 20 Three-point induction model

um. The third point of interaction comes from the π-stacking between the quinoline moiety and the pyridine ring in the substrate. It was believed that the π-stacking plays an important role in fixing the configuration of the reaction intermediate. The presence of a methoxy group enhances the π-stacking effect.

Therefore, any new mediator should have two chelating sites and an aromatic ring. Because of the difficulty in modifying quinine to examine further the key structural features needed for optimal activity in this reaction, Schering chemists decided to look for synthetic mediators, using the above three-point model as our starting point for mediator design. Norephedrine was selected as the starting material because of its easy derivatization. The ees of alkylated product with a variety of norephedrine-based tertiary amines only ranged from 13 to 28%. Therefore, a number of secondary amine derivatives were prepared with the best ee of 88% obtained with the trimethoxy benzylamine **51** (Fig. 3).

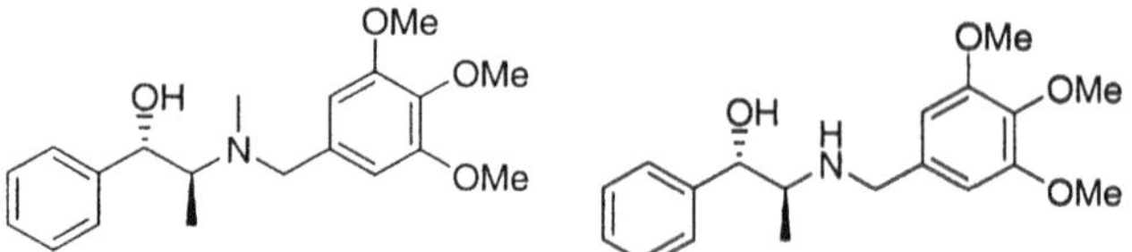

50 (28% ee) **51** (88% ee)

Fig. 3 Novel norephedrine-based ligands

One of the issues associated with the alkylation using both quinine and the trimethoxy benzylamine **51** was the inconsistency of ees from run to run. It was first suspected that water was detrimental to the LDA reaction. Counterintuitively, a consistently lower range of ees was observed when moisture was vigorously excluded from the system. To follow up on this unexpected observation, the reaction mixture was spiked with different amounts of water prior to the addition of LDA. Surprisingly, water has a pronounced positive effect on the enantioselectivity as shown in Table 1. The higher the water content the better the enantioselectivity with an optimum charge of one equiv. The same trend was also observed with the yield. With added water, 95% ee was consistently achieved using either quinine or **51**. Introduction of other additives such as LiOH, $B(OH)_3$, MeOH, and AcOH had no effect on the enantioselectivity. This novel alkylation works well for a variety of benzylic substrates and secondary electrophiles.

Table 1 Water effect on the selectivity of alkylation with **51**[a]

Entry	Water added	ee (%)	Yield(%)
1	0.0 eq	55–60	50–60
2	0.5 eq	72	N/A
3	0.7 eq	85	92
4	1.0 eq	95	95

[a] All experiments conducted with 1.8 eq. of the trimethoxy benzylamine**51**

To complete the reaction, the alkylation product was hydrolyzed in situ and crystallized as the *N*-acetyl-*L*-phenylalanine salt, further enhancing the ee to >98%. Both quinine and **51** can be readily recovered and recycled without loss of ee. After some practical modifications, this novel chiral alkylation was scaled up in the plant to a 33 kg batch size using the commercially available quinine to produce multi-hundred kilograms of Lonafarnib (Scheme 21).

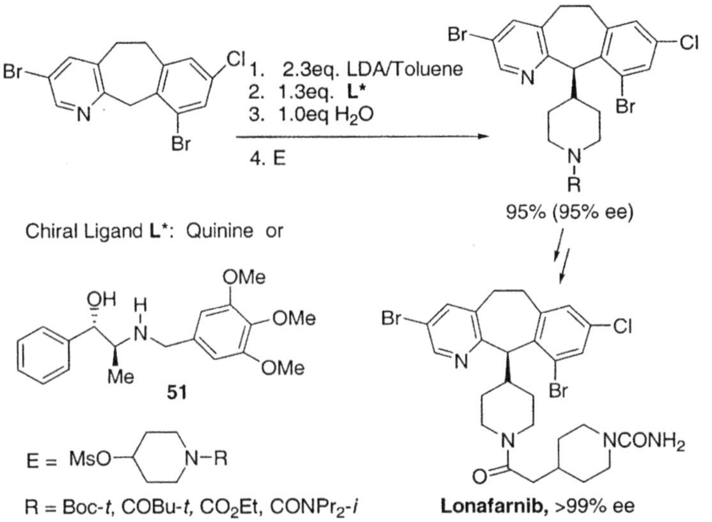

Scheme 21 Chiral ligand-mediated enantioselective alkylation

3
Chiral Nucleophilic Addition of Organolithium Reagents

Organometallic additions to carbonyls, including asymmetric additions, are ubiquitous in both industry and academia, and therefore this review will focus on additions to carbonyl derivatives. In particular, chiral nucleophilic addition to imines has been a focus of industrial research over the last decade since most drugs contain nitrogen. Thus, this section will discuss several examples of additions to imines.

3.1
Addition to C=N Bond

Addition of organolithium reagents to imines has been carefully studied in industry and academic research. There are several excellent reviews on this general topic by Enders [19], Bloch [20], and Kobayashi [21].

Pridgen et al. of SmithKline Beecham have described a stereoselective synthesis of both enantiomers of 2-(1′-amino-2′-methylpropyl)imidazole **52** through the nucleophilic addition of the lithium reagent to chiral 1,3-oxazolidines

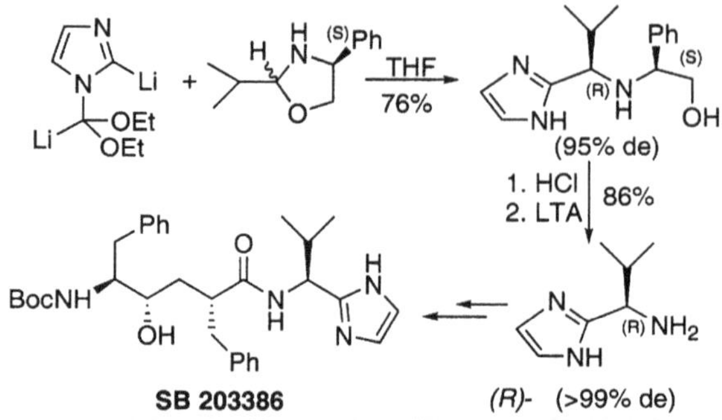

Scheme 22 Asymmetric nucleophilic addition for a key synthon of SB203386

(Scheme 22) [22]. The oxazolidine is presumably serving as a masked imine in this reaction. Thus, reaction of the dilithio anion of the ortho ester-protected imidazole **53** with nonracemic 2-isopropyloxazolidine **54** gave (*R*,*S*)-product **55** in 76% yield and 95% de. The (*S*,*R*)-enantiomer was prepared in 84% yield using the (*R*)-2-phenylglycinol as the auxiliary. An oxidative removal of the chiral auxiliary yielded the chiral 2-(1′-amino-2′-methylpropyl)imidazole **52**, a key synthon in the synthesis of a potent protease inhibitor **SB 203386.**

Senanayake et al. of Sepracor have reported the application of imine addition to the first asymmetric synthesis of (*R*)-desmethylsibutramine (DMS), a metabolite of an antiobesity compound [23]. Addition of *i*-BuLi to aldimine **56** in the presence of 0.2 equiv. of chiral *bis*-oxazoline **57** gave 40% ee of (*R*)-desmethylsibutramine with 95% conversion. Increasing or decreasing the amount of the chiral ligand did not affect the ee. Other C_2 symmetric chiral bis-oxazolines or sparteine gave poorer results. The ee could be enriched to >99% with >90% recovery by a single crystallization with (*R*)-mandelic acid. This catalytic enantioselective addition was used as a key step in the asymmetric synthesis of (*R*)-desmethylsibutramine (DMS) (Scheme 23).

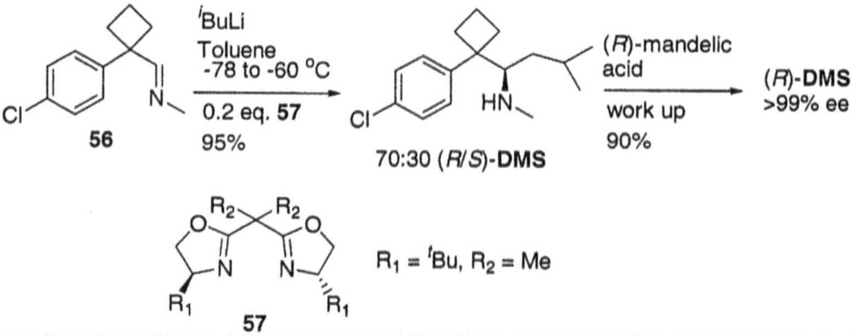

Scheme 23 Asymmetric addition of *i*-BuLi to imine catalyzed by bis-oxazoline

A novel one-step diastereo- and enantioselective formation of *trans*-β-lactams from a chiral aldol condensation of lithium dianions and its application to the synthesis of Ezetimibe, a cholesterol absorption inhibitor was reported by Schering-Plough's Process Development scientists (Scheme 24) [24, 28].

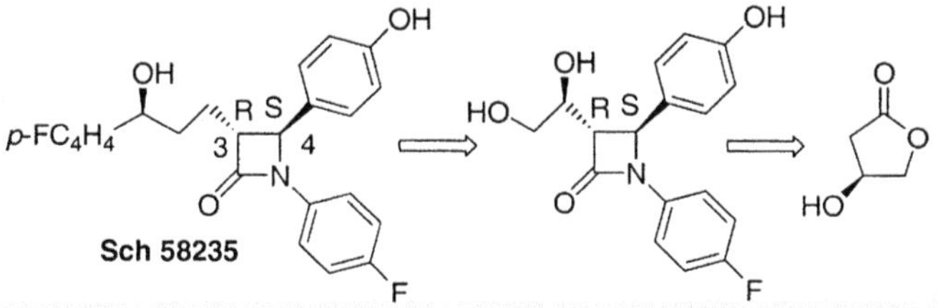

Scheme 24 Asymmetric synthesis of *trans*-β-lactams

Initially, addition of the lithium dianion of the hydroxy-γ-lactone to imine gave predominately the two adducts **58** and **59**. The low reactivity of the two intermediates are most likely due to the formation of stable lithium aggregates [25]. It was speculated that addition of a lithium salt may be able to form mixed aggregates and increase the reactivity [26]. As expected, addition of LiCl does accelerate the cyclization of these adducts to a mixture of two β-lactams in 35% isolated yield.

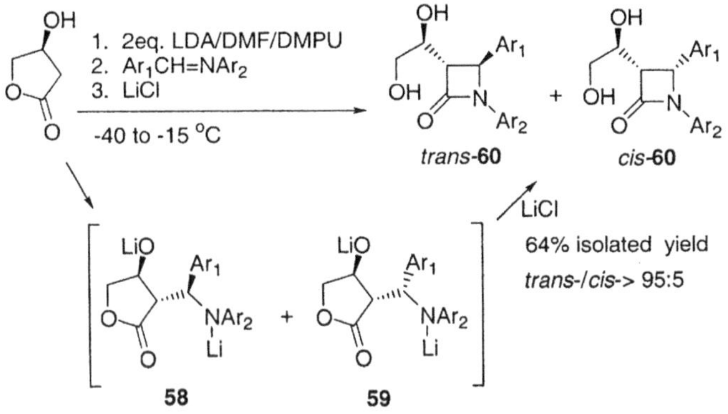

Scheme 25 Addition of chiral dianion to an imine

The imine condensation reaction gives exclusive control at C-3 but only moderate diastereoselectivity at the C-4 (or benzylic) position with an initial ratio of 73:27 (**58**/**59**) in favor of the desired *SSS*-**58** intermediate. It was also observed that the rate of cyclization of intermediate **58** to *trans*-**60** product is about four times as fast as that of **59** to *cis*-**60** product, enhancing the ratio of *trans* to *cis* from 73:27 (**58**/**59**) to 87:13 (*trans*/*cis*). Examination of the cation effect on the diastereoselectivity indicated that the weaker the coordination ability, the better the selectivity. For example, one equivalent each of Et_2Zn and LDA gave an 11:89

ratio of **58/59** while equal equivalents of NaHMDA and LiHMDA reversed the ratio to 86:14. Hence, either the *trans* or the *cis* isomer can be obtained from this condensation reaction depending upon which metal is used. A simple precipitation procedure was developed to further enhance the diastereomeric ratio in isolated product from 90:10 to better than 95:5. Under the optimized conditions, the reaction was scaled up smoothly to 300 g to give a 64% isolated yield (Scheme 25). It is interesting to compare these results with a previous report where an open-chain 3-hydroxybutyrate was used as a starting material [27]. The predominant products in both cases have the same absolute stereochemistry for the β-lactam ring despite the opposite stereochemistry for the side-chain alcohols. This clearly demonstrates the effect of the butyrolactone ring on the dianion configuration and reactivity.

This reaction represents the first one-step enantio- and diastereoselective, high yielding, and practical synthesis of *trans* β-lactams starting from (*S*)-3-hydroxy-γ-lactone. This one-step azetidinone formation is quite general for arylimines and various *trans* β-lactams can be prepared. The diol produced from this reaction is a versatile synthon. It may be oxidatively cleaved to an aldehyde, as is done in the synthesis of Ezetimibe. Alternatively, the diol may be converted to an epoxide. For example, the opening of a chiral epoxide derived from this synthon with lithium phenoxide for the synthesis of an azetedinone-based cholesterol absorption inhibitor (**Sch 57939**) was reported [28]. Initially, displacement of a monotosylate with sodium 4-fluorophenoxide gave a very slow reaction at 65 °C with less than 10% of the desired ether. Use of calcium 4-fluorophenoxide gave only a trace of the desired product even at elevated temperatures. Reaction with lithium 4-fluorophenoxide improved the yield to 43% but a long reaction time was still required (95% conversion at 60 h). In all of the reactions carried out, a common intermediate was isolated and found to be the corresponding epoxide.

The fact that sodium, calcium, and lithium gave different results suggested a cation effect. It was speculated that metal cations such as Zn^{2+} with both Lewis acid and base properties would be best suited for this type of reaction. On the Lewis base side, $4\text{-}FC_6H_4OZnX$ could act as a nucleophile, while on the Lewis acid side, the Zn^{2+} could activate the epoxide. This type of combined "push-pull" effect should speed up the epoxide opening and give a better yield [29]. As predicted, transmetalation of $4\text{-}FC_6H_4ONa$ with $ZnBr_2$ to $4\text{-}FC_6H_4OZnBr$ followed

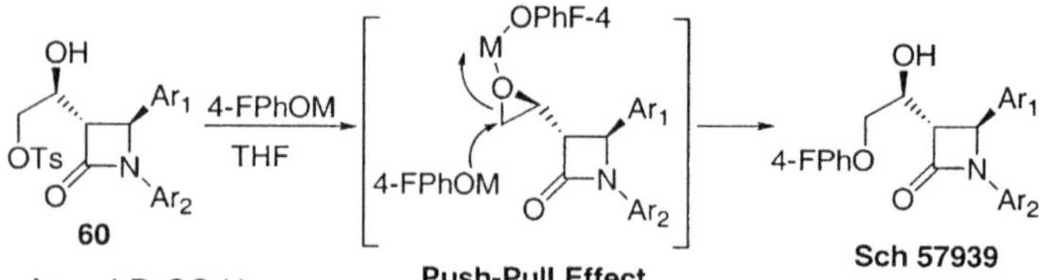

Ar_1 = 4-$BnOC_6H_4$-
Ar_2 = 4-FOC_6H_4-
M = Na (10%), Li (43%), ZnBr (67%)

Scheme 26 Push-pull effect in epoxide opening

by reaction with the monotosylate completed the reaction in less than 12 h, five times faster than its lithium counterpart. Accordingly, the isolated yield of the desired ether increased from 43% for lithium phenoxide to 67% for its zinc counterpart (Scheme 26).

3.2 Alkynylation of Imines and Ketones

Recently, investigators at Merck and DuPont developed several practical enantioselective syntheses for HIV-1 non-nucleoside reverse transcriptase inhibitors through addition of lithium acetylides to prochiral imines and ketones mediated by chiral lithium amino alkoxides [30–38].

Addition of Li-acetylide **61** to cyclic *N*-acyl ketimines **62** mediated by stoichiometric amounts of quinine Li-alkoxide gave the (*S*)-enantiomer **63** in 84% yield and 97% ee (Scheme 27) [31]. Nearly equal selectivity for the (*R*)-enantiomer could be obtained using quinidine as the mediator. The bulky 9-anthrylmethyl protecting group at the distal position on the imine was necessary for the high selectivity. The lithium salts of both acetylene and alcohol, generated with BuLi or LiHMDS, are better than sodium (NaHMDS) or magnesium (EtMgBr) salts. Homogeneous THF solutions of acetylide and alkoxide gave better selectivity than the suspensions obtained with toluene or diethyl ether.

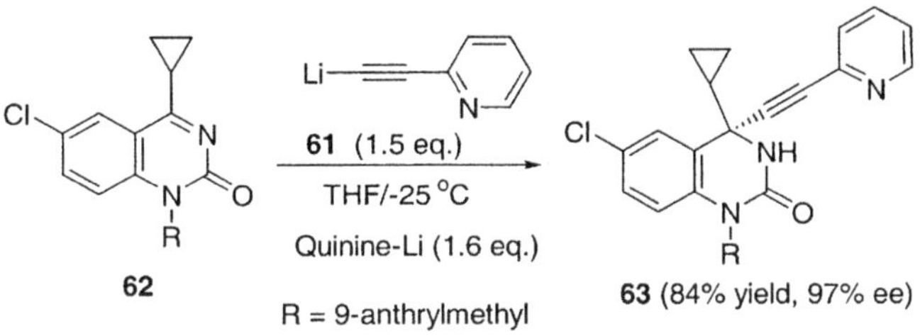

62 → **63** (84% yield, 97% ee)

61 (1.5 eq.), THF/-25 °C, Quinine-Li (1.6 eq.)

R = 9-anthrylmethyl

Scheme 27 Asymmetric alkynylation mediated by quinine Li-alkoxides

Quinine proved to be an unsatisfactory ligand for a conceptually similar asymmetric ketone addition in the synthesis of Efavirenz. (1*R*,2*S*)-*N*-Pyrrolidinylnorephedrine was chosen from a pool of chiral amino alcohols as the chiral mediator for the addition of lithium cyclopropylacetylide **65** to trifluoromethyl *p*-methoxybenzyl-protected ketoaniline **66a**. A 98% ee with 95% yield of **67** was achieved using THF as solvent at temperatures below -50 °C. The use of 2 equiv. of Li-acetylide, and 2 equiv. of Li-alkoxide followed by equilibration of the resulting acetylide-alkoxide solutions at temperatures above -40 °C prior to the addition of the ketoaniline were necessary to achieve high ees [32]. A cubic 2:2 tetramer formed from Li-acetylide and Li-alkoxide was proposed to be the active intermediate based on ^{6}Li, ^{13}C, and ^{15}N NMR spectroscopy studies in solution and X-ray crystallography in the solid state (Scheme 28) [32–38].

The success of alkynylation mediated by quinine or norephedrine relied on the *N*-protection of the ketone or ketimine substrates. However, a protection/de-

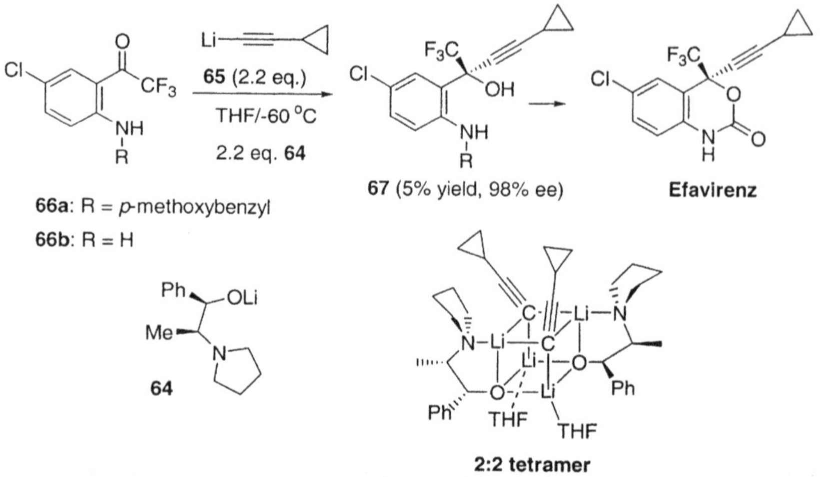

Scheme 28 Asymmetric alkynylation mediated by norephedrine Li-alkoxides

protection step was required. Direct alkynylation of unprotected ketoaniline or ketimines by the methods described above suffered from low conversion and low ee. Complexation of the lithium acetylide **65** with zinc alkoxide **68** lowers the basicity while maintaining the nucleophilicity of the acetylide. As a result, addition of lithium acetylide to the unprotected ketoaniline **66b** became possible in THF/toluene at 25 °C to give **69** in 83% yield and 83% ee. The optimized reaction has been carried out on a kilogram scale and is the basis of the most efficient synthesis of Efavirenz to date (Scheme 29) [33].

Parsons et al. of Merck have found that a (+)-3-carene-derived amino alcohol **70** is a uniquely effective chiral mediator for the addition of lithium cyclopropylacetylide to the unprotected *N*-acylketimines **71** [37]. Three equivalents of Li-

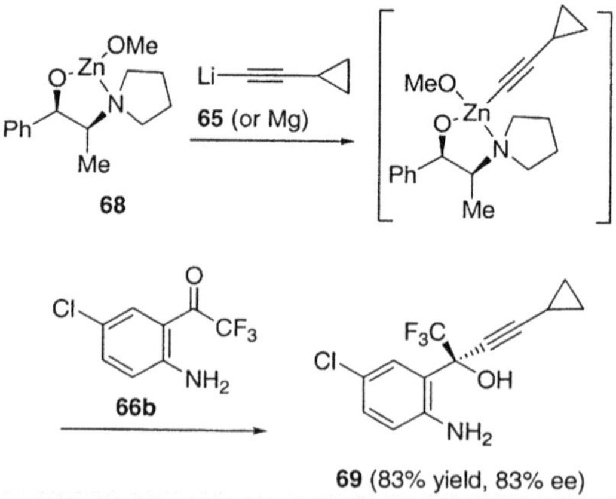

Scheme 29 Asymmetric alkynylation mediated by Zn-alkoxides

acetylide and 3.0 equiv. of the chiral mediator **70** were required to achieve high yield and high ee. Lithium bis(trimethylsily)amide (LiHMDS) proved superior to BuLi or other lithium amides as the strong base in this system. The amino alcohol ligand could be recovered in 92% yield by basification of the aqueous acid extracts. The recovered amino alcohol was of suitable purity to be recycled directly (Scheme 30) [30, 37, 38].

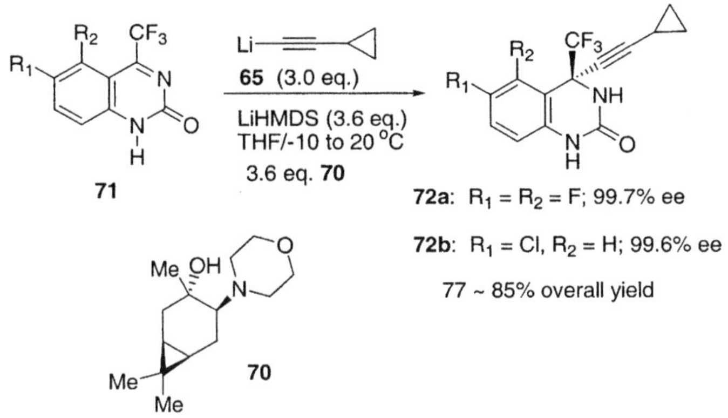

Scheme 30 Asymmetric alkynylation mediated by morpholinocaranol Li-alkoxides

Addition of lithium acetylide to a chiral epoxide provided an easy process for an asymmetric synthesis of chiral secondary alcohols (Scheme 31) [39].

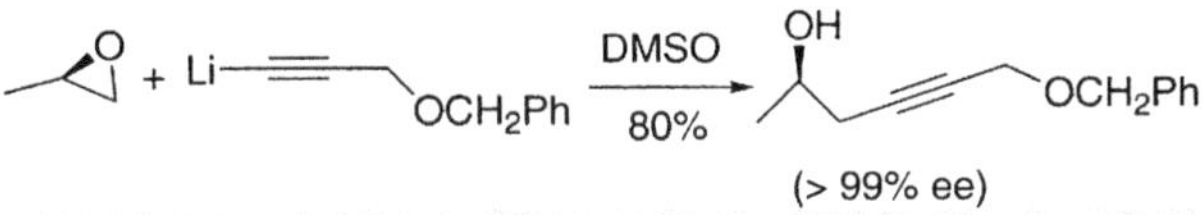

Scheme 31 Addition of lithium acetylide to a chiral epoxide

3.3 Addition to Esters

Nucleophilic addition of alkyllithiums or lithium amides to chiral esters has displayed a strong tendency to retain the configuration at α or β carbon. In general, this chiral retention could be attributed to the weak basicity of lithium reagents and the strong chelation tendency of lithium ion.

Chloromethyl lithium ($ClCH_2Li$) was used as a nucleophile in a number of additions to chiral esters. For example, a group of investigators at Roche reported the addition of chloromethyl lithium, generated in situ at -78 °C, to an *N*-protected α-amino ester **73** for the preparation of *N*-protected chloromethylketone **74** in 76% yield without racemization [40]. The *N*-protected chloromethylketone **74** was used in the synthesis of the first HIV-protease inhibitor, Saquinavir (Scheme 32).

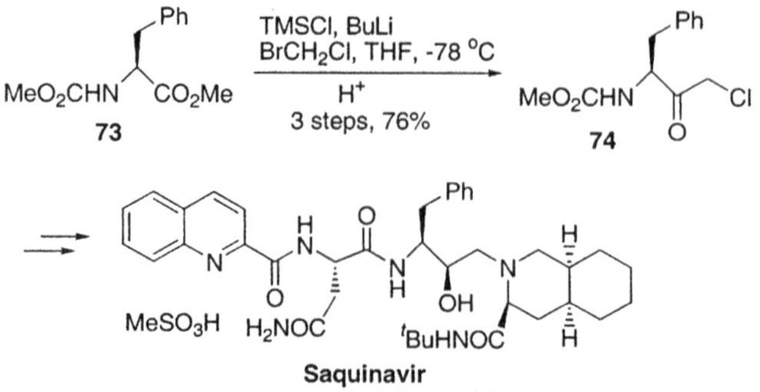

Scheme 32 Chiral retention in the chloromethylation of methoxycarbonyl-protected amino acid esters

Izawa et al. of Ajinomoto also reported that reaction of *N*-protected 3-oxazolidin-5-ones **75** with chloromethyl lithium afforded *N*-protected 5-chloromethyl-5-hydroxy-3-oxazolidines **76** without racemization [41]. The oxazolidines were easily hydrolyzed to give α-aminoalkyl-α′-chloromethyl ketones **77**. In this case, both the *N*, *O*-acetal and Boc groups were simultaneously deprotected by acid treatment to give α-aminoalkyl-α′-chloromethylketones (Scheme 33a). The *N*-Cbz-protected α-aminoalkyl-α′-chloromethylketones **80** could be prepared in a similar fashion (Scheme 33b).

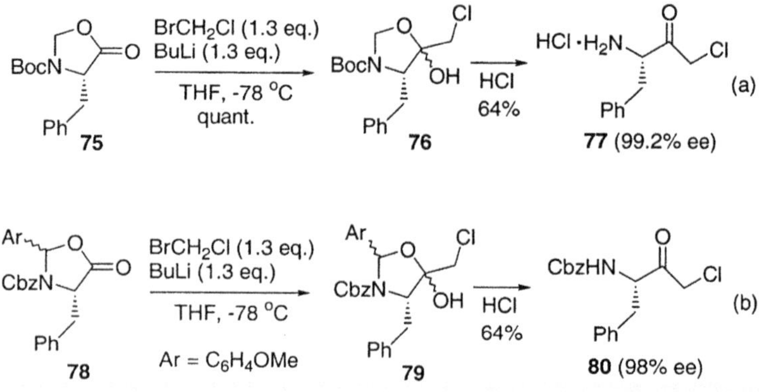

Scheme 33 Chiral retention in the chloromethylation of *N*-protected 3-oxazolidin-5-ones

Similarly, a chemoselective addition of $ClCH_2Li$ to the ester group of **81** was achieved in the presence of an imine group by Izawa et al. to give, after hydrolysis, an α-aminoalkyl-α′-chloromethylketones **82** in good yield without racemization. Surprisingly, both *N*-diphenylmethylene- and *N*-benzylidene-protected amino acid esters gave high optical purities (>98% ee) and good isolated yields (Scheme 34) [42].

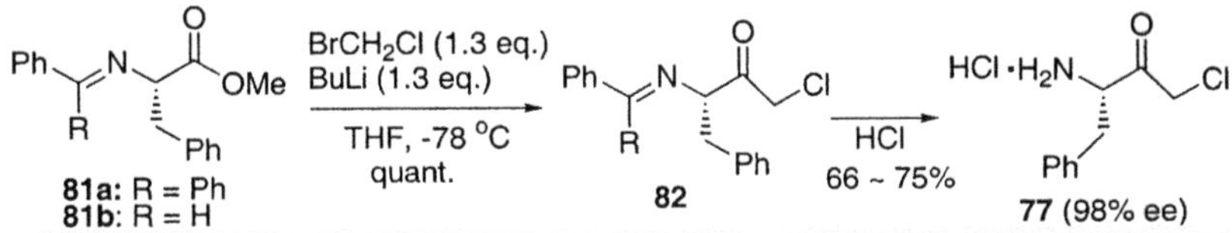

Scheme 34 Chiral retention in the chloromethylation of *N*-imine-protected amino acid esters

The full retention of the stereochemistry was attributed to the weak basicity of chloromethyllithium, generated in situ from $BrCH_2Cl$ and BuLi. In addition, a five-membered chelation of lithium metal with both the nitrogen and the carboxylic oxygen might have prevented the abstraction of the α-proton. α-Aminoalkyl-α′-chloromethylketones are useful intermediates for serine protease and hydroxyethyl isostere subunits found in many renin and HIV protease inhibitors.

Chang et al. of Abbott have shown that lithium amide gave superior chiral retention in the cyanomethylation of *N*, *N*-dibenzyl *L*-phenylalanine benzyl ester **83** in THF at room temperature than the corresponding sodium amide [43]. This was rationalized by two factors. First, deprotonation of acetonitrile with lithium amide is kinetically more favored than that of the α-proton, resulting in a better enantiomeric retention. Second, it is likely that there is a stronger chelation between lithium cation and carboxylic oxygen. This type of "push-pull" effect would increase the electrophilicity of the carboxylic carbon, hence favoring the nucleophilic addition of cyanomethyl anion to the carboxylic center rather than the deprotonation of the α-proton. The β-keto nitrile product **84** is a key intermediate in the synthesis of the protease inhibitor Ritonavir (Norvir) (Scheme 35).

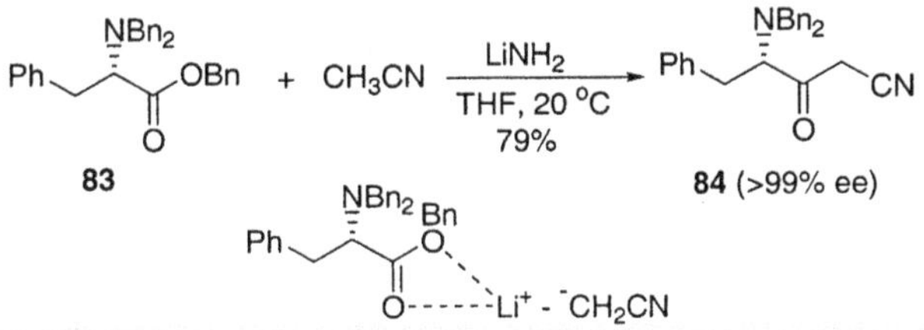

Scheme 35 Chiral retention in the cyanomethylation of *N*,*N*-dibenzyl *L*-phenylalanine benzyl ester

4 Organolithium in Chiral Aldol Condensations

Enantioselective aldol condensation is a very powerful tool for C-C bond formation. A general review of enantioselective aldol and Michael additions of chiral lithium amides and amines was published previously [44, 45].

4.1
Auxiliary-Induced Aldol Reaction

Iwanowicz et al. of Bristol-Meyers Squibb have described a new *anti*-selective aldol reaction utilizing 2,4-disubstituted-oxazolidine as a chiral auxiliary (Scheme 36) [46]. The reaction of the lithium enolate of **85** with various aldehydes gave predominantly a single *anti*-diastereomer **86** in good to excellent yields. The bulkier the substituent at the α-position to the carbonyl the better the diastereo- and enantioselectivity. In fact, the lithium enolate of a less sterically hindered glycine ester gave a mixture of all four possible diastereomers. The chiral auxiliary could be easily removed, allowing for the efficient preparation of chiral β-hydroxy-α-amino acids of erythro stereochemistry.

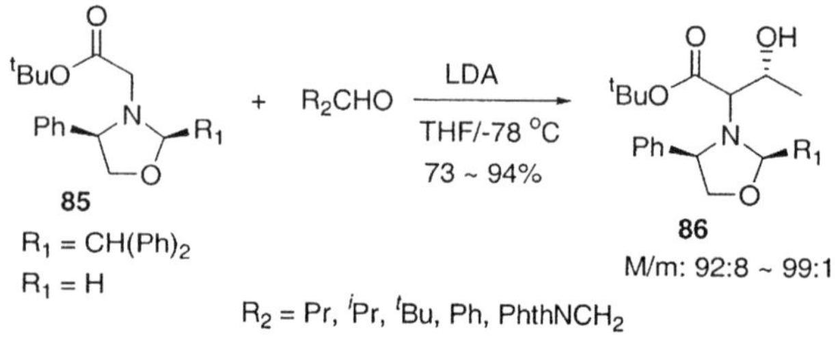

85 + R_2CHO → (LDA, THF/-78 °C, 73 ~ 94%) **86**

$R_1 = CH(Ph)_2$
$R_1 = H$

M/m: 92:8 ~ 99:1

R_2 = Pr, ^{i}Pr, ^{t}Bu, Ph, PhthNCH$_2$

Scheme 36 *anti*-Selective aldol reaction using chiral oxazolidine

Jacobson et al. of DuPont-Merck have developed a method for the synthesis of a variety of 2-hydroxy-2,3-disubstituted succinates **87** by the asymmetric additions of lithium, boron, or titanium enolates of Evans' chiral imides to α-keto esters (Scheme 37) [47]. The lithium enolate, prepared from the *S*-imide **88** and LDA in THF, reacted with a α-keto esters **89** to give the corresponding aldol-type adducts in good yield. The *Z*-enolate reacted preferentially from the *Si* face allowing for the control of the configuration at the C3-carbon. The control of stereoselectivity at the tertiary carbon C2 was found to be non-specific. The two isomers, *anti*- and *syn*-**87**, could be separated by flash chromatography in multigram quantities.

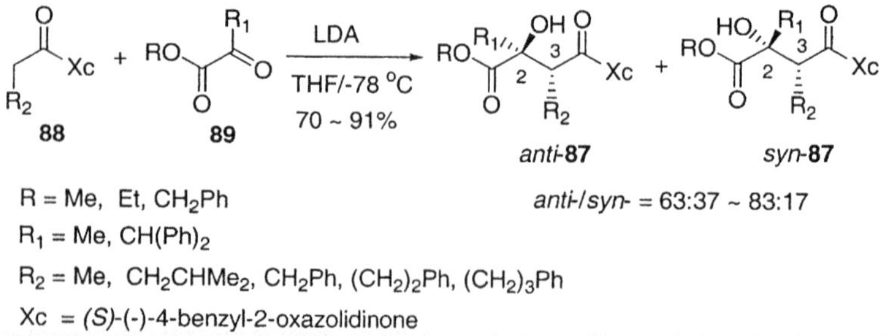

88 + **89** → (LDA, THF/-78 °C, 70 ~ 91%) *anti*-**87** + *syn*-**87**

anti-/syn- = 63:37 ~ 83:17

R = Me, Et, CH_2Ph
R_1 = Me, $CH(Ph)_2$
R_2 = Me, CH_2CHMe_2, CH_2Ph, $(CH_2)_2Ph$, $(CH_2)_3Ph$
Xc = *(S)*-(-)-4-benzyl-2-oxazolidinone

Scheme 37 Stereoselective aldol reactions of Evans' chiral imides

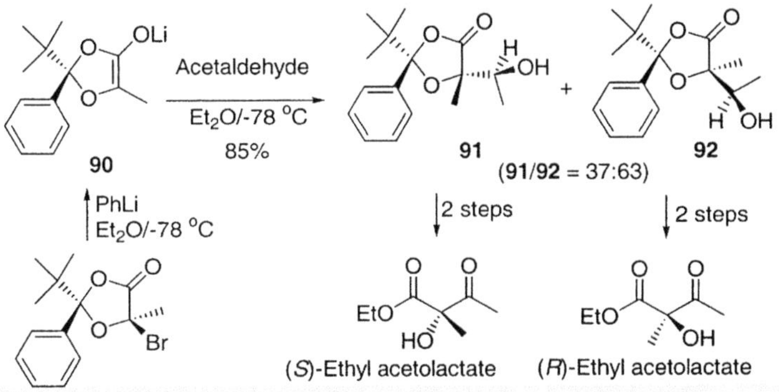

Scheme 38 Erythroselective aldol condensation of dioxolanone with aldehyde

Greiner et al. of Rhone Poulenc Agrochimie have developed a new method for the synthesis of the ethyl acetolactate enantiomers through the erythro-selective aldol condensation of sterically hindered dioxolanone with acetaldehyde (Scheme 38) [48]. The lithium enolate of 2-*tert*-butyl-5-methyl-2-phenyl-1,3-dioxolan-4-one **90**, generated by halogen metal exchange, reacted with acetaldehyde to yield two of the four possible diastereomers **91** and **92** in a 37:63 ratio. Separation of the resulting diastereomers, followed by alcoholysis led to the corresponding enantiomerically pure diols.

4.2 Lewis Acid-Mediated Aldol Reaction

Marumoto et al. of Sankyo Co. have shown that aldol addition of the lithium enolate of malonate ester to various α-alkoxy aldehydes **93** in the presence of zinc chloride gave preferentially *anti*-1,2-diols **94** in high yields (Scheme 39) [49]. In the absence of a Lewis acid, the aldol reaction gave product in low diastereoselectivity (*anti-/syn-*=60:40) and low yield (71%). Addition of $BF_3{\cdot}OEt_2$ gave high diastereoselectivity (*anti-/syn-*=91:9) but lower yield (52%). Other Lewis acids such as $MgBr_2$, $ZnBr_2$ gave lower selectivity. When the reaction was carried out at -98 °C in the presence of $ZnCl_2$, the *anti-* aldol adduct was obtained in excellent yield (89%) and high stereoselectivity (87:13). The stereoselectivity decreased when a bulky silyl group was used in place of the benzyl group as the protecting group. In contrast, when the alkyl group (R_1) was

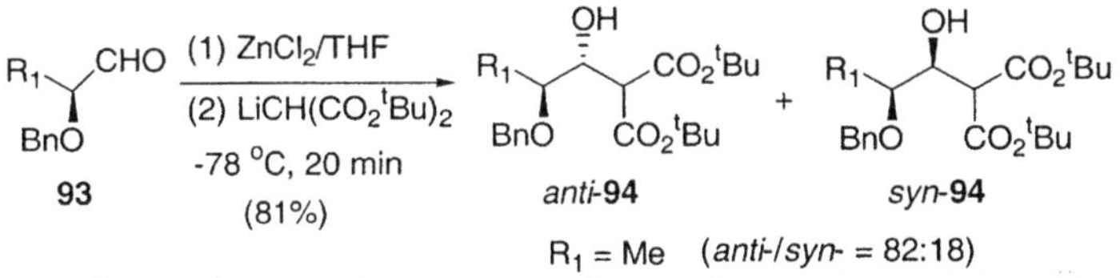

Scheme 39 Lewis acid-mediated aldol reactions (1)

changed from methyl to a more bulky isopropyl or phenyl group, high *anti*-selectivity was obtained.

Interestingly, the stereoselectivity was reversed from *anti*- to *syn*-aldol product with high selectivity (90:10) when the 2-trityloxypropanal **95** was used under the same conditions (Scheme 40).

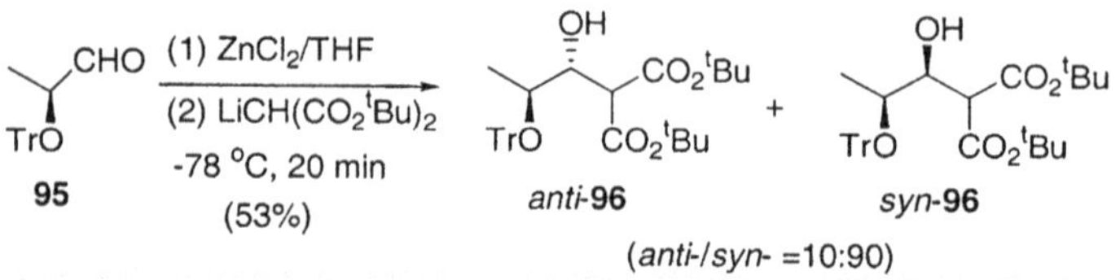

Scheme 40 Lewis acid-mediated aldol reactions (2)

4.3 Catalytic Asymmetric Aldol Condensation

A catalytic asymmetric aldol condensation between the TMS enolate and a benzaldehyde and its application to the synthesis of **SCH 58053**, a cholesterol absorption inhibitor has been reported by a Schering-Plough's Process Development group [50]. Enolization of the spiro ester **97** with LDA followed by trapping with TMSCl gave the TMS enolate **98.** Mukaiyama type of aldol condensation of the TMS enolate with a benzaldehyde in the presence of *D*-valine sulfonamide oxazaborolidine **99** as a chiral catalyst gave the β-hydroxy ester **100** in 90% yield and 95% ee. The hydroxy ester was then converted in high yield to the corresponding spiro-β-lactam, **SCH 58053** (Scheme 41).

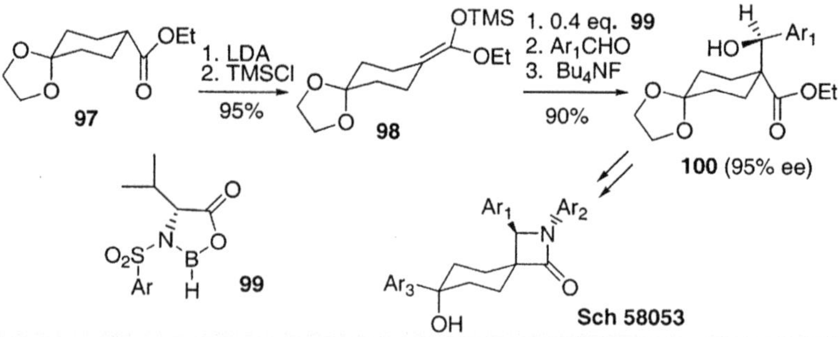

Scheme 41 Oxazaborolidine catalyzed aldol reaction

5 Organolithium in Chiral Conjugated Additions

A number of chemical processes employ chiral conjugated addition in their key steps. For a Michael addition, the chiral template can be placed either on the Michael donor or on the Michael acceptor. A general review of the Michael addition has been published previously [45].

5.1 Conjugated Addition to a Chiral Acceptor

Conjugated addition of a lithium reagent to a chiral Michael acceptor is one of the most effective ways for asymmetric C-C bond formations.

Ebata et al. of Japan Tobacco Inc. have disclosed a process for the synthesis of *trans*-3,4-disubstituted-γ-lactones **101** [51]. The key step was the conjugate addition of alkyl or alkenyl lithium to a chiral levoglucosenone **102** in the presence of copper salts, such as copper halides. Alkyl or alkenyl magnesium could also be used in the addition reaction (Scheme 42).

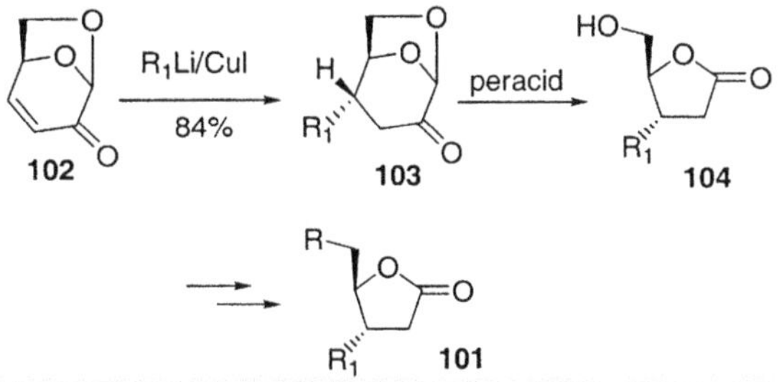

Scheme 42 Conjugate addition to a chiral acceptor

Dunn et al. of Pfizer England have reported an asymmetric synthesis of β-amino acid derivatives by Michael addition of lithium enolate to chiral 2-aminomethylacrylates [52]. The addition of the lithium enolate of cyclopentanecarboxylic acid **105a** to chiral aminomethylacrylates **106** produced the (*S*, *S*, *S*)-glutarates **107a** and **107b** in 83% to 86% yield and 94 to 98% de. The Michael addition of the lithium enolate of methyl cyclopentanecarboxylate **105b** to acrylate

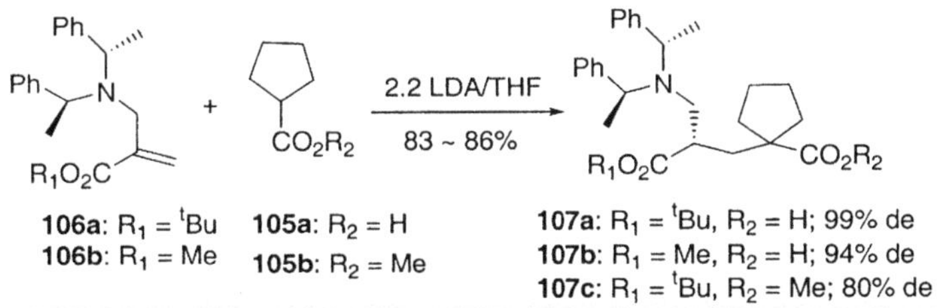

Scheme 43 Michael addition to chiral 2-aminomethylacrylates

gave a glutarate diester **107c** in 50% yield and 80% de. The chiral β-amino acid esters are useful intermediates in the construction of neutral endopeptidase inhibitors (Scheme 43).

5.2 Auxiliary-Induced Asymmetric Conjugate Additions

The use of chiral auxiliaries has been established as an effective method for the control of asymmetric conjugated additions. The asymmetric induction can be achieved by using either a chiral acceptor or a chiral donor.

Song et al. reported that aryllithium reagents react rapidly and efficiently with β-aryl-α, β-unsaturated *tert*-butyl esters bearing a chiral imidazolidine or oxazolidine **108** to give 1,4-addition products **109** in high yield and optical purity [8, 53]. The levels of stereoselectivity in these reactions depend on the structure of the chiral auxiliary, on the substitution in the Michael acceptor/nucleophile, and on the solvent used. For any given combination of aryllithium rea-

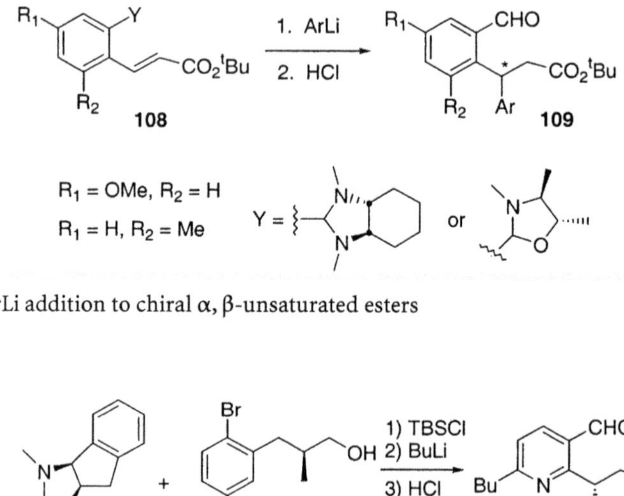

Scheme 44 ArLi addition to chiral α, β-unsaturated esters

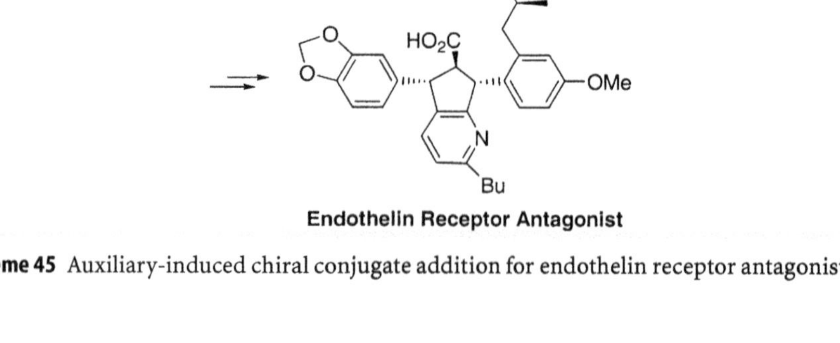

Endothelin Receptor Antagonist

Scheme 45 Auxiliary-induced chiral conjugate addition for endothelin receptor antagonist (1)

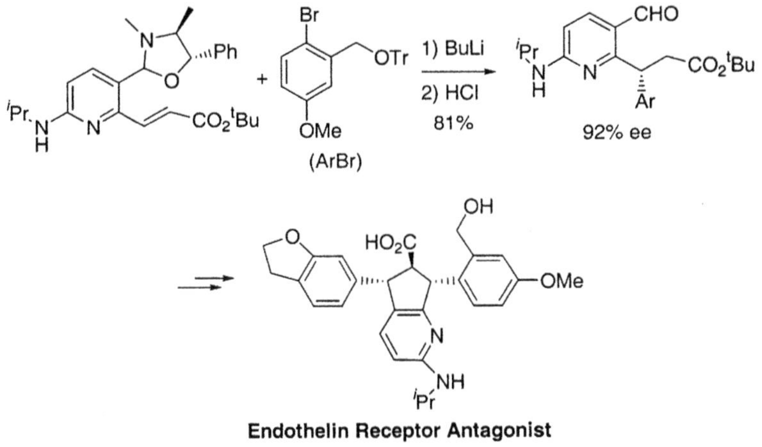

Endothelin Receptor Antagonist

Scheme 46 Auxiliary-induced chiral conjugate addition for endothelin receptor antagonist (2)

gent and acceptor, optimal selectivities could be obtained by matching the auxiliary and solvent effects. This practical chemistry utilizes readily available and inexpensive chiral auxiliaries, and is applicable to prepare a wide variety of structurally complex chiral β, β-diaryl propanoates on large scale (Scheme 44). Successful examples presented by investigators at Merck include the synthesis of key intermediates of several endothelin receptor antagonists (Schemes 45 and 46) [8, 53b].

The alternative strategy of using an auxiliary on the Michael donor has also been developed by Smitrovich et al. of Merck for the asymmetric synthesis of 3-aryl-δ-lactones (Scheme 47) [54].

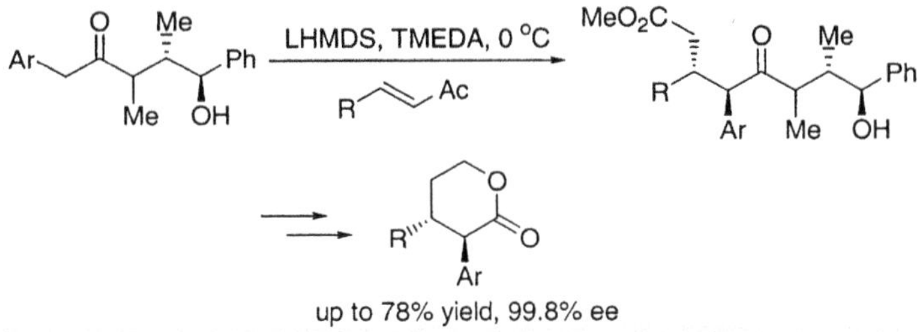

up to 78% yield, 99.8% ee

Scheme 47 Asymmetric Michael reactions in the synthesis of 3-aryl-δ-lactones

5.3
Chiral Additive-Mediated Conjugate Addition

Aryllithium reagents generated from the reaction of arylbromides with *t*-BuLi react efficiently and highly chemoselectively with α,β-unsaturated *tert*-butyl esters in the presence of a range of chiral ligands such as diamines, diethers, or amino ethers to give the β,β-diaryl propanoates with moderate enantioselectivity. Optimal selectivities (up to 88% ee) were obtained using either (-)-sparteine or a C_2-symmetry ligand, 1,2-diphenylethylene dimethylether as a chiral ligand and toluene as a solvent (Scheme 48) [55].

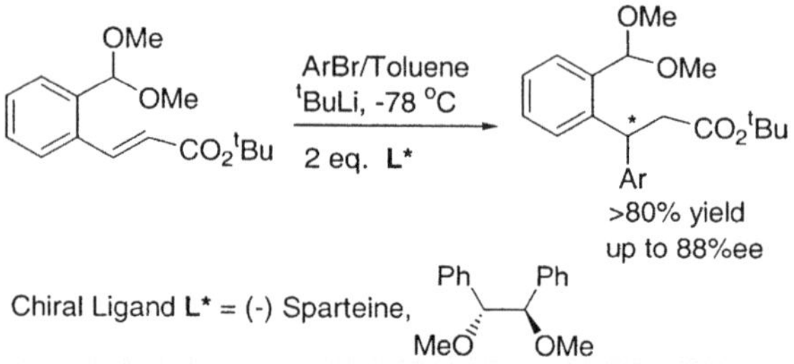

ArBr/Toluene
^{t}BuLi, -78 °C
2 eq. L*
>80% yield
up to 88%ee
Chiral Ligand L* = (-) Sparteine,

Scheme 48 External ligand-mediated asymmetric conjugate addition

5.4
Catalytic Asymmetric Conjugate Addition

A $ZnCl_2$-MAEP complex was demonstrated to catalyze the diastereoselective Michael reaction of the lithium enolate of dioxolane **110** with 2-cyclopenten-1-one for the preparation of contiguous quaternary-tertiary chiral centers [56]. The adduct **112** was isolated out of three possible isomers in 74% yield and 99.0% de. The unsymmetrical triamine ligand 1-(2-dimethylaminoethyl)-4-methylpiperazine (MAEP) itself is inactive to the Michael reaction but it does stabilize the lithium enolate. The formation of the homogeneous reaction mixture prior to the addition of cyclopentenone is critical for high selectivity. The Michael addition product was used to prepare multi-kilograms of a muscarinic receptor antagonist (Scheme 49).

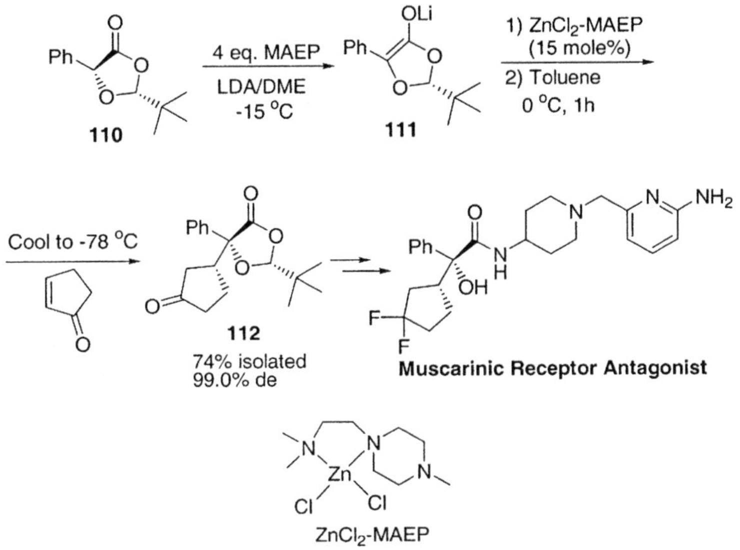

Scheme 49 Catalytic asymmetric conjugate addition

6 Conclusion

Organolithium reagents are one of the most powerful, versatile, and widely used reagents. In addition to the commercially available reagents, such as BuLi, LDA, PhLi, and LiHMDS, a variety of other lithium species can be generated in situ in excellent yield. These lithium species play an important role in chemical processes, particularly in asymmetric syntheses and, as a result, have been the subject of extensive research in academia and industry. In the past decade, many of the asymmetric processes involving organolithium reagents have been developed and used in multi-kilo to multi-hundred kilo processes to produce a number of pharmaceutical products for clinical studies. Some of these chemical processes have made their way to commercial production. Some applications are drawn directly from literature precedents, but as we have attempted to show in this review there are also a number of industry examples of pioneering work in this field.

Application of organometallic chemistry particularly organolithium chemistry to asymmetric syntheses will continue to be the focus of industrial research. With a better understanding of the reaction mechanisms, there will be more new discoveries of practical and efficient asymmetric processes.

Acknowledgements We thank Drs. Michael Mitchell and Doris Schumacher for their support and proof-reading of the manuscript.

References

1. (a) O'Brien P (1998) J Chem Soc Perkin Trans 1 8:1439; (b) Reed F (1991) Speciality Chemicals 11:148; (c) Fehr C (1996) Angew Chem Intl Ed 35:2566
2. Rathman TL (1993) Chimica Oggi 11:15
3. (a) Senanayake CH (1998) Aldrichimica Acta 31:3; (b) Meyers AI (1998) J Heterocyclic Chem 35:991; (c) Evans DA (1982) Aldrichimica Acta 15:23
4. Holla EW, Napierski B, Rebenstock HP (1994) Synlett 333
5. Hilpert H (2001) Tetrahedron 57:7675
6. (a) Askin D, Wallace MA, Vacca JP, Reamer RA, Volante RP, Shinkai I (1992) J Org Chem 57:2771; (b) Askin D, Eng KK, Rossen K, Purick RM, Wells KM, Volante RP, Reider PJ (1994) Tetrahedron Lett 35:673; (c) Maligres PE, Upadhyay V, Rossen K, Cianciosi SJ, Purick RM, Eng KK, Reamer RA, Askin D, Volante RP, Reider PJ (1995) Tetrahedron Lett 36:2195; (d) Maligres PE, Weissman SA, Upadhyay V, Cianciosi SJ, Reamer RA, Purick RM, Sager J, Rossen K, Eng KK, Askin D, Volante RP, Reider PJ (1996) Tetrahedron 52:3327; (e) Askin D, Volante RP, Ryan KM, Reamer RA, Shinkai I (1988) Tetrahedron Lett 34:4245
7. Lee J, Choi WB, Lynch JE, Volante RP, Reider PJ (1998) Tetrahedron Lett 39:3679
8. Song ZJ, Zhao MZ, Desmond R, Devine P, Tschaen DM, Tillyer R, Frey L, Heid R, Xu F, Foster B, Li J, Reamer R, Volante R, Grabowski EJJ, Dolling UH, Reider, PJ, Okada S, Kato Y, Mano E (1999) J Org Chem 64:9658
9. Kress MH, Yang C, Yasuda, N, Grabowski EJJ (1997) Tetrahedron Lett 38:2633
10. (a) Dragovich PS, Prins TJ, Zhou R (1997) J Org Chem 62:7872; (b) Sandham D, Taylor R, Carey JS, Fassler A (2000) Tetrahedron Lett 41:10,091; (c) Myers AG, Yang BH, Chen H, McKinstry L, Kopecky DJ, Gleason JL (1997) J Am Chem Soc 119:6496
11. Armstrong JD III, McWilliams C (1997) US Patent 5 977 371
12. Hanessian S, Margarita R (1998) Tetrahedron Lett 39:5887
13. Tian Q, Nayyar NK, Babu S, Chen L, Tao J, Lee S, Tibbetts A, Moran T, Liou J, Guo M, Kennedy TP (2001) Tetrahedron Lett 42:6807
14. (a) Frutos RP, Stehle S, Nummy L, Yee N (2001) Tetrahedron: Asymmetry 12:101; (b) Yee N (2000) Org Lett 2:2781
15. Napolitano E, Farina V (2001) Tetrahedron Lett 42:3231
16. Chen L, Zaks A, Chackalamannil S, Dugar S (1996) J Org Chem 61:8341
17. (a) Dolling UH, Davis P, Grabowski EJJ (1984) J Am Chem Soc 106:446; (b) O'Donnell MJ (2001) Aldrichimica Acta 34:3; (c) Lim SH, Ma S, Beak P (2001) J Org Chem 66:9056; (d) Serino C, Stehle N, Park YS, Florio S, Beak P (1999) J Org Chem 64:1160; (e) Thayumanavan S, Basu A, Beak P (1997) J Am Chem Soc 119:8209; (f) Matsuo J, Odashima K, Kobayashi S (1999) Org Lett 1:345. (g) Yamashita Y, Odashima K, Koga K (1999) Tetrahedron Lett 40:2803; (h) Murakata M, Yasukata T, Aoki T, Nakajima M, Koga K (1998) Tetrahedron 54:2449; (i) Sato D, Kawasaki H, Shimado I, Arata Y, Okamura K, Date T, Koga K (1997) Tetrahedron 53:7191; (j) Koga K (1994) Pure Appl Chem 66:1487; (k) Imai M, Hagihara A, Kawasaki H, Manabe K, Koga K (1994) J Am Chem Soc 116:8829; (l) Berrisford DJ (1995) Angew Chem Int Ed Engl 34:178; (m) Hoppe I, Marsch M, Harms K, Boche G, Hoppe D (1995) Angew Chem Int Ed Engl 34:3419
18. (a) Kuo S-C, Chen F, Hou D, Kim-Meade A, Bernard C, Liu J, Wu G (2002) Org Chem 68:4984 (b) Kuo, S-C, Bernard C, Chen F, Hou D, Kim-Meade A, Wu G (2001) US Patent 6,307,048
19. Enders D, Reinhold U (1997) Tetrahedron Asymmetry 8:1895
20. Bloch R (1998) Chem Rev 98:1407
21. Kobayashi S, Ishitani H (1999) Chem Rev 99:1069
22. Pridgen LN, Mokhallalati MK, Mcguire MA (1997) Tetrahedron Lett 38:1275
23. Krishnamurthy D, Han Z, Wald SA, Senanayake CH (2002) Tetrahedron Lett 43:2331
24. Wu G, Wong Y, Chen X, Ding Z (1999) J Org Chem 64:3714
25. Gilchrist JH, Collum DB (1992) J Am Chem Soc 114:794

26. McGarrity JF, Ogle CA, Brich Z, Loosli HR (1985) J Am Chem Soc 107:1810
27. Georg GI, Kant J, Gill HS (1987) 109:1129
28. Wu G (2002) Org Pro Res Dev 4:298
29. Wu G, Schumacher DP, Tormos W, Clark JE, Murphy BL (1997) J Org Chem 62:2996
30. Parsons RL Jr (2000) Curr Opin Drug Discov Dev 3:783
31. Huffman MA, Yasuda N, DeCamp AE, Grabowski EJJ (1999) J Org Chem 60:1590
32. Xu F, Reamer RA, Tillyer R, Cummins JM, Grabowski EJJ, Reider PJ, Collum DB, Huffman J (2000) J Am Chem Soc 122:11,212
33. Tan L, Chen C, Tillyer RD, Grabowski EJJ, Reider PJ (1999) Angew Chem Int Ed 38:711
34. Thompson A, Corley EG, Huntington M, Grabowski EJJ, Reamer JF, Collum DB (1998) J Am Chem Soc 120:2028
35. Pierce ME, Parsons RL, Radesca LA, Lo YS, Silverman S, Moore JR, Islam Q, Choudhury A, Fortunak JMD, Nguyen D, Luo C, Morgan SJ, Davis WP, Confalone PN, Chen C, Tillyer R, Frey L, Tan L, Xu F, Zhao D, Thompson A, Corley EG, Grabowski EJJ, Reamer JF, Reider PJ (1998) J Org Chem 63:8536
36. Thompson A, Corley EG, Huntington M, Grabowski EJJ (1995) Tetrahedron Lett 36:8937
37. Parsons RL, Fortunak JM, Dorow RL, Harris GD, Kauffman GS, Nugent WA, Winemiller MD, Briggs TF, Xiang B, Collum DB (2001) J Am Chem Soc 123:9135
38. (a) Kauffman GS, Harris GD, Dorow RL, Stone BRP, Parsons RL, Pesti JA Jr, Magnus NA, Fortunak JM, Confalone PN, Nugent WA (2000) Org Lett 2:3119; (b) Parsons RL, Dorow RL, Davulcu AH, Fortunak JM, Harris GD, Kauffman GS, Nugent WA, Radesca LA (2000) PCT Int Appl WO 01/70707
39. Klein JP, Leigh AJ, Michnick J, Kumar AM, Underiner GE (1995) PCT Int Appl WO 9531450
40. Goehring W, Gokhale S, Hilpert H, Roessler F, Schlageter M, Vogt P (1996) Chimia 50:532
41. Onishi T, Hirose T, Nakano T, Nakazawa M, Izawa K (2001) Tetrahedron Lett 42:5883
42. Onishi T, Nakano T, Hirose T, Nakazawa M, Izawa K (2001) Tetrahedron Lett 42:5887
43. Chang S-J, Stuk TL (2001) Synth Commun 30:955
44. Juaristi E, Beck AK, Hansen J, Matt T, Mukhopadhyay T, Simson M, Seebach D (1993) Synthesis 1271
45. Palomo C, Oiarbide M, Garcia J (2002) Chem Eur J 8:37
46. Iwanowicz EJ, Blomgren P, Cheng PTW, Smith K, Lau WF, Pan YY, Gu HH, Malley MF, Gougoutas JZ (1998) Synlett 664
47. Jacobson IC, Reddy GP (1996) Tetrahedron Lett 37:8263
48. Greiner A, Ortholand J-Y (1992) Tetrahedron Lett 33:1897
49. Marumoto S, Kogen H, Naruto S (1998) Chem Commun 2253
50. Wu G, Tormos W (1997) J Org Chem 62:6412
51. Ebata T, Matsushita H, Kawakami H, Koseki K (1991) Eur Pat Appl EP 460,413
52. Barnish IT, Corless M, Dunn PJ, Ellis D, Finn PW, Hardstone JD, James K (1993) Tetrahedron Lett 34:1323
53. (a) Frey LF, Tillyer RD, Caille A-S, Tschaen DM, Dolling U-H, Grabowski EJJ, Reider PJ (1998) J Org Chem 63:3120; (b) Song ZJ, Zhao MZ, Frey L, Li J, Tan L, Chen CY, Tschaen DM, Tillyer R, Grabowski EJJ, Volante R, Reider PJ, Kato Y, Okada S, Nemoto T, Sato H, Akao A, Mase T (2001) Org Lett 3:3357
54. Smitrovich JH, Boice GN, Qu C, DiMichele L, Nelson TD, Huffman MA, Murry J, McNamara J, Reider PJ (2002) Org Lett 4:1963
55. Xu F, Tillyer RD, Tschaen DM, Grabowski EJJ, Reider PJ (1998) Tetrahedron Asymmetry 9:1651
56. Mase T, Houpis I, Akao A, Dorziotis I, Emerson K, Hoang T, Iida T, Itoh T, Kamei K, Kato S, Kato Y, Kawasaki M, Lang F, Lee J, Lynch J, Maligres P, Molina A, Nemoto T, Okada S, Reamer R, Song ZJ, Tschaen DM, Wada T, Zewge D, Volante R, Reider, PJ, Tomimoto K (2001) J Org Chem 66:6775

Topics Organomet Chem (2004) 6: 37–61
DOI 10.1007/978-3-540-36966-0

Applications of Organotitanium Reagents

David L. Hughes

Department of Process Research, Merck and Co. Inc., Rahway, NJ 07065, USA
E-mail: Dave_Hughes@Merck.com

Abstract Organometallic transformations involving titanium are employed both catalytically and stoichiometrically in the pharmaceutical and fine chemical industry. The current review outlines a number of examples of the industrial applications of organotitanium reagents over the past decade. In addition, several reactions are briefly reviewed for which no industrial applications have been reported, but for which commercial utility may be envisaged. The review is organized by reaction type: carbon-carbon bond formation, carbon-heteroatom bond formation, oxidations, reductions, and hydrolysis/ester formation.

Keywords Organotitanium · Asymmetric catalysis · Pharmaceutical applications · Oxidation · Reduction

List of Abbreviations

BINOL 1,1-Binaphth-2-ol
Cp Cyclopentadienyl
de Diasteromeric excess
ee Enantiomeric excess
TADDOL Tetraaryl-1,3-dioxolane-4,5-dimethanol
TMEDA Tetramethylethylenediamine

1 Introduction

Titanium is the first member of the d-block transition metals and has four valence electrons. Titanium(II), (III), and (IV) are the readily accessible oxidation states, with the (IV) oxidation state being most common for organometallic complexes. Titanium(IV) accomodates up to six ligands, and many of these complexes form dimeric structures or aggregates via bridging ligands. Titanium is highly oxophilic, a characteristic that is nearly always an important factor in reactions in which it is involved. Being relatively inexpensive and generally having low toxicity, titanium compounds are of utility in both stoichiometric as well as catalytic reactions.

The use of titanium in industrial organic chemistry dates back to the 1950s when Ziegler discovered $TiCl_4$ in combination with Me_3Al polymerized ethylene under mild conditions (room temperature and 1 atm) [1]. This spurred a massive development effort, leading to the commercialization of the new polymerization technique and the birth of the polymer industry. Ziegler and Natta were awarded the Nobel Prize in 1963 in recognition of their pioneering contributions to the field. Over the past four decades, an intense research effort has been directed toward the understanding of the Ziegler-Natta polymerization process, which in turn has contributed to the development of other applications of organotitanium chemistry, both academically and industrially.

Titanium was also involved in what may be considered the beginning of modern asymmetric catalysis, when Sharpless and Katsuki reported in 1980 the asymmetric epoxidation of allylic alcohols catalyzed by a titanium-tartrate complex [2]. Up to this time, it was usual to rely upon asymmetry already present in the substrate to control the stereochemistry of asymmetric reactions. The significant aspect of the Sharpless-Katsuki discovery was that the stereochemistry of the epoxidation product could be predicted and controlled based on which enantiomer of the reagent (tartrate) was used. Thus, the stereochemical information inherent in the reagent could be transferred to a pro-chiral substrate, and either product enantiomer could be generated depending on which reagent enantiomer was used in the reaction. This work led in part to the awarding of the Nobel Prize to Sharpless in 2001.

The use of titanium reagents in organic synthesis is widespread; the Dictionary of Organometallic Compounds has 73 pages devoted to a listing of thousands of organotitanium structures reported in the literature [3]. Several specific and general reviews of organotitanium chemistry have appeared over the past two decades [4–12]. The aim of this review is to describe a number of examples of the use of organotitanium reagents in the pharmaceutical and fine chemical industry over the past decade, emphasizing not only the scientific aspects, but also some of the developmental issues (safety, environmental, economical) resulting from implementation on a larger scale. In addition, several reactions are briefly covered for which no industrial applications have been reported, but for which commercial utility may be envisaged. The review is organized by reaction type: carbon-carbon bond formation, carbon-heteroatom bond formation, oxidations, reductions, and hydrolysis/ester formation, and covers the literature through 2001.

2 Carbon-Carbon Bond Formation

2.1 Titanium Carbanions

2.1.1 *Titanium Enolate Reactions*

Stereoselective aldol reactions have been achieved through the use of metal enolates that provide a rigid framework conducive to high selectivity. In 1991, Evans and co-workers reported that titanium enolates participated in highly diastereoselective aldol reactions with selectivities comparable with those achieved via boron-mediated processes, and generally in higher yield [13]. The syn stereochemistry of the product was rationalized based on a six-membered chair transition state with the oxygens of the enolate and aldehyde coordinated to titanium.

Cooke and co-workers at Glaxo-Wellcome used titanium enolate chemistry in the preparation of an elastase inhibitor aimed at treating respiratory diseases (Scheme 1) [14]. In the original synthesis, a 3:1 diastereoselectivity was obtained via coupling of a silyl ketene acetal **1** with an acyl iminium ion generated by BF_3-etherate [15]. To improve selectivity and avoid the use of an additional protect-

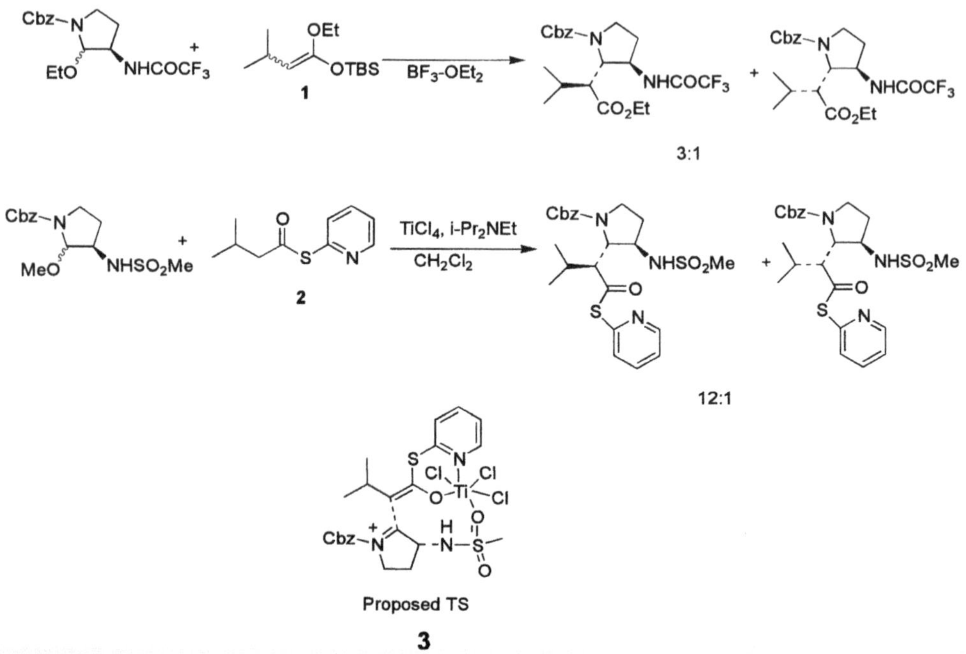

Scheme 1

ing group (trifluoroacetamide), the Glaxo group turned to metal enolate chemistry, with titanium providing the best results. The titanium enolate of the pyridylthioester **2** was prepared at -20 to -10 °C using titanium tetrachloride and *i*-Pr_2NEt. The enolate geometry was determined by NMR experiments, and found to be an 88.5:11.5 mixture of *Z*/*E* enolates at -10 °C. The enolate was unstable at temperatures above -10 °C, leading to lower *Z*/*E* ratios and regenerating the pyridyl ester. On the other hand, the aminal was unreactive at -10 °C, and instead required a temperature of 0–5 °C for the reaction to occur at a reasonable rate. Thus, the reaction temperature for the acyliminium ion reaction was a compromise between the instability of the enolate and its reactivity with the aminal. Due to instability, 2 equivalents of the enolate were required for a rapid reaction rate and optimum yield. At 0–5 °C a 12:1 mixture of product diastereomers was produced. The authors proposed a transition state structure (**3**) wherein titanium is coordinated with the enolate oxygen, the sulfonamide oxygen, and the pyridine of the thioester, thus providing a rigid geometry for the observed selectivity. It would have been of interest to investigate if the pyridine was important for high selectivity, as proposed. The product was isolated by work up with citric acid (to remove titanium from the organic layer) and crystallized from EtOAc/cyclohexane.

Another example where a titanium enolate was used in a stereoselective acyl iminium coupling is the work of Pye and Rossen, from the Merck Process Research group, in the synthesis of an anti-MRSA beta-methyl carbapenem (Scheme 2) [16]. To make use of the readily available and inexpensive acetoxy-

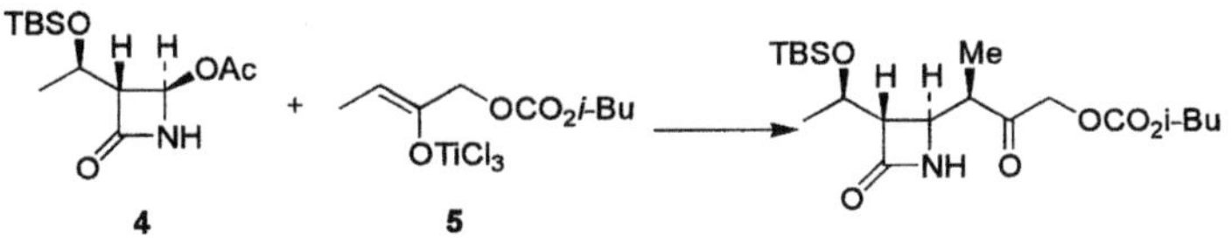

Scheme 2

azetidinone **4**, a 4-carbon coupling partner was required along with a method for installing the methyl group stereoselectively. Extensive studies had been reported in the literature on introduction of the beta-methyl group as part of a 3-carbon synthon, but none for the 4-carbon group required in this case. Enolate or enolate equivalents of the inexpensive 1-hydroxy-2-butanone would afford the desired framework and also provide a handle for further elaboration of the side chain. Somewhat surprisingly, no reports could be found in the literature regarding the enolate chemistry of 1-hydroxy-2-butanone. Probe experiments using the lithium enolate demonstrated coupling was feasible, but a roughly 1:1 mixture of α:β isomers was obtained. After considerable experimentation, the titanium enolate of the hydroxy-butanone, protected as the *iso*-butyl carbonate **5**, provided the desired beta-methyl isomer with excellent 95:5 selectivity in 82% yield. NMR experiments indicated the titanium enolate existed as a 10:1 mixture of *Z*:*E* isomers. This geometry preference is consistent with a cyclic transition state leading to the β-methyl orientation. In the extensive field of β-methyl carbapenem chemistry, which is strewn with lengthy processes having poor selectivity, the elegance of the Pye/Rossen method for introducing the beta-methyl group with high selectivity is especially noteworthy.

The Schering-Plough Process group used titanium enolate chemistry in two approaches to the cholesterol absorption inhibitor, Sch 48461 (Scheme 3) [17]. In the first approach, the titanium enolate of the chiral *N*-acyloxazolidinone **6**,

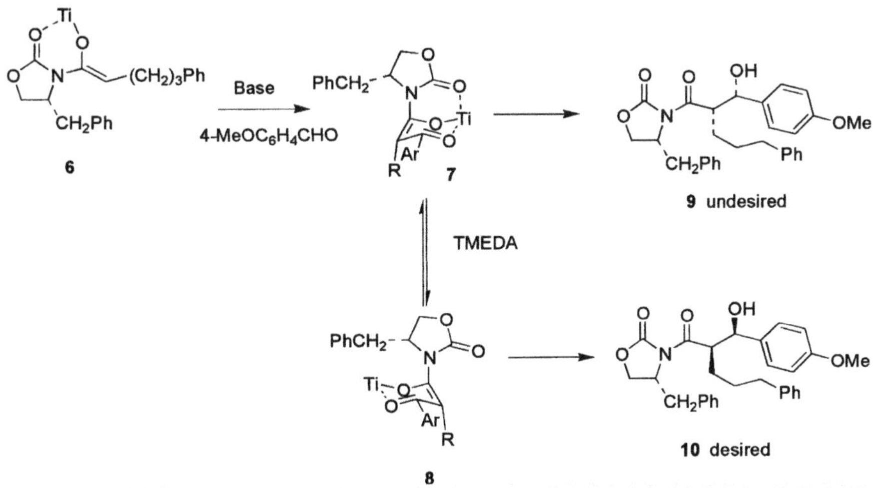

Scheme 3

generated with 2 equiv. i-Pr_2NEt and $TiCl_4$, was reacted with p-anisaldehyde to provide a 1:2 mixture of the two syn aldols **9** and **10**, with the major isomer being the undesired one. When tetramethylethylenediamine (TMEDA) replaced i-Pr_2NEt, a complete reversal was obtained, with the desired isomer generated with >99:1 selectivity. Two equivalents of TMEDA were required to obtain high selectivity, which led the authors to speculate the first equivalent acts as a base to deprotonate the ketone to form the chelated enolate **7**, while the second equivalent serves to disrupt the chelation to provide the open form of the enolate **8**. In line with this hypothesis, use of 1 equiv of Et_3N for deprotonation followed by 1 equiv. of TMEDA also gave high selectivity. On larger scale these were the preferred conditions, since reactions with Et_3N gave better conversion (>97%) compared with the all-TMEDA reactions which proceeded to only 85–90% conversion. Hydrolysis, formation of the (4-methoxyphenyl) anilide, and cyclization provided the beta-lactam drug candidate.

Since the first approach outlined above required use of the unnatural (expensive) enantiomer of the chiral auxiliary and chromatographic purification, a second approach was designed and developed based on an enolate addition to an imine rather than an aldehyde (Scheme 4). Since the imine already contains the p-methoxyphenyl group, this route is more convergent than the aldol route. The titanium enolate generated from the chiral imide using 2 equiv. of TMEDA in dichloromethane gave a 95:5 ratio of product isomers, with the major one being the anti isomer. Thus, the configuration at C-2 is opposite to that in the aldol product, so the desired isomer **11** was obtained via the chiral auxiliary with the opposite and natural (S) configuration. Unlike the aldol reaction, the stereoselectivity was unaffected by the nature of the tertiary amine base, with Et_3N, i-Pr_2NEt, and TMEDA all giving similar selectivities. The observed stereochemistry can be explained by invoking a chelated transition state. Although the selectivity was high (95:5), removal of the minor diastereomer via crystallization was not successful, so the chiral auxiliary was replaced with the commercially available (S)-phenyloxazolidinone, which also provided a 95:5 ratio of anti:syn diastereomers in the titanium imine chemistry. In this case, however, the minor diastereomer could be purged by a single crystallization.

Inconsistent yields and selectivities in these reactions were found to be a function of the workup protocol, as apparently retro-aldol reaction was occurring. Quenching at -20 °C with HOAc, presumably to protonate the Ti-N bond, gave >90% yields with anti:syn ratio of 87:13. The higher selectivity obtained

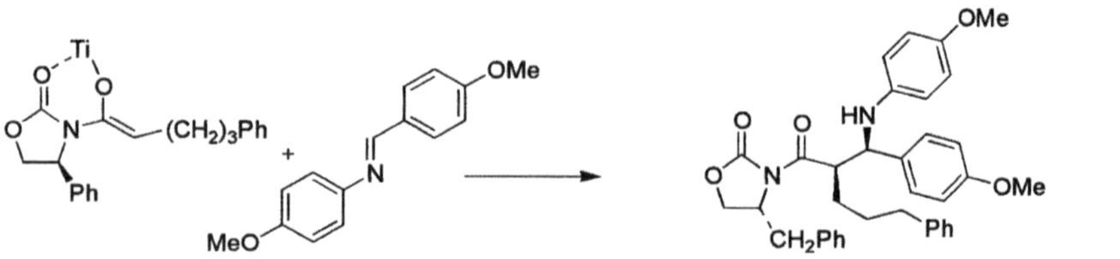

11

Scheme 4

with the room temperature quench is believed to be due to retro-aldol of the minor isomer.

Overall, the final process for Sch 48461 involved just three steps from the chiral auxiliary with a yield of 55–60%.

Using the same chemistry, the titanium enolate reaction with the appropriate imine has also been used for the preparation of Schering-Plough's hypocholesterolemic compound Sch 58235, (ezetimibe, Zetia) for which a new drug application (NDA) was filed in Dec 2001 [18].

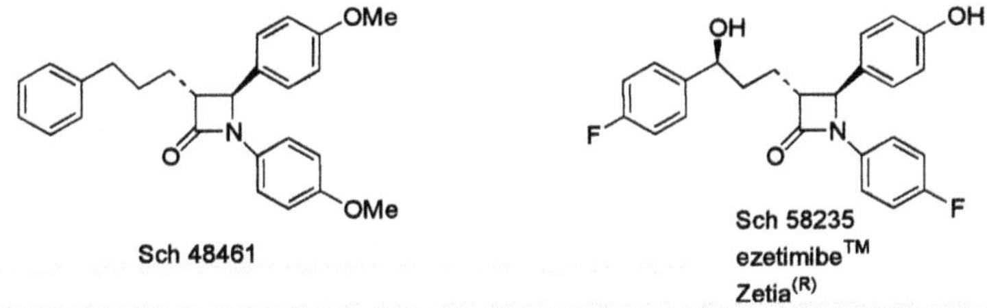

Sch 48461

Sch 58235
ezetimibe™
Zetia®

The Schering-Plough group has also used titanium enolate chemistry in the preparation of the antifungal drug candidate, Sch 56592, which was in Phase III studies in 2001. In this case, alkylation of the imide enolate provided the benzyl ether in 84% yield and 98% de. Triethylamine was used as the base in the generation of the titanium enolate, which was carried out in dichloromethane at 0 °C (Scheme 5) [19].

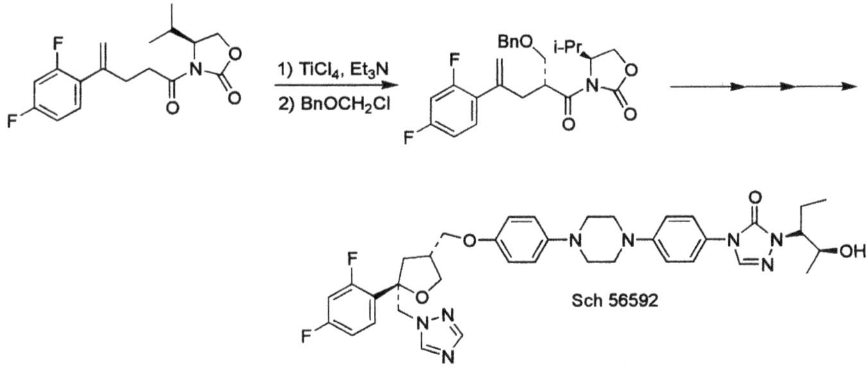

1) $TiCl_4$, Et_3N
2) $BnOCH_2Cl$

Sch 56592

Scheme 5

2.1.2
Titanium-Mediated Homo-Enolate Reactions

Titanium homo-enolates can be generated from 3-iodo-propionates by metalation with Zn-Cu couple followed by transmetalation with titanium, or via opening of silyloxycyclopropanes. The resulting homo-enolates can then be reacted with aldehydes to provide the 3-carbon homologated product [20].

The homo-aldol reaction has been used by Merck Process chemists in the synthesis of HIV-protease inhibitors. In the initial process, the titanium homo-enolate **12** was generated via the iodopropionate with Zn-Cu, followed by transmetalation with trichlorotitanium isopropoxide (Scheme 6). Reaction with the chiral aldehyde provided hydroxy-ester **13** in 82% yield and 16:1 diastereoselectivity. The homo-enolates derived from transmetalation with chlorotitanium triisopropoxide or dichlorotitanium diisopropoxide afforded product with significantly lower selectivities. The diastereoselectivity can be explained via a chelated transition state [21]. Incorporating *cis*-aminoindanol as a chiral auxiliary into the titanium homo-enolate (**14**) afforded the opposite stereochemistry at the 4-position of the product (**15**), apparently as a result of non-chelation control in the transition state. Simple achiral amides were used as substrate probes to determine the effect of the chiral auxiliary on diastereoselectivity. In these cases, the selectivities were very low, indicating that the indanol provides a powerful influence over the stereocontrol in these reactions [22].

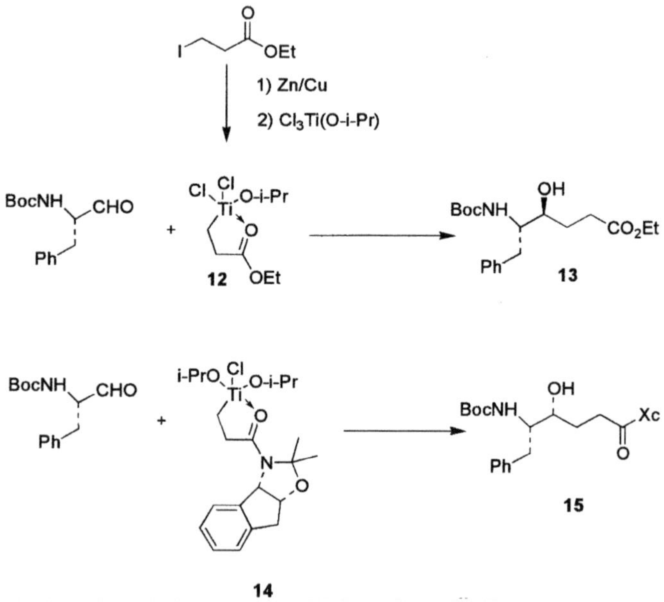

OEt
1) Zn/Cu
2) $Cl_3Ti(O\text{-}i\text{-}Pr)$
BocNH CHO
Ph
Cl Cl O-i-Pr Ti O
12 OEt
BocNH OH CO_2Et
Ph 13
BocNH CHO
Ph
i-PrO Cl O-i-Pr Ti O
N O
14
BocNH OH Xc O
Ph
15

Scheme 6

A major drawback to the above protocol was the use of the Zn-Cu couple for generation of the homo-enolates. An elegant one-pot process was developed in which stereoselective 1,2-migration and the homo-aldol reactions were carried out in tandem [23]. The sequence (Scheme 7) starts with the amide incorporating the chiral auxiliary (**16**), which is deprotonated with BuLi, then transmetalated with bis(iodomethyl)zinc to form **17**. This intermediate rearranges with the aid of added lithium alkoxide to afford the homo-enolate **18** without loss of the stereochemistry at the alpha position. This homoenolate then is transmeta-

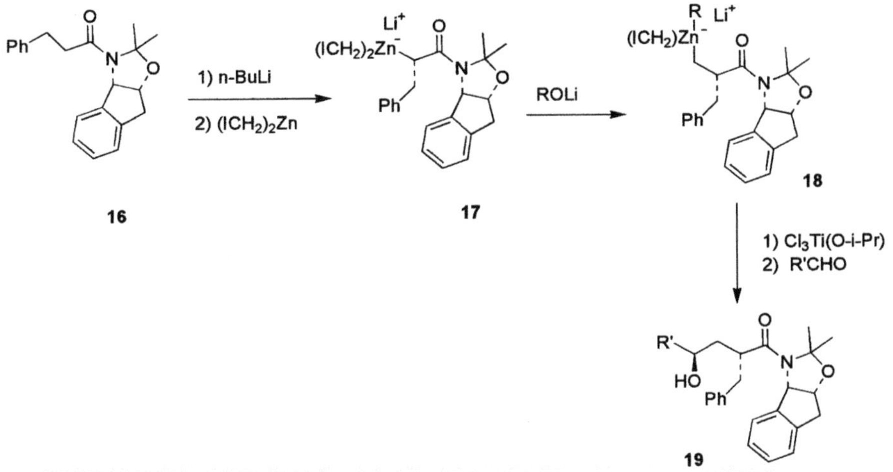

Scheme 7

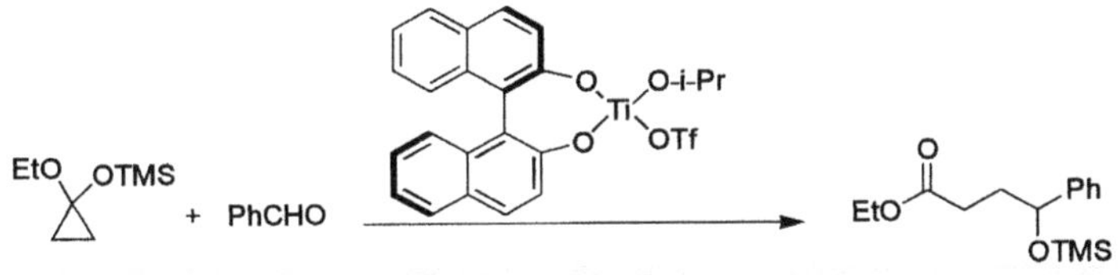

Scheme 8

lated to provide the titanium homoenolate as above, which reacts with a variety of aldehydes to provide the hydroxy-amide **19** with high diastereoselectivity.

The above examples use a chiral auxiliary to control asymmetry in the product and require a stoichiometric quantity of the auxiliary as well as titanium. An asymmetric homo-aldol reaction catalytic (10 mol%) in titanium and ligand has been developed by Martins and Gleason [24]. A chiral ligand (BINOL) was used to promote the reaction, but enantioselectivities were low (15–20%) (Scheme 8). Development of this reaction to provide a catalytic, asymmetric homo-aldol reaction would be an attractive route for preparing asymmetric 1,4-hydroxyacids.

2.2 Titanium Carbenes/Carbenoids

2.2.1 *Tebbe/Petasis Reaction*

Esters, ketones, amides, and other carbonyl compounds can be methylenated using four titanium-based reagents that generate an active titanium carbene: (1) the Tebbe reagent derived from titanocene dichloride and trimethylaluminum [25]; (2) the Grubbs' titanacyclobutanes derived from the Tebbe reagent and an

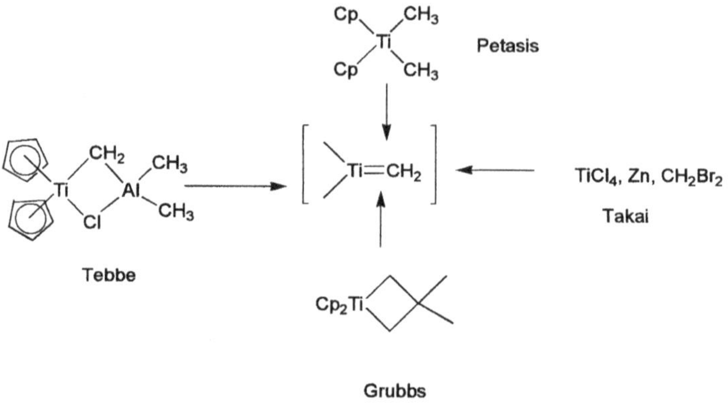

$Cp_2Ti(CH_3)_2$ Petasis

$[Cp_2Ti{=}CH_2]$

$TiCl_4$, Zn, CH_2Br_2 Takai

Tebbe

Cp_2Ti

Grubbs

Scheme 9

appropriate olefin [26]; (3) the Takai reaction using methylene bromide, zinc, and titanium tetrachloride [27]; and (4) dimethyltitanocene, recently pioneered for olefinations by Petasis [28] (Scheme 9).

Of these methods, olefinations using dimethyltitanocene are particularly attractive for larger scale use since the reagent is nonpyrophoric, stable to air and water, and therefore much easier to prepare and use as compared with the Tebbe and Grubbs reagents. In addition, the Tebbe reagent and its reaction byproducts contain Lewis-acidic aluminum which is absent in reactions involving dimethyltitanocene. Consequently, dimethyltitanocene has been used successfully in a wide variety of olefinations where delicate functionalities such as silyl esters, anhydrides, carbonates, imides, and acylsilanes are present in the substrate [28].

The Petasis olefination of ester **20** using dimethyltitanocene was a key step in the synthesis of Merck's Substance P antagonist aprepitant (Scheme 10) [29]. The olefination was used to insert a methylene group which was then stereoselectively reduced to a methyl group (**21**). A number of scientific, technical and safety issues had to be addressed in order to develop a reaction capable of being scaled in a safe and efficient manner, and provides an insight into the process by which a sensitive and potentially unsafe reaction is modified to allow scale up from the lab to pilot plant.

First, a safe way to prepare and use dimethyltitanocene needed to be developed. The literature preparation of dimethyltitanocene requires use of methyllithium, a pyrophoric reagent undesirable for large scale use [30]. Fortunately, methyl Grignard was found to be an effective replacement. The straightforward preparation of the reagent involved reaction of methyl magnesium chloride (3 mol/l in THF) with titanocene dichloride in toluene, followed by an aqueous workup, azeotropic drying, and concentration [31]. Toluene was chosen as the solvent since the methylenation reaction is run in toluene, and this would allow the reagent to be formed and used in toluene without having to carry out a solvent switch or isolation. Although Cp_2TiMe_2 is a crystalline solid, it is unstable as a solid [32], and it is best stored and used as a 20% (ca. 1 mol/l) solution in toluene. In the dark and under a nitrogen atmosphere, the reagent is stable in

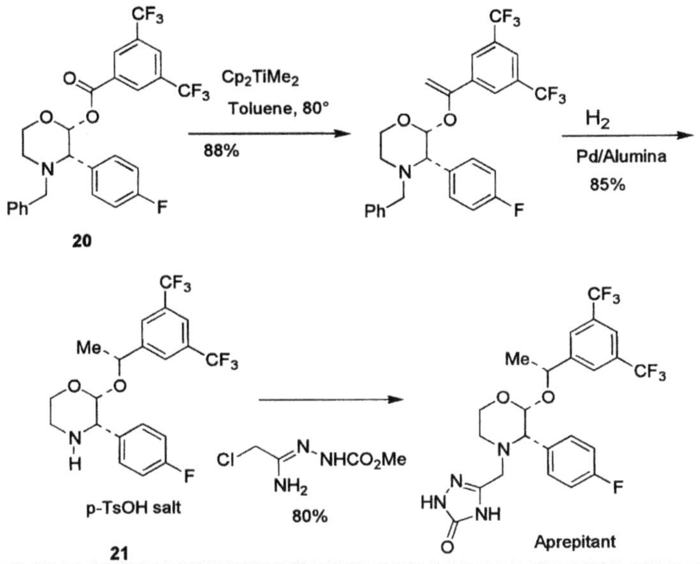

Scheme 10

toluene for weeks at 5 °C. However, a rapid, autocatalytic, and highly exothermic decomposition accompanied by a rapid pressure increase can initiate as low as 40 °C, or even at ambient temperature after several days. This autocatalytic decomposition could be hazardous during the vacuum distillation required to produce the azeotropically dried and concentrated reagent. Use of the in situ formed dimethyltitanocene without an aqueous workup was studied, but no olefination occurred in the presence of the salts, so an aqueous workup was essential to wash out salts. To avoid the potential decomposition of the reagent, an additive was needed to provide stabilization. In the first iteration of the process, the additive best suited for this turned out to be the ester substrate itself. By adding the ester substrate before concentration, the potential decomposition pathway is replaced by a controlled reaction to form product, so the solution can be safely distilled to produce the dried and concentrated solution.

A second issue that arose was that the vinyl ether product began to decompose at the end of the methylenation reaction. Dimethyltitanocene reacts preferentially with the ester when present, but when the ester is depleted, the dimethyltitanocene begins attacking the vinyl ether product. Since heat up and cool down cycles increase on scale up, significant yield loss and impurity generation was occurring during the time required to assay for complete reaction and during the subsequent cool down. The solution to this problem was to add a hindered ester to scavenge unreacted dimethytitanocene [33]. The hindered ester needed to be less reactive than the substrate ester, but more reactive than the vinyl ether product. The ideal hindered ester would not react when the substrate ester was present, but would react once substrate ester had been depleted, sparing the product from decomposition. After experimentation with a number of esters, the readily available 1,1-dimethyl-2-phenylethyl acetate was found to be the best candidate [33]. The small amount of vinyl ether product produced from the hindered ester did not impact work-up and crystallization of the desired product.

In the second iteration of the process, the hindered ester was added to the dimethyltitanocene preparation before concentration instead of the ester substrate itself. This allowed the dimethyltitanocene to be prepared independently of the ester substrate, thus providing a safe process for a contract manufacturer.

A third issue was determining how to deal with the titanium waste. Since the reaction is stoichiometric in titanium, a large amount of titanium byproduct is produced and must be not only separated from the product, but either disposed or recycled. In early versions of the process, the titanium byproducts were oxidized and hydrolyzed to undetermined inorganic titanium solids and filtered from the product mixture. However, given the cost of titanocene dichloride and the expense and environmental concerns with titanium disposal, the titanium reagent needed to be recycled to develop an overall economical and environmentally-sound process. Key to developing a recycle process was the isolation and characterization of the titanium byproducts from the reaction mixture, which revealed that the major product was a titanium dimer, $(Cp_2TiMe)_2O$ [35]. The work-up of the olefination process was modified such that the titanium dimer could be crystallized and isolated. A simple process for converting the dimer to titanocene dichloride was developed which involved treatment of the dimer with HCl in THF and filtration of the resulting dichloride (Scheme 11). Thus, about 70% of the titanocene dichloride could be recovered and recycled by this process [34].

$$Cp_2Ti(Me)O(Me)TiCp_2 + HCl \longrightarrow Cp_2TiCl_2$$

Scheme 11

2.2.2
Asymmetric Cyclopropanation

Charette and co-workers have recently demonstrated the enantioselective cyclopropanation of allylic alcohols mediated by Ti-TADDOLS [36]. The optimized reaction utilizes 1 eq of $Zn(CH_2I)_2$ and 25 mol% of the titanium TADDOL in dichloromethane solvent with molecular sieves, providing the cyclopropane product in 85% yield and 92% ee (Scheme 12). The zinc reagent is prepared from diethylzinc and diiodomethane. The method appears capable of development for larger scale implementation.

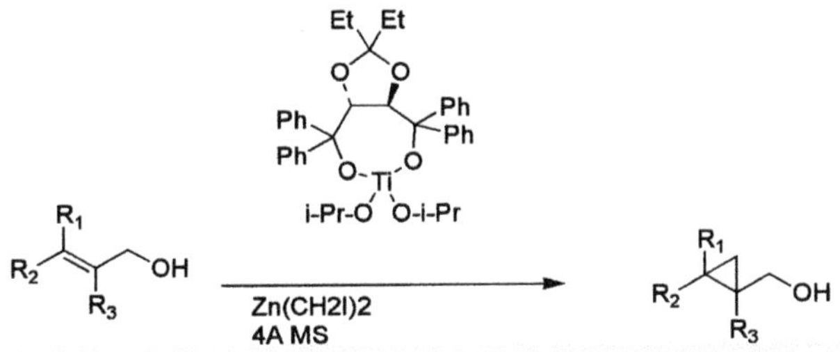

Scheme 12

2.3 Titanium-Mediated Radical and Radical Anion Processes

2.3.1 *Pinacol Couplings*

The pinacol reaction is a reductive coupling of aldehydes mediated by a one-electron reductant, commonly Ti(III) [37]. The reaction proceeds via the coupling of the radical anions formed by reduction of the aldehyde. The asymmetric, catalytic pinacol coupling has been an area of active research in recent years. While progress has been made, most of the reported methods suffer from the use of exotic catalysts at high loadings and by poor product enantioselectivity. One of the most promising catalysts has been reported from the laboratory of Riant (Scheme 13) [38]. The asymmetric ligands are salen Schiff bases derived from readily available aminoalcohols. Asymmetric pinacol couplings of aromatic aldehydes using the low valent Ti-salen catalysts afford the chiral diols with high diastereoselectivity and ee's up to 91% with stoichiometric catalyst. When used in a catalytic mode (2–10%, with TMS-Cl as co-reactant), the ee's drop by 20–40%. Another drawback is the use of 3 equiv. of manganese metal as the terminal reductant, which would be undesirable for commercial applications.

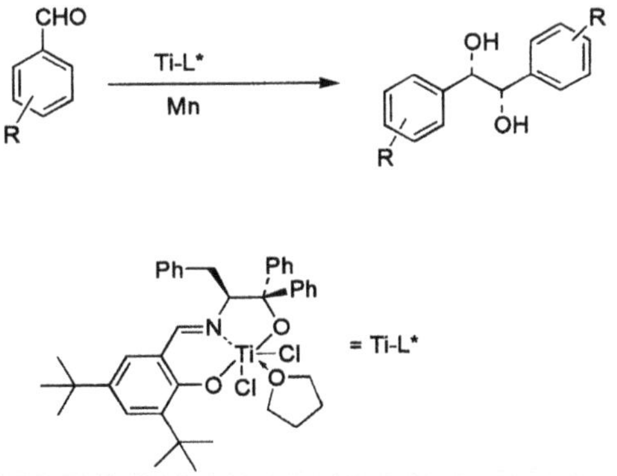

Scheme 13

The asymmetric-variant of the pinacol reaction continues to see rapid improvement and is likely to be further developed into a reaction amenable to scale up.

2.4 Organotitanium Lewis Acid-Mediated Transformations

2.4.1 *Asymmetric Catalytic Cyanohydrin Formation*

Catalytic asymmetric cyanohydrin formation employing transition metals and a variety of chiral ligands have been reported in a number of laboratories over the past 15 years [39].

In conjunction with Professors M. North and Y. Belokon [40–44], the fine chemical company Avecia is commercializing compounds made via asymmetric cyanohydrin chemistry based on catalysts derived from chiral salens. Both Ti and V salen catalysts were screened, with the vanadium catalysts being somewhat more selective but far less active. With the titanium catalysts, catalyst loadings of only 0.1% are needed for effective catalysis. Inconsistent results obtained initially were traced to the amount of water in the reactions, with the best ee's obtained when some water was present [42]. Further investigation revealed that a titanium dimer (**22**) was being formed in the presence of water, and this dimer was isolated and a crystal structure obtained (Scheme 14). The isolated dimers

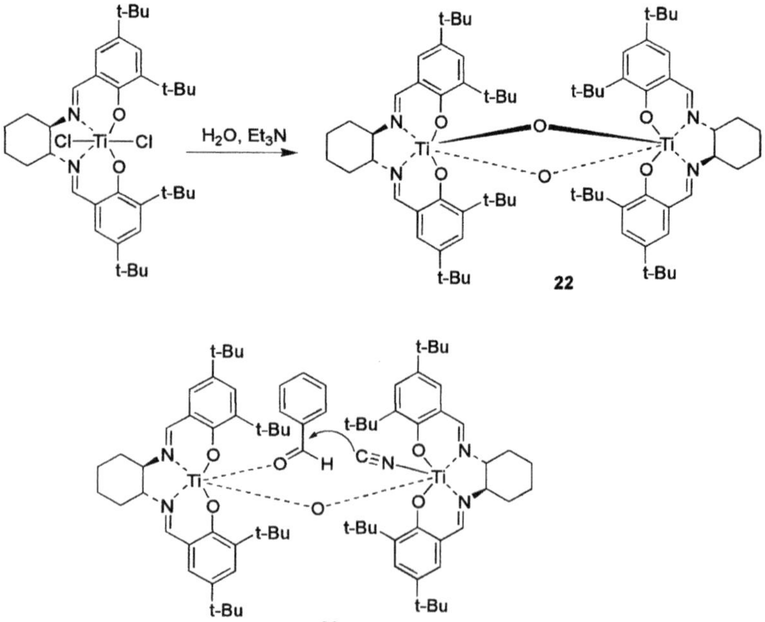

Scheme 14

are themselves active catalysts, and kinetics experiments suggested the dimer is the true catalyst. The authors suggest a transition state (**23**) in which one half of the dimer activates the aldehyde while the other half coordinates cyanide, and that an intramolecular transfer of the cyanide occurs (Scheme 14).

As with all asymmetric cyanohydrin catalysts to date, aldehydes give much better ees than ketones. With the North and Belokon catalysts, however, respectable ee's of 70% were obtained with acetophenone [41].

According to a report at the ChiraSource Symposium in September 2001, and abstracted in C&EN, Avecia is producing (*R*)-2-chloromandelic acid (**24**) via this chemistry as well as a nicotine-derived aminoalcohol **25** (Scheme 15) [45].

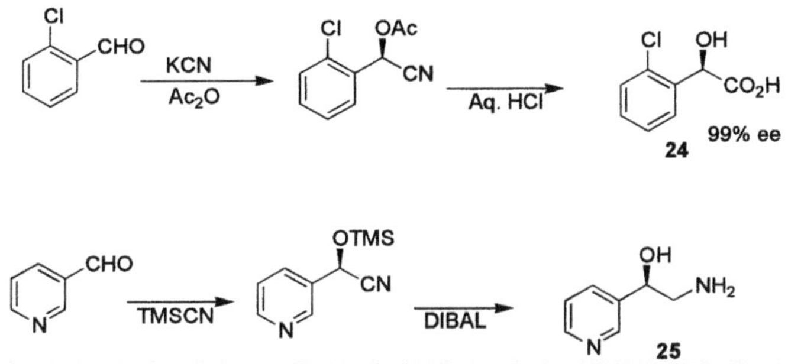

Cl CHO; KCN, Ac_2O → Cl OAc CN; Aq. HCl → Cl OH CO_2H **24** 99% ee

CHO (N); TMSCN → OTMS CN (N); DIBAL → OH NH_2 (N) **25**

Scheme 15

2.4.2 *Catalytic Asymmetric Ene Reactions*

Mikami and co-workers have developed a chiral titanium catalyst based on chiral 1,1′-binaphth-2-ol for the asymmetric carbonyl-ene reaction. With methyl glyoxylate, ees >95% were obtained with several 1,1-disubstituted olefins. For the best results, the catalyst is prepared in situ in the presence of molecular sieves from diisopropoxy titanium dihalides (Cl or Br) and optically pure BINOL or 6-Br-BINOL [46].

The ene-reaction was applied by Hoffman-La Roche chemists as the first step in the synthesis of the matrix metalloproteinase inhibitor Trocade (Scheme 16) [47, 48]. They found the enantioselectivity of the reaction was highly sensitive to the water level in the molecular sieves used in the catalyst preparation. With wet sieves (6–10 wt%), ee's in the 95–98% range were obtained compared to only 20–30% when activated, dry sieves were used. While this observation was not pursued further, the effect of water on the reaction is similar to that observed in the asymmetric cyanohydrin reactions studied by North and Belokon [40–44]. As noted above, these workers determined that titanium dimers were involved, and the same is likely in this case.

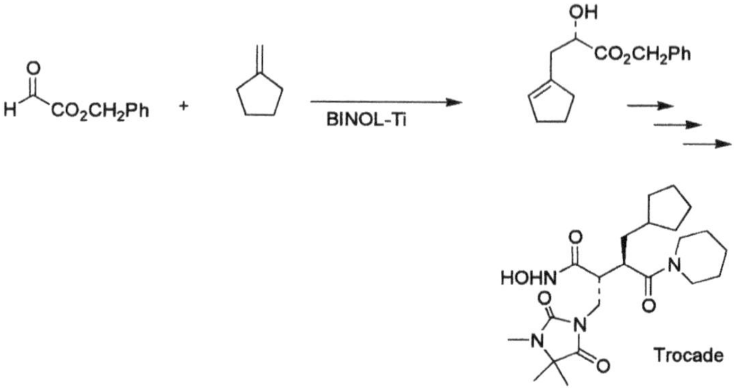

Scheme 16

2.4.3
Mukaiyama-Type Aldol Reactions

In 1990 Mukaiyama and co-workers reported the titanium-BINOL mediated asymmetric aldol reaction of silyl ketene thioacetals with aromatic aldehydes, affording the aldol product in up to 85% ee [49] (Scheme 17). Mikami and Keck have extensively developed and expanded the scope of the Ti-BINOL aldol reactions over the past decade [50]. The ready availability of both enantiomers of BINOL make this reaction one that should find application in the pharmaceutical and fine chemical industry.

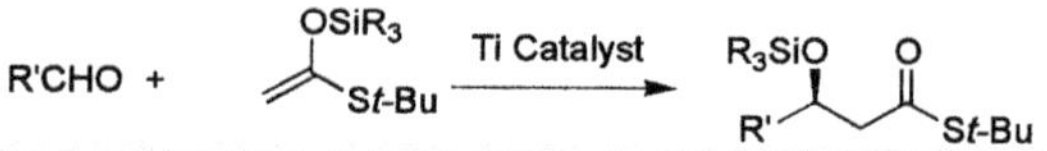

Scheme 17

2.4.4
Diels-Alder Reactions

Titanium TADDOL complexes have been used extensively to promote asymmetric Diels-Alder reactions with a wide range of reacting partners in both intra- and inter-molecular reactions [12, 51]. This reaction has been used by Rhone-Poulenc in the synthesis of the Substance P antagonist RPR 107880 (Scheme 18) [52] Reaction of *N*-benzylmaleimide with 1-acetoxyisoprene with a stoichiometric amount of Ti-TADDOL (prepared from 2 equiv. $Ti(O\text{-}i\text{-}Pr)_4$ and 1 equiv. TADDOL) provided the Diels-Alder product with 90% ee (using the Np-TADDOL) and 90% yield. Interestingly, the major product (**26**) has lost the acetoxy group. The minor product (**27**), which contains the acetoxy group, is present at the 5–10% level depending on the conditions and is nearly racemic. This is strong evidence that pre-coordination of the two reacting partners is vital for high selectivity. In an attempt to make the reaction catalytic, 20 mol% NpTAD-

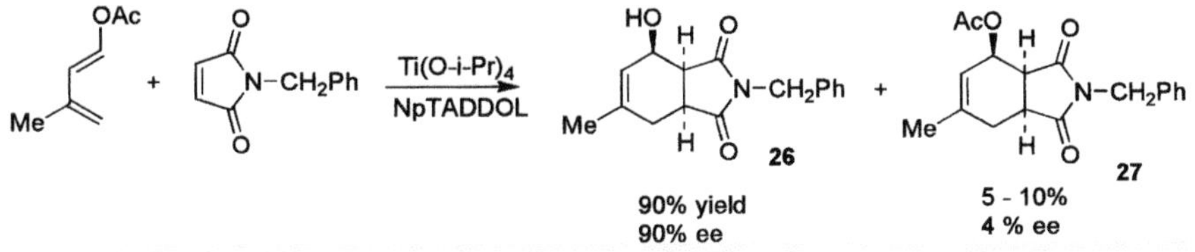

Scheme 18

DOL was tried, but the ee dropped to 30%, suggesting the TADDOL ligand remains strongly bound to the product and does not exchange with excess Ti(O-*i*-Pr)$_4$ to regenerate the chiral catalyst.

3 Formation of Carbon-Heteroatom Bonds

3.1 Hetero Diels-Alder Reaction

Hetero Diels Alder reactions provide an efficient route to dihydropyran derivatives. Asymmetric variants of the hetero Diels Alder reaction catalyzed by titanium diol complexes have been pioneered by the Mikami group [53] and the fine chemical company Takasago holds patents to this chemistry [53]. In a recent report, Chinese scientists have optimized the reaction such that catalyst loadings of <0.1% afford dihydropyrans with high ee's under solvent free conditions at room temperature (Scheme 19) [54]. This highly practical reaction should be readily scalable.

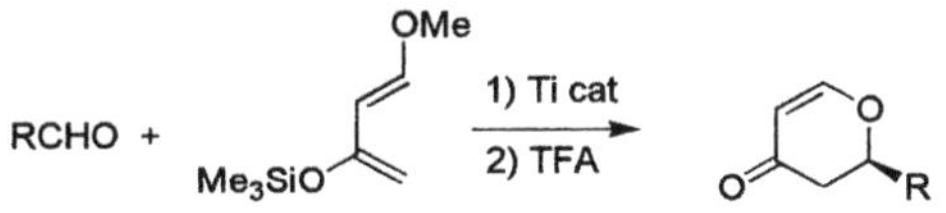

Scheme 19

4 Oxidation Reactions

4.1 Sharpless-Katsuki Asymmetric Epoxidation

Discovered in the late 1970s and reported in 1980 by Katsuki and Sharpless, the epoxidation of primary allylic alcohols was a breakthrough in modern catalytic asymmetric synthesis in that the product chirality is controlled by the asymmetry of the reagent (tartrate) [3]. Thus, either enantiomer of the product can be obtained depending on which enantiomer of the reagent is used. The original conditions used stoichiometric reagents, but the reaction was made catalytic when water was scavenged with molecular sieves. Epoxidations on unsubstituted or *trans*-allylic alcohols generally give product ee's >90%, but poor results are obtained with *cis*-allylic alcohols. The titanium-tartrate complex was shown to be dimeric by X-ray crystallography, which has led Sharpless to propose a transition state model **28**, with titanium coordinated to the peroxide, the hydroxyl of the allylic alcohol, the two hydroxyls of the tartrate, and one of the hydroxyls of the tartrate from the second titanium.

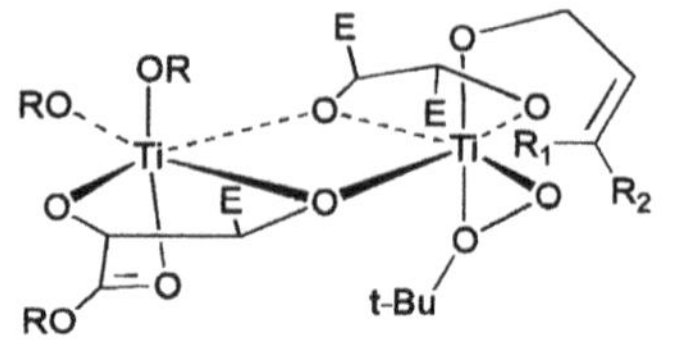

28

Thiamphenicol and florfenicol are broad spectrum antibiotics used for both Gram negative and Gram positive bacterial infections. The original commercial synthesis employed a resolution of an aminoalcohol intermediate using the unnatural isomer of tartaric acid. Wu and co-workers at Schering-Plough have recently devised and developed an asymmetric synthesis of florfenicol wherein the asymmetry is introduced using the Sharpless-Katsuki oxidation [55]. The overall sequence is shown in Scheme 20. The epoxidation of the *E*-allylic alcohol **29** using *L*-diisopropyl tartrate, titanium isopropoxide, *tert*-butylhydroperoxide and sieves provided a 97% ee when 20 mol% catalyst was used, and 90% ee with 10 mol% catalyst. The epoxyalcohol **30** was opened using a sequential addition of sodium hydride, zinc chloride, and dichloroacetonitrile to afford the oxazoline **31**. Since the epoxidation was only effective with the *trans*-alkene, the epoxide which was produced had the opposite stereochemistry from the desired

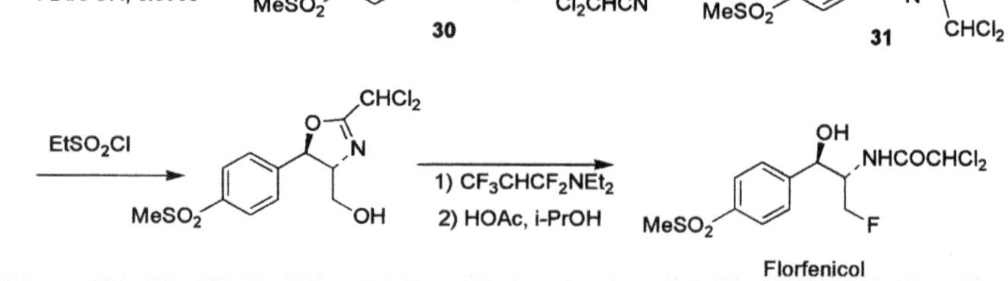

Scheme 20

product. Therefore, the oxazoline **31** was inverted using ethanesulfonyl chloride in a nifty intramolecular reaction. This product was fluorinated and hydrolyzed to provide florfenicol.

Schering-Plough chemists have used the Katsuki-Sharpless epoxidation in the synthesis of an antifungal agent denoted Sch 42427 [56]. The epoxidation of the *cis*-olefin **32** affords epoxide **33** with an ee of 76%, which is in line with other examples for *cis*-olefins, where the ee's are generally lower than for the *trans* analogs. In the Sumitomo synthesis of anti-fungal agent SM-8668 [57], the *trans* analog **34** was epoxidized with an ee >95% (Scheme 21).

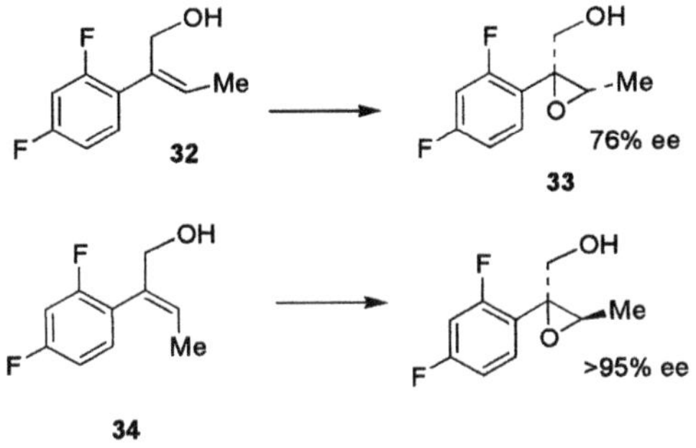

Scheme 21

4.2 Asymmetric Sulfide Oxidation

In 1984 Kagan reported the stoichiometric asymmetric oxidation of sulfides to sulfoxides using a modification of the titanium reagent used by Sharpless for allylic alcohol epoxidations [58]. The reagent was generated from titanium tetraisopropoxide, diethyl tartrate, and water in a 1:2:1 ratio, and *tert*-butyl hydroperoxide was the terminal oxidant in methylene chloride as solvent. In 1987 the same group reported that ee's were improved with a number of substrates by using cumene hydroperoxide as oxidant [59], with an ee of 96% achieved for oxidation of methyl *p*-tolyl sulfide. Obtaining reproducible results required that a carefully optimized protocol be followed, including appropriate addition order and strict age times [60]. Making the reaction catalytic involved the addition of molecular sieves to help control water, with ee's up to 88% obtained with 20 mol% catalyst [61]. The catalyst loading was further reduced to 10 mol% by replacing water with isopropyl alcohol.

The catalytic species is not known, but it is speculated that the active catalyst involves a tridendate tartrate and an η2-coordinated hydroperoxide [58].

The most important industrial application of the titanium-mediated asymmetric sulfoxidation is the production of the proton-pump inhibitor esomeprazole, (Nexium), the single enantiomer of omeprazole (Prilosec) [62]. (Scheme 22) As an effective treatment for acid-reflux and the prevention of ulcers, Prilosec was one of the top selling drugs in the world in 2001 (>$6 billion).

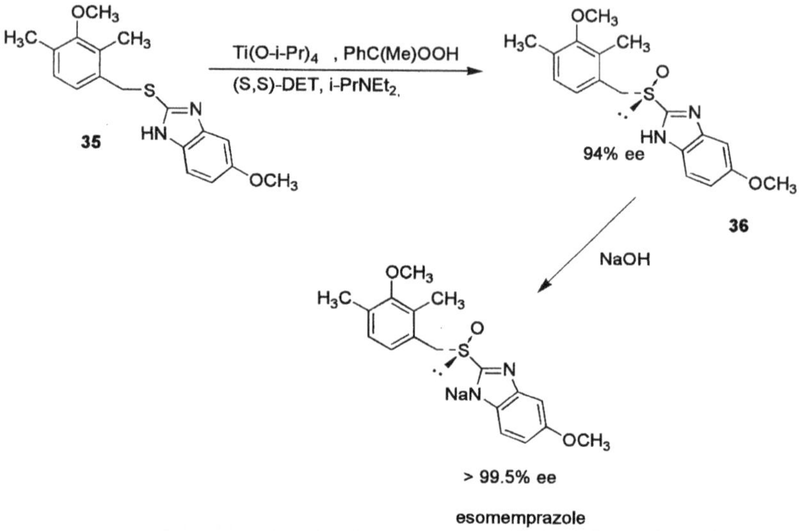

Scheme 22

The asymmetric oxidation of omeprazole sulfide **35** required significant modifications of the Kagan procedure in order to obtain the chiral sulfoxide **36** in high yield and ee in a process amenable for scale up. In general, the Kagan procedure works best with sulfides having groups quite different in size, and it is thought this size differentiation is key to obtaining high ee's. However, in the case of the omeprazole sulfide, the two groups are of comparable size. Two modifications of the Kagan procedure were key for obtaining high enantioselectivities, and it is not well-understood how each modification is operating. First, the preparation of the titanium complex, involving titanium tetraisopropoxide, (*S*,*S*)-diethyl tartrate, and water, was performed in the presence of the sulfide and this mixture was equilibrated at an elevated temperature (54 °C for 50 min). This suggests that the active catalyst is a complex involving not only the tartrate and titanium, but also the substrate. Coordination of the sulfide may be occurring via the imidazole NH (or potentially the deprotonated imidazole) since alkylated imidazoles are not effectively oxidized with high ee. A possible catalytic intermediate is shown as **37**.

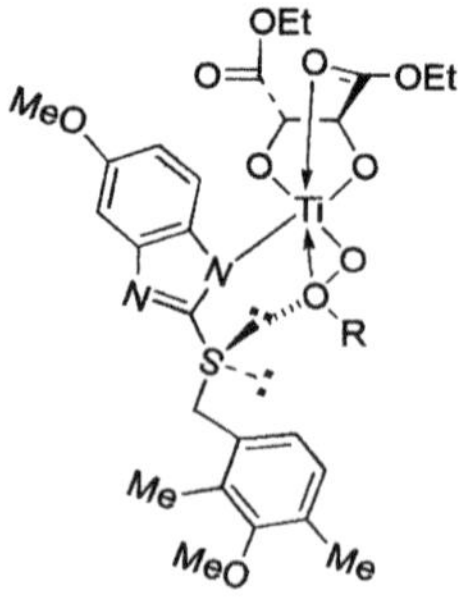

The second major modification was the use of diisopropylethylamine in the oxidation mixture. Amines such as triethylamine and *N*-methylmorpholine could also be used but gave lower ee's. If stronger bases were used, such as DBU and tetramethylguanidine, the ee's were dramatically reduced, which led the authors to conclude that imidazole deprotonation was probably not accounting for the increased ee with diisopropylethylamine. However, it is quite possible that the hindered base deprotonates the imidazole and that the deprotonated imidazole coordinates to titanium, while the more basic amines which are not hindered are themselves involved in coordination, disrupting the complex involving the substrate.

An important consequence of these improved conditions was that the reaction no longer needed to be carried out in a chlorinated solvent at low temperature. Environmentally acceptable solvents such as toluene and ethyl acetate were suitable, and the reaction was normally carried out for 1 h at 30 °C. Crystallization as the sodium salt provided an upgrade in ee from 94% in the reaction mixture to >99.5% in the isolated product (Scheme 22).

The Astra-Zeneca group has extended this chemistry to the preparation of a number of omeprazole analogs with success [63].

Oxidation of similar substrates was carried out prior to the Astra-Zeneca work by Pitchen and co-workers at Rhone-Poulenc Rorer in the preparation of the ACAT inhibitor RP 73163 [64]. The penultimate sulfide was readily oxidized, but with no enantioselectivity, so the oxidation was carried out at an earlier stage where the two groups on sulfur are very different in size (**38**). In these cases, with the imidazole nitrogen protected with a benzyl group, ee's >98% were achieved (Scheme 23). Interestingly, the group at Astra-Zeneca applied their improved oxidation conditions to the RP 73163 sulfide precursor, and were able to obtain 92% ee in the sulfide oxidation [62, 63].

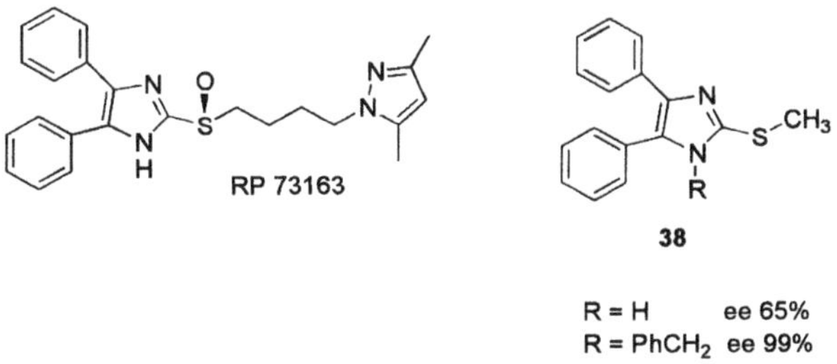

RP 73163

38

R = H ee 65%
R = $PhCH_2$ ee 99%

Scheme 23

The platelet adhesion drug candidate, OPC-29030, from Otsuka Pharmaceutical also has a chiral sulfoxide as a part of the molecule. The original synthesis used the Kagan oxidation procedure, which provided the sulfoxide **39** with only 54% ee. A survey of a wide variety of ligands other than tartrate was undertaken, with mandelic acid providing the highest ee (76%) (Scheme 24) [65]. The free hydroxyl group was found to be essential for high ee, suggesting it is bound to Ti in the transition state. Replacing the aryl group in the mandelic acid with *i*-Pr or

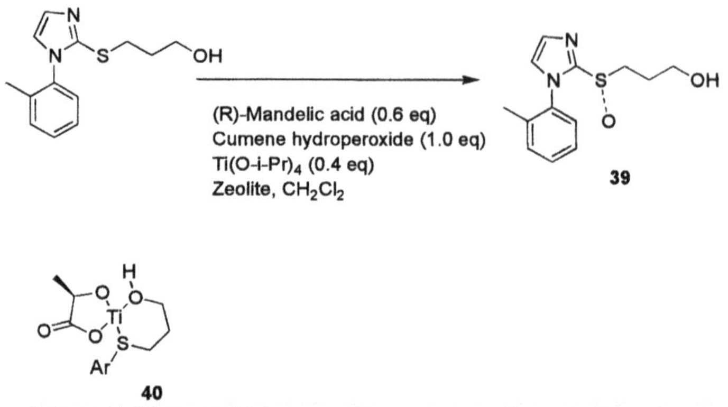

(R)-Mandelic acid (0.6 eq)
Cumene hydroperoxide (1.0 eq)
$Ti(O\text{-}i\text{-}Pr)_4$ (0.4 eq)
Zeolite, CH_2Cl_2

39

40

Scheme 24

cyclohexyl led to a reduction in ee, leading the authors to suggest π-stacking was also playing a role in the transition state. Based on the survey of ligands and substrates, a working model for the transition state was postulated as the rigid structure **40**.

5 Reductions

5.1 Hydride Reductions of Ketones and Imines

Titanium-based reagents have been used for asymmetric reductions of ketones and imines. Imine reductions using titanium are likely to have industrial applications in the future based on the recent work of the Buchwald group [66] who have carried out asymmetric imine reductions and hydrosilations using the Britzinger type ansa-titanocenes **41** (Scheme 25) [67]. The original synthesis of the chiral titanocene dichloride required resolution of the racemic compound using optically pure 2,2′-binaphth-1-ol in the presence of finely divided sodium metal. The product was purified by chromatography. The Buchwald group vastly

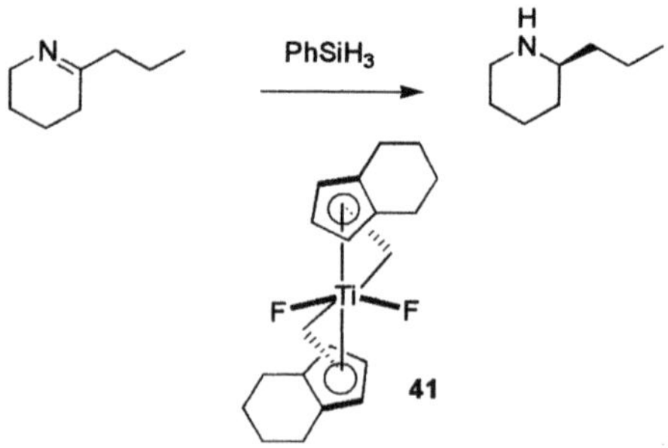

$PhSiH_3$

41

Scheme 25

simplified this procedure by running the resolution with 4-aminobenzoic acid, triethylamine, and 2,2′-binaphth-1-ol. No chromatography or sodium metal was required, and both enantiomers could be recovered. The process was demonstrated on a 10 g scale and should be amenable to further scale up. Conversion to the active difluoride catalyst is accomplished in one step. An example of the use of this catalyst in asymmetric reduction is shown in Scheme 25.

6 Hydrolysis/Transesterification

Seebach and his group have demonstrated the enantioselective desymmetrization of meso esters, anhydrides, and sulfonylimides using Ti-TADDOL reagents (Scheme 26) [68]. In the case of the anhydrides, a catalytic quantity (20 mol%) of the titanium reagent could be used along with a stoichiometric amount of aluminum triisopropoxide. The method should be competitive with enzymatic methods, and may be applicable for substrates where enzymes are ineffective.

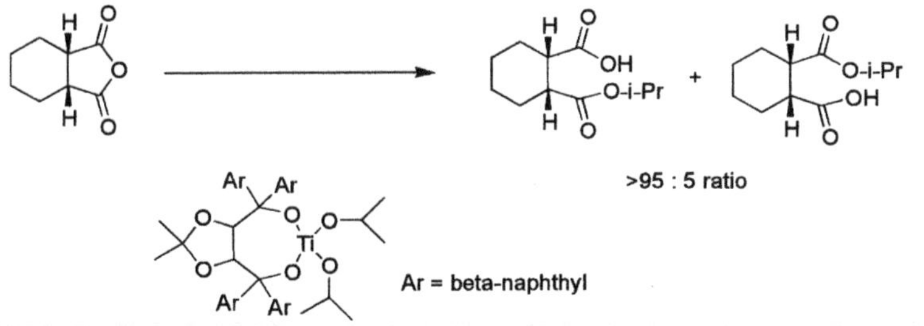

>95 : 5 ratio

Ar = beta-naphthyl

Scheme 26

References

1. Yamamoto A (1986) Organotransition metal chemistry. Wiley, New York, p 307
2. Katsuki T, Sharpless KB (1980) J Am Chem Soc 102:5974
3. McIntyre JE (ed) (1995) Dictionary of organometallic compounds, 2nd edn, vol 4. Chapman and Hall, New York, p 4313
4. Reetz MT (1982) Organotitanium reagents in organic synthesis. Springer, Berlin Heidelberg New York
5. Reetz MT (1986) Organotitanium reagents in organic synthesis. Springer, Berlin Heidelberg New York
6. Seebach D, Beck AK, Schiess M, Widler L, Wonnacott A (1983) Pure Appl Chem 55:1807
7. Ferreri C, Palumbo G, Caputo R (1991) Organotitanium and organozirconium reagents. In: Comprehensive organic synthesis, vol 1. Pergamon Press, Oxford
8. Reetz MT (ed) (1992) Tetrahedron 48:5557 (Tetrahedron Symposia-in-Print Number 47)
9. Duthaler RO, Hafner A (1992) Chem Rev 92:807
10. Narasaka K, Iwasawa N (1993) Asymmetric reactions promoted by titanium reagents. In: organic synthesis: theory and applications, vol 2. JAI, London, p 93
11. Kulinkovich OG, de Meijere A (2000) Chem Rev 100:2789
12. Seebach D, Beck AK, Heckel A (2001) Angew Chem Int Ed Engl 40:92

13. Evans DA, Rieger DL, Bilodeau MT, Urpi F (1991) J Am Chem Soc 113:1047
14. Cooke JWB, Berry MB, Caine DM, Cardwell KS, Cook JS, Hodgson A (2001) J Org Chem 66:334
15. Macdonald SJF, Clarke GDE, Dowle MD, Harrison LA, Hodgson ST, Inglis GGA, Johnson MR, Shah P, Upton RJ, Walls SB (1999) J Org Chem 64:5166
16. Humphrey GR, Miller RA, Pye PJ, Rossen K, Reamer RA, Maliakal A, Ceglia SS, Grabowski EJJ, Volante RP, Reider PJ (1999) J Am Chem Soc 121:11,261
17. Thiruvengadam TK, Sudhakar AR, Wu G (1999) Practical enantio- and diastereo-selective processes for azetidinones. In: Gadamasetti KC (ed) Process chemistry in the pharmaceutical industry. Dekker, New York, p 221
18. Thiruvengadam TK, Fu X, Tann CH, McAllister TL, Chiu, JS, Colon C (2001) US Patent 6,207,822
19. Saksena AK, Girijavallabhan VM, Wang H, Liu YT, Pike RE, Ganguly AK (1996) Tetrahedron Lett 37:5657
20. Crimmins MT, Nantermet PG (1993) Org Prep Proc Int 25:43
21. DeCamp AE, Kawaguchi AT, Volante RP, Shinkai I (1991) Tetrahedron Lett 32:1867
22. Armstrong JD, Hartner FW, DeCamp AE, Volante RP, Shinkai I (1992) Tetrahedron Lett 33:6599
23. McWilliams JC, Armstrong JD, Zheng N, Bhupathy M, Volante RP, Reider PJ (1996) J Am Chem Soc 118:11,970
24. Martins EO, Gleason JL (1999) Org Lett 1:1643
25. Tebbe FN, Parshall GW, Reddy GS (1978) J Am Chem Soc 100:3611; Pine SH (1993) Org React (NY) 43:1
26. Brown-Wensley KA, Buchwald SL, Cannizzo L, Clawson L, Ho S, Meinhardt D, Stille JR, Straus D, Grubbs RH (1983) Pure Appl Chem 55:1733
27. Takai K, Hotta Y, Oshima K, Nozaki H (1978) Tetrahedron Lett 2417; Hibino J, Okazoe K, Takai K, Nozaki H (1985) Tetrahedron Lett 26:5579
28. Petasis NA, Bzowej EI (1990) J Am Chem Soc 112:6392; Petasis NA, Lu S, Bzowej EI, Fu D, Staszewski JP, Akritopoulou-Zanze I, Patane MA, Hu Y (1996) Pure Appl Chem 68:667
29. Hale JJ, Mills SG, MacCoss M, Finke PE, Cascieri MA, Sadowski S, Ber E, Chicchi GG, Kurtz M, Metzger J, Eiermann G, Tsou NN, Tattersall D, Rupniak NMJ, Williams AR, Rycroft W, Hargreaves R, MacIntyre DE (1998) J Med Chem 41:4607
30. Claus K, Bestian H (1962) Annalen 654:8
31. Payack JF, Hughes DL, Cai D, Cottrell I, Verhoeven TR, Reider PJ (1995) Org Prep Proced Int 27:715
32. Erskine GJ, Hartgerink J, Weinberg EL, McCowen JD (1979) J Organmet Chem 170:51
33. Huffman M (2001) US Patent 6,255, 550
34. Huffman M, Payack J (2000) US Patent 6,063,950
35. Hughes DL, Payack JF, Cai D, Verhoeven TR, Reider PJ (1996) Organometallics 15:663
36. Charette AB, Molinaro C, Brochu C (2001) J Am Chem Soc 123:12,168
37. Robertson GM (1991) Pinacol coupling reactions In: Trost BM (ed) Comprehensive organic synthesis, vol 3. Pergamon Press, NY, p 563
38. Bensari A, Renaud JL, Riant O (2001) Org Lett 3:3863
39. Mori A, Inoue S (1999) Cyanation of carbonyl and imino groups In: Jacobsen E, Pfaltz A, Yamamoto H (eds) Comprehensive asymmetric catalysis. Springer, Berlin Heidelberg New York, p 983
40. Tararov VI, Hibbs DE, Hursthourse MB, Ikonnikov NS, Malik KMA, North M, Orizu C, Belokon YN (1998) Chem Commun 387
41. Belokon YN, Green B, Ikonnikov NS, North M, Tararov VI (1999) Tetrahedron Lett 40:8147
42. Belokon YN, Caveda-Cepas S, Green B, Ikonnikov NS, Khrustalev VN, Larichev VS, Moscalenko MA, North M, Orizu C, Tararov VI, Tasinazzo M, Timofeeva GI, Yashkina LV (1999) J Am Chem So 121:3968

43. Belokon YN, Green B, Ikonnikov NS, Larichev VS, Lokshin BV, Moscalenko MA, North M, Orizu C, Peregudov AS, Timofeeva GI (2000) Eur J Org Chem 2655
44. Belokon YN, Green B, Ikonnikov NS, North M, Parsons T, Tararov VI (2001) Tetrahedron 57:771
45. Stinson SC (2001) Chem Eng News 79:35
46. Mikami K, Terada M (1999) Ene-type reactions. In: Jacobsen E, Pfaltz A, Yamamoto H (eds) Comprehensive asymmetric catalysis. Springer, Berlin Heidelberg New York, p 1143
47. Hilpert H (2001) Tetrahedron 57:7675
48. Brown PA, Hilpert H (2001) US Patent 6,172,238
49. Mukaiyama T, Inubushi A, Suda S, Hara R, Kobayashi S (1990) Chem Lett 1015
50. Carreira EM (1999) Mukaiyama aldol reaction. In: Jacobsen E, Pfaltz A, Yamamoto H (eds) Comprehensive asymmetric catalysis. Springer, Berlin Heidelberg New York, p 997
51. Evans DA, Johnson JS (1999) Diels-Alder reactions. In: Jacobsen E, Pfaltz A, Yamamoto H (eds) Comprehensive asymmetric catalysis. Springer, Berlin Heidelberg New York, p 1177
52. Bienayme H (1997) Angew Chem Int Ed 36:2670
53. Jorgensen KA (2000) Angew Chem Int Ed Engl 39:3558
54. Long J, Hu J, Shen X, Ji B, Ding K (2002) J Am Chem Soc 124:10
55. Wu G, Schumacher DP, Tormos W, Clark JE, Murphy BL (1997) J Org Chem 62:2996
56. Bennett F, Ganguly AK, Girijavallabhan VM, Pinto PA (1995) Synlett 1110
57. Miyauchi H, Nakamura T, Ohashi N (1996) Bull Chem Soc Jpn 69:2625
58. Pitchen P, Deshmukh M, Dunach E, Kagan HB (1984) J Am Chem Soc 106:8188
59. Zhao S, Samuel O, Kagan HB (1987) Tetrahedron 43:5135
60. Brunel JM, Diter P, Duetsch M, Kagan HB (1995) J Org Chem 60:8086
61. Brunel M, Kagan HB (1996) Synlett 404
62. Cotton H, Elebring T, Larsson M, Li L, Sorensen H, von Unge S (2000) Tetrahedron Asymm 11:3819
63. Larrson EM, Stenhede UJ, Sorensen H, von Unge PO, Cotton HK (1996) PCT WO 9602535A1
64. Pitchen P, France CJ, McFarlane IM, Newton CG, Thompson DM (1994) Tetrahedron Lett 35:485
65. Matsugi M, Fukuda N, Muguruma Y, Yamaguchi T, Minamikawa J, Otsuka S (2001) Tetrahedron 57:2739
66. Chin B, Buchwald SL (1996) J Org Chem 61:5650
67. Wild FRWP, Zsolnai L, Huttner G, Britzinger HH (1982) J Organomet Chem 232:233
68. Jaeschke G, Seebach D (1998) J Org Chem 63:1190

Topics Organomet Chem (2004) 6: 63–95
DOI 10.1007/978-3-540-36966-0

Rhodium/Ruthenium Applications

Kenzo Sumi · Hidenori Kumobayashi

Takasago International Corporation, Central Research Laboratory, 1-4-11 Nishiyawata, Hiratsuka-City, Kanagawa 254-0073, Japan
E-mail: kenzo_sumi@takasago.com, hidenori_kumobayashi@takasago.com

Dedicated to Professor Ryoji Noyori in honor
of his being awarded the Nobel prize in Chemistry for 2001
for the development of catalytic asymmetric synthesis

Abstract Asymmetric catalytic hydrogenation and hydrogen transfer reactions of various substrates such as prochiral ketones and olefins using Rh(I) and Ru(II) complexes will be reviewed. These asymmetric transformations have provided organic chemists with efficient routes to a wide variety of optically active compounds with important biological activities. Based on the success and the marvelous abilities of the BINAP ligand, new analogues of BINAP and SEGPHOS ligands have been developed. This review surveys recent advances in industrial chiral technology using rhodium and ruthenium catalysts.

Keywords Rh · Ru · Ligand · Asymmetric · Catalytic

List of Abbreviations and Symbols

ADHD	Attention-deficit hyperactivity disorder
BDPP	2,4-Bis(diphenylphosphino)pentane
BINAP	2,2′-Bis(diphenylphosphino)-1,1′-binaphthyl
(*R*,*S*)-BINAPHOS	(*R*)-2-(Diphenylphosphino)-1,1′-binaphthalen-2′-yl (*S*)-1,1′-binaphthalene-2,2′-diyl phosphite
BIPHEMP	(6,6′-Dimethylbiphenyl-2,2′-diyl)bis(diphenylphosphine)
BIPNOR	2,2′,3,3′-Tetraphenyl-4,4′,5,5′-tetramethyl-6,6′-bis-1-phosphanorborna-2,5-dienyl
BPPM	*N*-(*tert*-Butoxycarbonyl)-4-(diphenylphosphino)-2-[(diphenylphosphino)methyl]pyrrolidine
c	cyclo
DAIPEN	1-Isopropyl-2,2-bis(*p*-methoxyphenyl)-1,2-ethylenediamine
DBT	Dibenzoyl tartrate
DIOP	2,3-*O*-Isopropylidene-2,3-dihydroxy-1,4-bis(diphenylphosphino)butane
DIPAMP	1,2-Bis(*o*-anisylphenylphosphino)ethane
DPEN	1,2-Diphenylethylenediamine
H_8-BINAP	2,2′-Bis(diphenylphosphino)-5,5′,6,6′,7,7′,8,8′-octahydro-1,1′-binaphthyl
HMG-CoA	3-Hydroxy-3-methylglutaryl coenzyme-A
(*R*,*R*)-Me-DuPHOS	(-)-1,2-Bis[(2*R*,5*R*)-2,5-dimethylphospholano]benzene
MeO-BIPHEP	(6,6′-Dimethoxybiphenyl-2,2′-diyl)bis(diphenylphosphine)
2-Nap-BIPNITE-*p*-F	2-Di(2-naphthyl)phosphino-2′-bis(4-fluorophenyl)phosphinoxy-1,1′-binaphthyl
[2.2]PHANEPHOS	4,12-Bis(diphenylphosphino)-[2.2]-paracyclophane
S/C	Substrate-to-catalyst ratio
SEGPHOS	(4,4′-Bi-1,3-benzodioxole)-5,5′-diylbis(diarylphosphine)
TON	Turnover number
TBDMS (TBS)	*tert*-Butyldimethylsilyl
θ	Dihedral angle

1 Introduction

An increasing number of pharmaceutical, agrochemical, flavor, and fragrance products rely on enantiomerically pure intermediates. The world pharmaceutical market of chiral drugs grew by more than 13% to $133 billion in 2000, while 40% of all drug sales in that year were of single enantiomers [1].

Drug companies are increasingly developing chiral drugs as single enantiomers. Recently, chirality has been used as a tool for drug life-cycle management, whereby racemic mixtures are reformulated as single enantiomers. In this matter of racemic switches, drug companies manage the life cycles of their own drugs by patenting the individual enantiomers and then switching the drugs as a means of prolonging the patent life.

Enantiomerically pure compounds can be prepared by biocatalysis, asymmetric chemocatalysis, crystallization, or modification of natural resources. Among these methods, asymmetric chemocatalysis is one of the most efficient and versatile methods for the preparation of a wide range of chiral target compounds. Asymmetric hydrogenation catalyzed by transition metal complexes containing optically active phosphine ligands has attracted significant interest in industry and academia for its synthetic utility. Recent catalytic asymmetric syntheses have utilized atropisomeric diphosphine ligands, among which BINAP (Scheme 1) is preeminent. The 1,1′-binaphthyl ring system is a key component of a number of chiral ligands that has been used very successfully for asymmetric synthesis. The new chiral biaryl ligand, SEGPHOS [(4,4′-bi-1,3-benzodioxole)-5,5′-diylbis(diarylphosphine)], is also a very promising candidate (Scheme 1). In this chapter, we will survey the effectiveness of the new optically active bisphosphine ligands, such as the SEGPHOSes, in asymmetric hydrogenation.

PR_2 PR_2 BINAPs

PR_2 PR_2 SEGPHOSes

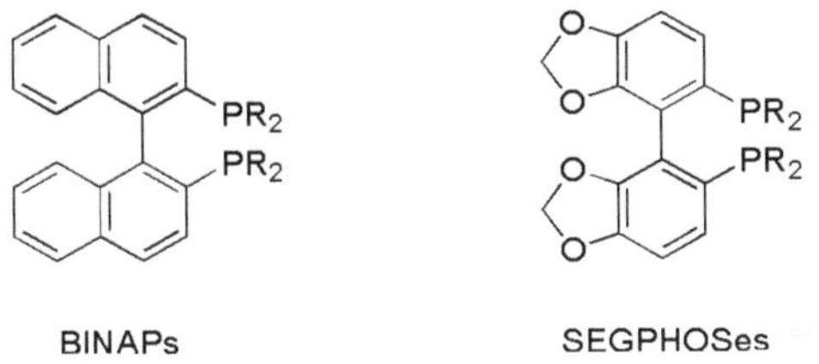

Scheme 1

2 BINAP

Since Kagan [2] and Knowles [3, 4] demonstrated that DIOP- and DIPAMP-Rh(I) complexes were highly effective for asymmetric hydrogenation of 2-(acylamino)acrylic acids, vast numbers of chiral ligands [5] and catalysts [6] have been developed by many researchers in academia and industry. Professor Noyori and the late professor Takaya subsequently designed and synthesized a bidentate phosphine that contained an atropisomeric 1,1′-binaphthyl structure as a chiral element for use in transition metal-catalyzed asymmetric reactions. From these efforts the axially dissymmetric ligand, BINAP [2,2′-bis(diphenylphosphino)-1,1′-binaphthyl] [7] has become one of the most effective ligands in asymmetric synthesis. Since the practical method for the synthesis of BINAPs was reported in 1986 [8], a series of BINAP ligands has been prepared at TAKASAGO (Scheme 2).

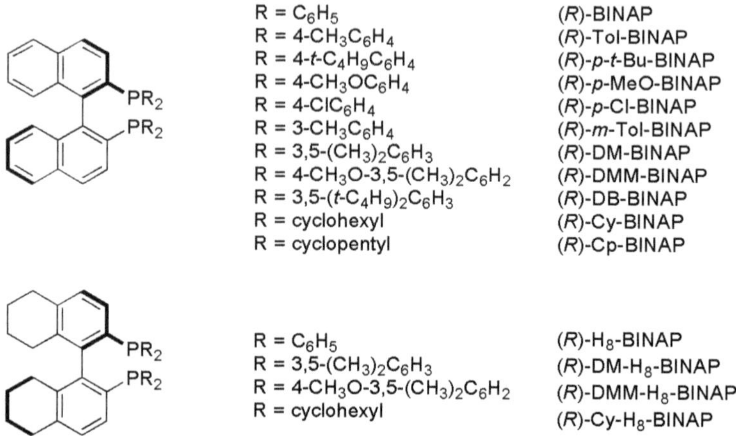

Scheme 2 Lineup of BINAPs

Now, most chiral BINAP ligands are synthesized by using a metal-catalyzed coupling reaction of easily accessible chiral 2,2′-bis[(trifluoromethanesulfonyl)oxy]-1,1′-binaphthyl with diarylphosphine oxides [9, 10]. Similarly, 2,2′-bis(diphenylphosphino)-5,5′,6,6′,7,7′,8,8′-octahydro-1,1′-binaphthyl (H_8-BINAP) is obtained from 2,2′-bis[(trifluoromethanesulfonyl)oxy]-5,5′,6,6′,7,7′,8,8′-octahydro-1,1′-binaphthyl [9–12]. The BINAP–Rh and –Ru complexes act as very efficient catalysts for the enantioselective isomerization of allylamines and the asymmetric hydrogenation of ketones and olefins.

3 SEGPHOS

A new chiral phosphine ligand, (4,4′-bi-1,3-benzodioxole)-5,5′-diyldiphosphine (hereafter abbreviated SEGPHOS), in Scheme 3 designed with the assistance of CAChe MM2 computations has been synthesized [13]. SEGPHOS ligands show

R = C_6H_5	(*R*)-SEGPHOS
R = 4-$CH_3C_6H_4$	(*R*)-Tol-SEGPHOS
R = 3,5-$(CH_3)_2C_6H_3$	(*R*)-DM-SEGPHOS
R = 4-CH_3O-3,5-(*t*-$C_4H_9)_2C_6H_2$	(*R*)-DTBM-SEGPHOS
R = cyclohexyl	(*R*)-Cy-SEGPHOS

Scheme 3 Lineup of SEGPHOSes

high crystallinity, which is convenient for industrial use. SEGPHOS differs from BINAP in that the outer ring of the naphthyl group of BINAP has been replaced with a methylenedioxy moiety. As a result, the electron-density at the phosphorus atoms is significantly increased. The biaryl backbone of SEGPHOS metal complexes has a narrower dihedral angle.

At the beginning, the following working hypothesis was made (as in Scheme 4). A biarylphosphine–Ru complex can be presumed to coordinate to a bidentate substrate. Then, by narrowing the dihedral angle (θ) in the biaryl backbone, the steric hindrance between the substrate and a phenyl group on the phosphorus atom can be increased. This steric repulsion would increase the stereoselectivity.

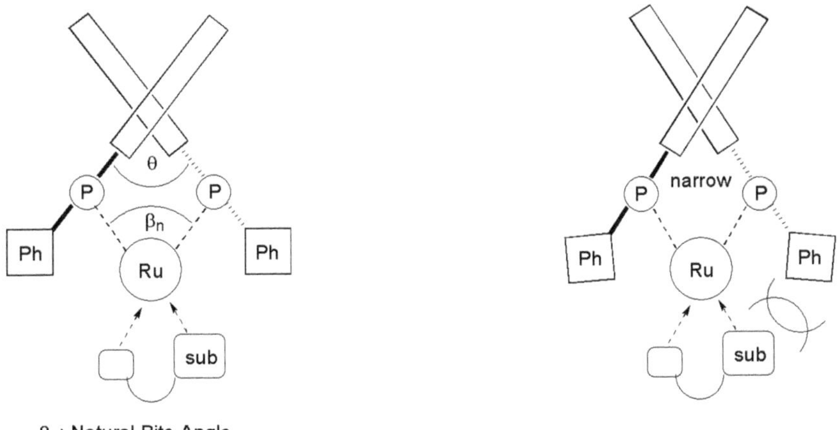

Scheme 4 Steric consideration based on dihedral angle (θ) in biaryl backbone

Hydroxyacetone was selected as the substrate for running the basic experiment to check catalytic activity and enantioselectivity (Scheme 5). The respective dihedral angles of the biaryl ligands, BINAP, BIPHEMP [14], and MeO-BIPHEP [15] were calculated. Finally, the reaction conditions were standardized; hydrogen pressure was 30 atm, the substrate-to-catalyst ratio (S/C) was 2000, the solvent was methanol, and the reaction temperature was 65 °C. The results corroborated the hypothesis that the narrower the dihedral angle in the biaryl backbone, the higher the observed optical yield. As a result of these experiments, MeO-BIPHEP was determined to be the best ligand for the reduction of hydroxyacetone. Consequently, we started to design a new ligand that was close to MeO-BIPHEP.

In the general structure shown in Scheme 6, MeO-BIPHEP has methoxy groups in the X positions and hydrogen atoms in the Y positions. After looking at many candidates, SEGPHOS which has a methylenedioxy group connecting the X and Y positions was tested. The calculated value of the dihedral angle in SEGPHOS is smaller (64.99°) than those in BINAP, BIPHEMP, and MeO-BIPHEP predicting that SEGPHOS would give higher enantioselectivity.

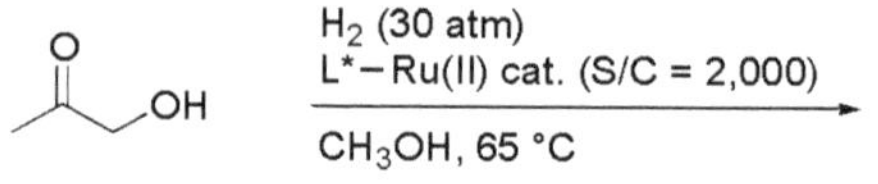

H_2 (30 atm)
L*–Ru(II) cat. (S/C = 2,000)
CH_3OH, 65 °C

L*	Dihedral Angle (θ)	% ee
(*R*)-BINAP	73.49°	89.0
(*R*)-BIPHEMP	72.07°	92.5
(*R*)-MeO-BIPHEP	68.56°	96.0

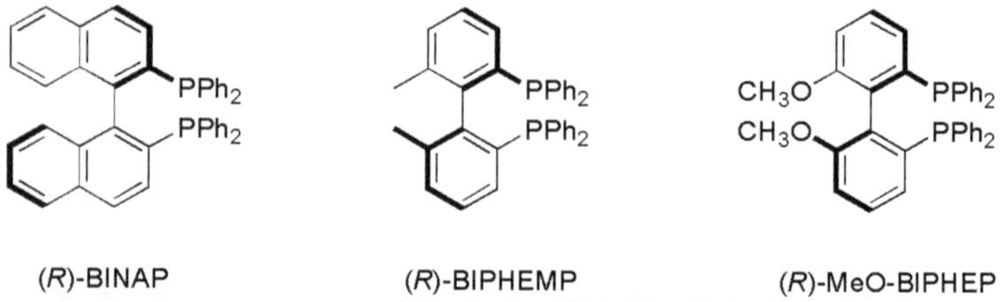

(*R*)-BINAP (*R*)-BIPHEMP (*R*)-MeO-BIPHEP

Scheme 5 Dihedral angles (θ) and enantioselectivities

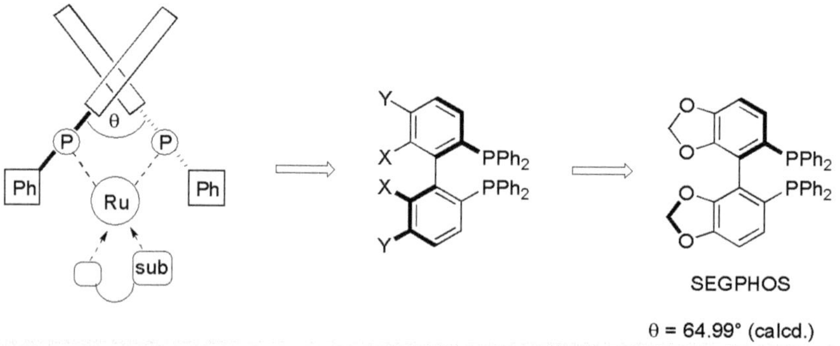

SEGPHOS

θ = 64.99° (calcd.)

Scheme 6 New chiral ligand

The preparation of SEGPHOS was achieved using, as the key reaction, an oxidative homo-coupling [16, 17] of 5-diarylphosphinyl-1,3-benzodioxoles **1** at the C4 position (Scheme 7). The phosphine oxide is obtained from its corresponding bromide and is then lithiated with LDA (lithium diisopropylamide); an oxidative coupling of the anion **2** affords the racemic biaryl phosphine oxide **3**. Optically active (*R*)-SEGPHOSO is prepared by optical resolution with (+)-dibenzoyl tartrate (DBT) followed by liberation of the DBT by treatment with alkaline solution. Next, the phosphine oxide is reduced with trichlorosilane to afford the expected biaryl phosphine, (*R*)-SEGPHOS.

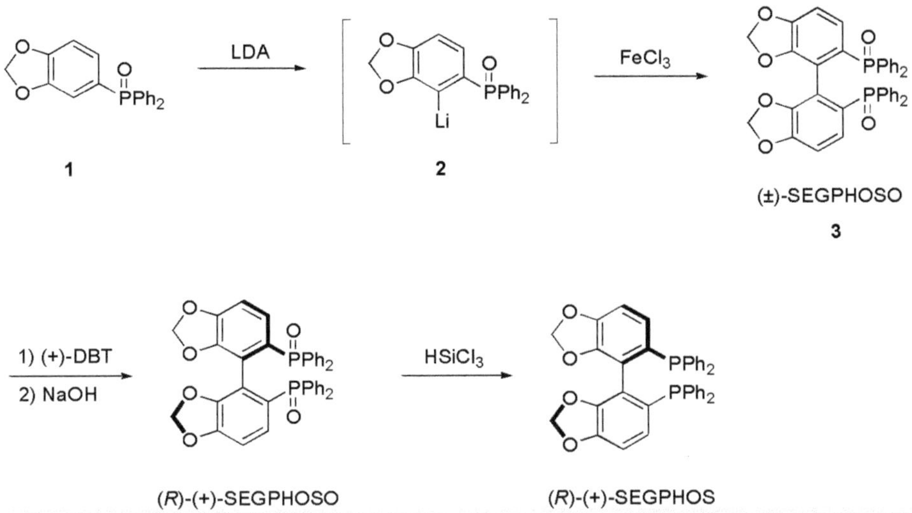

LDA
FeCl$_3$
1
2
(±)-SEGPHOSO
3
1) (+)-DBT
2) NaOH
HSiCl$_3$
(*R*)-(+)-SEGPHOSO
(*R*)-(+)-SEGPHOS

Scheme 7 Synthesis of SEGPHOS

4 Rh

Rhodium catalysts can perform enantioselective double-bond migration, asymmetric hydrogenation, asymmetric hydroformylation, and intramolecular N–H insertion (see later).

4.1 Asymmetric Isomerization

In 1973, TAKASAGO started to develop the asymmetric isomerization of *N*,*N*-diethylgeranylamine to citronellalenamine catalyzed by transition-metal complexes [18]. Subsequently, in 1980, a cationic BINAP–Rh(I) complex developed by Takaya and Noyori was reported to act as a very efficient catalyst for such asymmetric isomerizations [19, 20].

4.1.1 *Menthol*

The ability to conduct asymmetric synthesis with an organometallic catalyst realized the long-sought economic production of artificial *l*-menthol (Scheme 8). This process is sometimes cited as a typical example of an industrial application of homogeneous asymmetric synthesis. It is also the first successful industrial asymmetric synthesis of *l*-menthol. The starting material myrcene, which is readily available and cheap, is converted to diethylgeranylamine with lithium/diethylamine. Using a cationic bis(Tol-BINAP)–Rh(I) complex, the intermediate allylamine is isomerized to (*R*)-citronellalenamine. The catalytic activity is

myrcene —Li, $(C_2H_5)_2NH$→ N,N-diethylgeranylamine —(S)-Tol-BINAP–Rh(I), TON = 400,000→ (R)-citronellalenamine

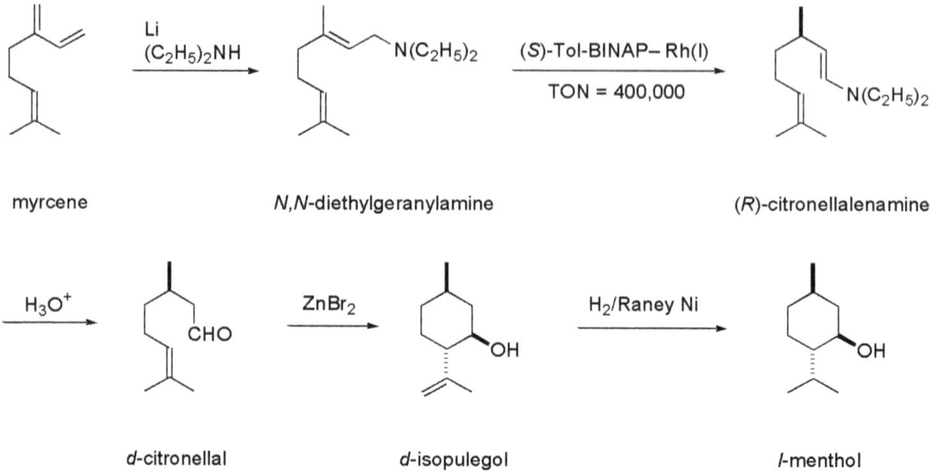

—H_3O^+→ d-citronellal —$ZnBr_2$→ d-isopulegol —H_2/Raney Ni→ l-menthol

Scheme 8 Asymmetric synthesis of *l*-menthol

extremely high and the turnover number (TON) reaches 400,000. The resultant enamine is hydrolyzed to *d*-citronellal under acidic conditions. Next, the intramolecular ene reaction is catalyzed by zinc bromide to give *d*-isopulegol with high stereoselectivity. Finally, the olefin is hydrogenated to give the desired *l*-menthol.

This process also allows the synthesis of (+)-7-hydroxydihydrocitronellal [21, 22] (a perfumery agent with specific olfactory properties), a side-chain of vitamin E [23], and the juvenile hormone methoprene [24].

4.2 Asymmetric Hydrogenation

Hydrogenation is a core technology in industry and academia because hydrogen is the simplest molecule and a clean resource which is available in abundance at a low cost. The asymmetric hydrogenation of C=C, C=O, and C=N bonds in organic substrates using organometallic complexes has been achieved with considerable success owing to significant synthetic advances in the development of chiral ligands over the past twenty years, primarily using optically active phosphines.

4.2.1 *Ketone*

Ketones are the most common unsaturated substrates. Optically active secondary alcohols derived from asymmetric hydrogenation of functionalized ketones are important intermediates for syntheses of various biologically active substances. So far, rhodium complexes as catalysts were applied to the hydrogenation of such substrates. Hayashi [25] accomplished the high optical yield of 95%

ee in the hydrogenation of α-aminoacetophenones with a rhodium complex of a ferrocene ligand. High enantioselectivities were achieved in asymmetric hydrogenations when ligands such as DIOP [26] or BPPM [27] were used.

Sakuraba and Achiwa [28] reported a practical asymmetric synthesis of Eli Lilly's drug (*R*)-fluoxetine hydrochloride [29] by catalytic asymmetric hydrogenation of β-amino ketone **4** as a key step with (2*S*,4*S*)-*N*-(methylcarbamoyl)-4-(dicyclohexylphosphino)-2-[(diphenylphosphino)methyl]-pyrrolidine [(2*S*,4*S*)-MCCPM]-rhodium complex (Scheme 9).

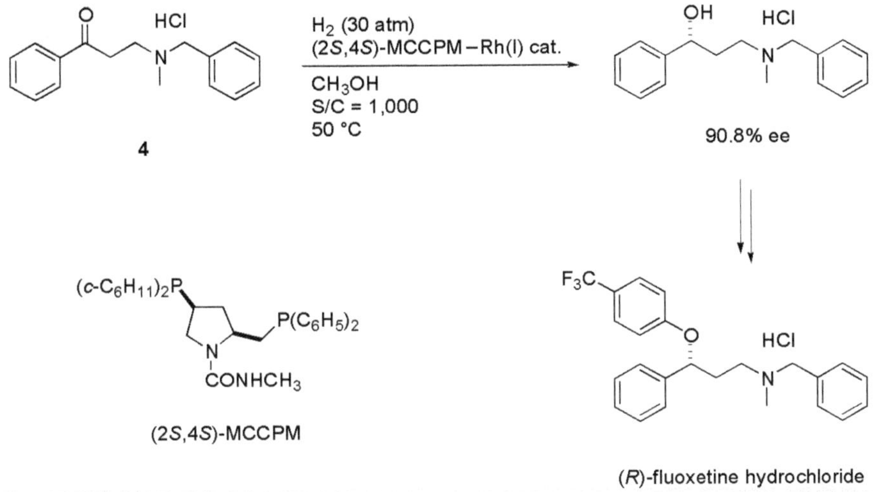

HCl
4
H_2 (30 atm)
(2*S*,4*S*)-MCCPM–Rh(I) cat.
CH_3OH
S/C = 1,000
50 °C
OH HCl
90.8% ee
$(c\text{-}C_6H_{11})_2P$
$P(C_6H_5)_2$
N
$CONHCH_3$
(2*S*,4*S*)-MCCPM
F_3C
O
HCl
(*R*)-fluoxetine hydrochloride

Scheme 9 Asymmetric synthesis of (*R*)-fluoxetine hydrochloride

Until recently, in spite of extensive studies, only a limited number of transition metal catalysts were known to exhibit satisfactory activity in the hydrogenation of simple ketones that have no functionality close to the carbonyl group [30–41]. Schrock and Osborn found that $[RhH_2\{P(C_6H_5)(CH_3)_2\}_2L_2]X$ (L=solvent, X=PF_6 or ClO_4) effectively reduces acetone under 1 atm of H_2 in the presence of a small amount of water [42]. Additionally, Tani et al. achieved hydrogenation of ketonic substrates using a cationic rhodium complex with a fully alkylated bidentate diphosphine [43, 44]. In those cases, the high basicity of the ligands [45] increased the electron density of the rhodium center so that the oxidative addition of H_2 might be accelerated [43, 44, 46].

Asymmetric hydrogenation of simple ketones was even more difficult. In 1985, Markó et al. [47] reported that a BDPP-Rh(I) [BDPP=2,4-bis(diphenylphosphino)pentane] complex in methanol and triethylamine at a substrate to catalyst ratio (S/C) of 100:1 catalyzed the hydrogenation of acetophenone under 69 atm of H_2 at 50°C to afford 1-phenylethanol in 82% ee. Various chiral phosphine-rhodium [48, 49], -iridium [50], and -ruthenium complexes [51] were devised for asymmetric hydrogenation. However, the number of substrates giving reasonable enantioselectivity was limited and the reactivity and selectivity remained unsatisfactory.

4.2.2
Olefin

Burk developed an improved process for the synthesis of key intermediate **6** required for Pfizer's drug candoxatril (Scheme 10) [52]. A cationic (*R*,*R*)-Me-DuPHOS–Rh catalyst was found to allow efficient and enantioselective hydrogenation of unsaturated carboxylate substrate **5** to afford the desired product **6** in greater than 99% ee.

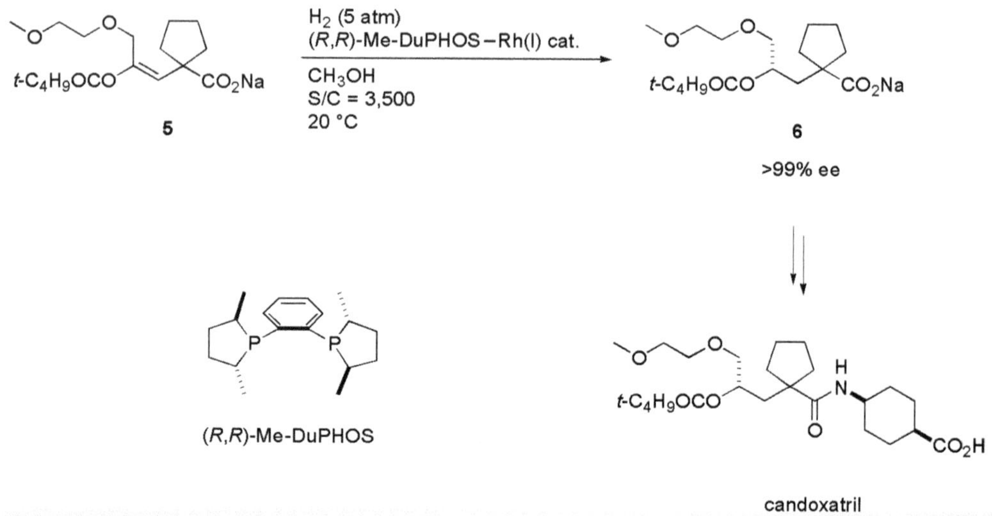

H_2 (5 atm)
(*R*,*R*)-Me-DuPHOS–Rh(I) cat.
CH_3OH
S/C = 3,500
20 °C

5 **6** >99% ee

(*R*,*R*)-Me-DuPHOS

candoxatril

Scheme 10 Synthesis of candoxatril with the Me-DuPHOS/Rh-catalyzed process: Pfizer

The cationic (*R*,*R*)-Me-DuPHOS-Rh catalyst could also be used for the synthesis of key intermediate **7** for an HIV protease inhibitor, tipranavir (Scheme 11) [53, 54]. The novel asymmetric process to the key intermediate was developed in a collaboration between Pharmacia & Upjohn and Chirotech.

A Merck group developed the planar chiral C_2-symmetric bisphosphine [2.2]PHANEPHOS [4,12-bis(diphenylphosphino)-[2.2]-paracyclophane] (Scheme 12) [55, 56]. [[2.2]phanephos–Rh]$^+$OTf$^-$ proved to be a highly enantioselective catalyst for the hydrogenation of tetrahydropyrazine **8** to afford **9**, a precursor to the HIV protease inhibitor indinavir, in 86% ee.

Rhodia [57] has investigated the catalytic properties of BIPNOR [2,2′,3,3′-tetraphenyl-4,4′,5,5′-tetramethyl-6,6′-bis-1-phosphanorborna-2,5-dienyl], incorporated as a Rh(I) catalyst for the hydrogenation of olefins, in collaboration with Mathey (Scheme 13) [51]. An enantiomeric excess of 98% was obtained for the asymmetric reduction of 2-(6′-methoxy-2′-naphthyl)acrylic acid (**10**) to the commercially important anti-inflammatory agent naproxen.

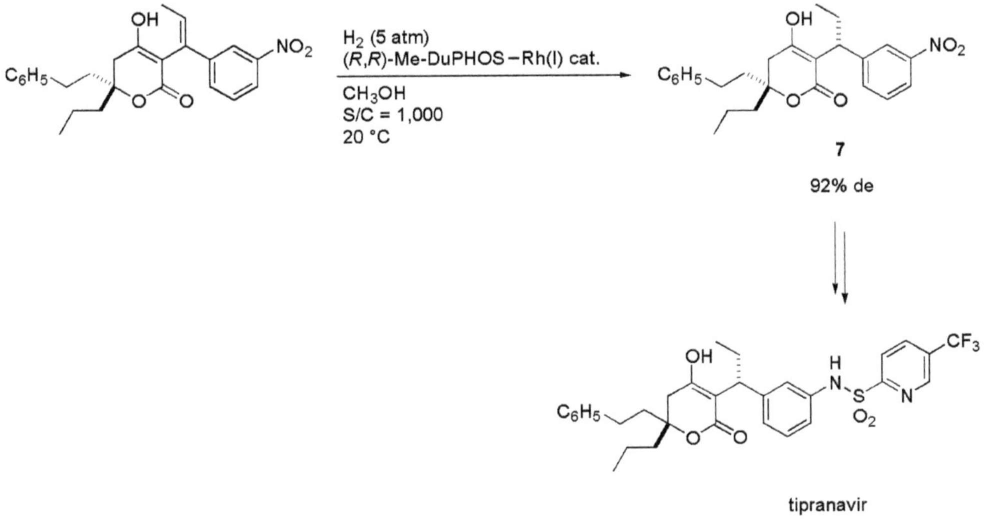
OH
NO_2
C_6H_5
O
H_2 (5 atm)
(R,R)-Me-DuPHOS–Rh(I) cat.
CH_3OH
S/C = 1,000
20 °C
7
92% de
H
N
S
O_2
N
CF_3
tipranavir

Scheme 11 Synthesis of tipranavir intermediate precursor: Pharmacia & Upjohn

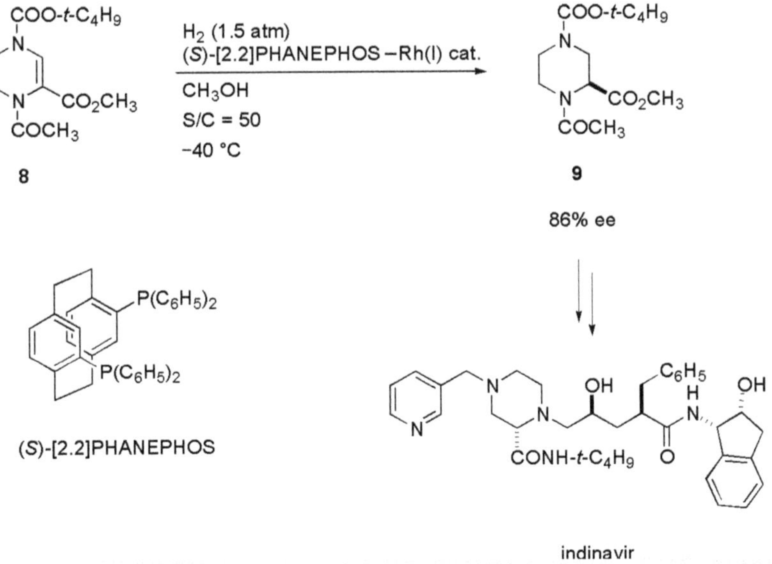
COO-t-C_4H_9
N
CO_2CH_3
$COCH_3$
8
H_2 (1.5 atm)
(S)-[2.2]PHANEPHOS–Rh(I) cat.
CH_3OH
S/C = 50
-40 °C
9
86% ee
$P(C_6H_5)_2$
$P(C_6H_5)_2$
(S)-[2.2]PHANEPHOS
OH
C_6H_5
H
CONH-t-C_4H_9
O
indinavir

Scheme 12 Synthesis of indinavir intermediate precursor: Merck

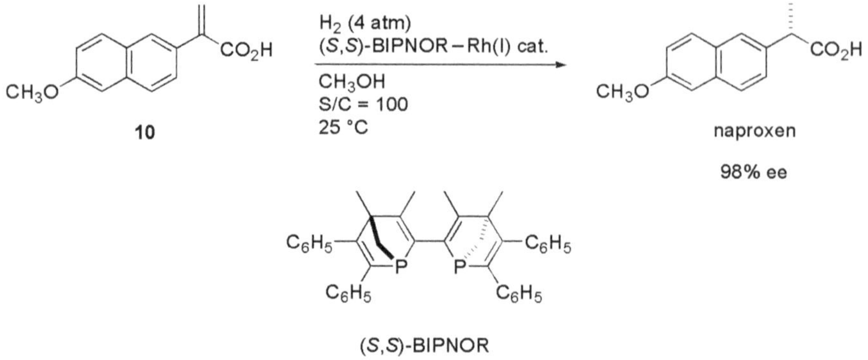

Scheme 13 Asymmetric hydrogenation using BIPNOR complex: Rhodia

4.3
Asymmetric Hydroformylation

Takaya et al. [58] developed a highly enantioselective hydroformylation of various olefins using a rhodium complex of a chiral phosphine–phosphite ligand, (*R*,*S*)-BINAPHOS [(*R*)-2-(diphenylphosphino)-1,1′-binaphthalen-2′-yl (*S*)-1,1′-binaphthalene-2,2′-diyl phosphite], as a catalyst.

4.3.1
1β-Methylcarbapenem

The high stereoselectivity and the versatility obtained by this hydroformylation prompted TAKASAGO to apply to the synthesis of 1β-methylcarbapenem (Scheme 14). When the phosphine–phosphite ligand, (*R*,*S*)-BINAPHOS, was employed, a satisfactory β/α selectivity of 93:7 was obtained, but the *iso*/*normal* ratio was very poor. Similarly, hydroformylation of 4-vinyl β-lactam **11** catalyzed by a rhodium complex of a chiral phosphine–phosphinite ligand, (*R*)-2-Nap-BIPNITE-*p*-F, gave *iso*/*normal* and β/α selectivities of 74:26 and 96:4, respectively. The oxidation of the aldehyde proceeded without epimerization to produce the corresponding carboxylic acid **12**. Thus, a new synthetic route toward 1β-methylcarbapenem antibiotics has been developed [59].

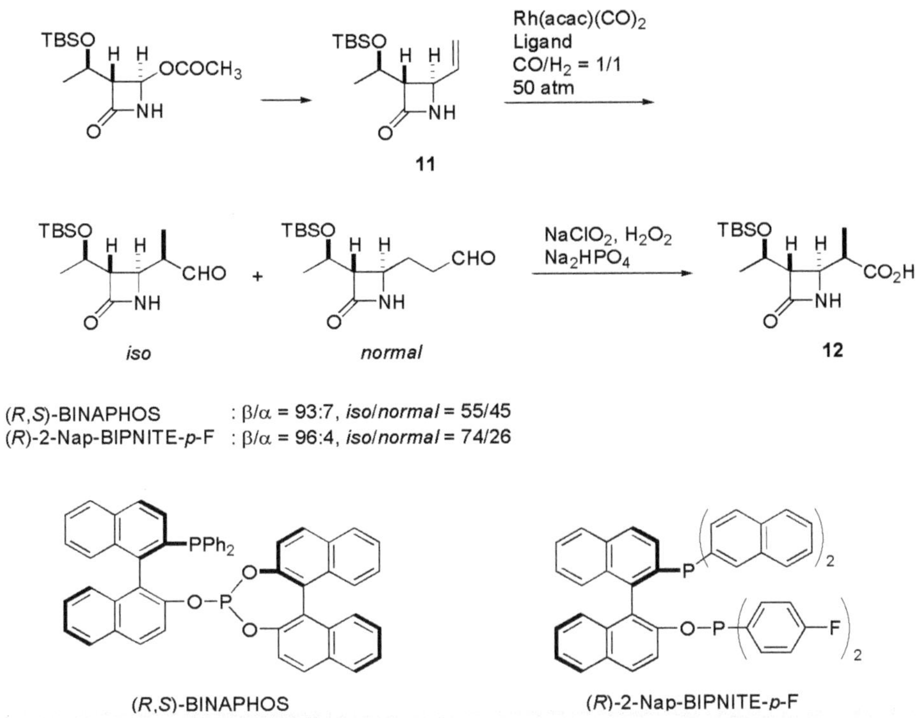

(*R*,*S*)-BINAPHOS : β/α = 93:7, *iso*/*normal* = 55/45
(*R*)-2-Nap-BIPNITE-*p*-F : β/α = 96:4, *iso*/*normal* = 74/26

Scheme 14 Asymmetric hydroformylation of 4-vinyl β-lactam

5
Ru

Triarylphosphine–Rh(I) catalysts have shown unsatisfactory catalytic efficiency, catalytic activity, and optical yields in many hydrogenations. High catalytic activity is necessary from an industrial point of view since Rh metal is very expensive. In contrast, diphosphine–Ru(II) complexes are effective catalysts exhibiting high catalytic activities for a wide range of substrates.

5.1
BINAP–Ru Complexes

The BINAP–Ru(II) complex, $Ru_2Cl_4(binap)_2{\cdot}N(C_2H_5)_3$ (**13**), was synthesized by Ikariya [60] for the first time as shown in Scheme 15. The structure was later published by Takaya et al. [61]. Subsequently, a mononuclear complex [62] which had acyloxy groups **14** instead of the halides in Ikariya's complex and a cationic ruthenium complex [63] derived from $[RuX_2(arene)]_2$ and BINAP were prepared.

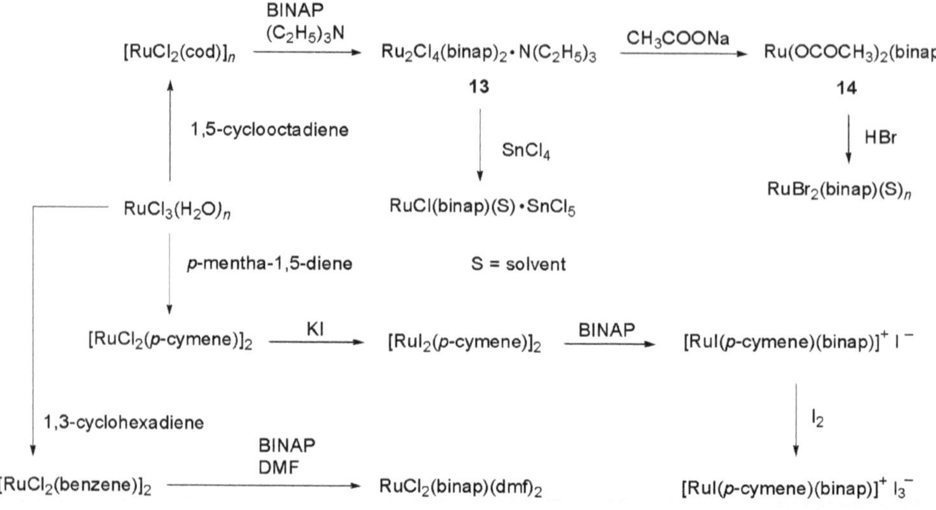

Scheme 15 BINAP–Ru(II) complexes

5.2 SEGPHOS–Ru Complexes

The highly active SEGPHOS–Ru(II) complex **17** [13] was developed at TAKASAGO by Takaya's method [61] for the preparation of the BINAP–Ru(II) analogue (Scheme 16). Mononuclear metal complex **16** was obtained from *p*-cymene–Ru chloride (**15**) and (*R*)-SEGPHOS, which was subsequently treated with dimethylamine hydrochloride to produce the binuclear SEGPHOS–Ru complex **17**. Its structure is similar to a BINAP–ruthenium complex.

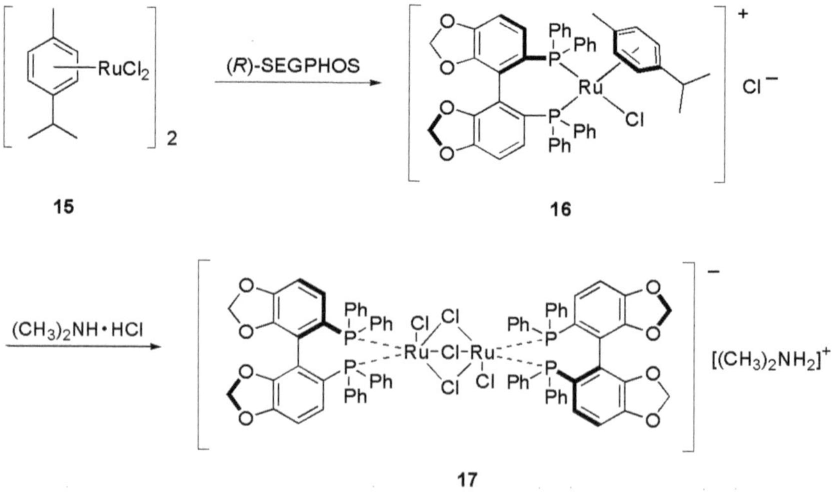

(*R*)-SEGPHOS

15 **16** Cl^-

$(CH_3)_2NH \cdot HCl$

17 $[(CH_3)_2NH_2]^+$

Scheme 16 Preparation of catalyst **17**

5.3
Asymmetric Hydrogenation

Ruthenium catalysts containing phosphine ligands are uniquely capable of hydrogenating a range of substrate types with extremely high enantioselectivities and high efficiencies. The commercial potential of this technology is vast.

5.3.1
Ketone

Enantioselective reduction of carbonyl compounds is important for the synthesis of optically active secondary alcohols. Hydrogenation, among the various reduction methods, is obviously the most desirable for this purpose. This section concentrates on this class of reactions using chiral ruthenium–phosphine complexes.

5.3.1.1
Functionalized Ketone

BINAP–Ru(II) complexes containing halide ligands effect the highly enantioselective hydrogenation of a wide range of functionalized ketones including *o*-bromoacetophenone and ketones that have functional groups, such as NR_2, OH, OR, $OSiR_3$, C=O, COOR, COSR, $CONR_2$, and COOH, at the β-position [64–66]. The SEGPHOS–Ru catalyst was used for the hydrogenation of hydroxyacetone (**18**) (Scheme 17). The reaction was run under 30 atm of hydrogen pressure at 65 °C. The substrate-to-catalyst ratio of the catalyst was more than 10,000, and the optical yield using SEGPHOS was 98.5% ee. In contrast, BINAP and BIPHEMP gave 89.0% and 92.5% ee, respectively. This result supports our hypothesis (refer to

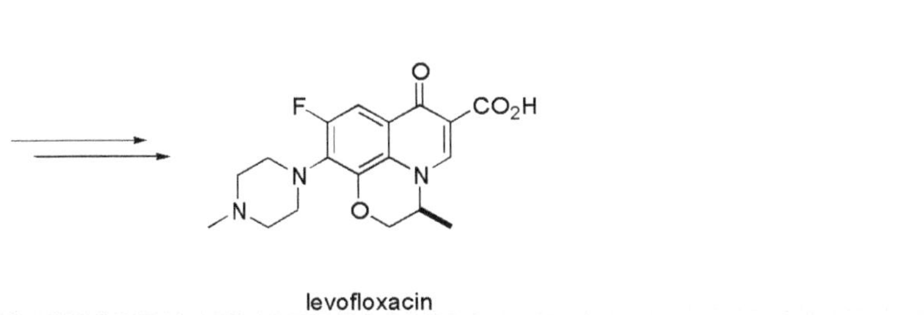

18 → **19**: H_2 (30 atm), (*R*)-SEGPHOS–Ru(II) cat., CH_3OH, S/C = 10,000, 65 °C

19: 98.5% ee

BINAP: 89.0% ee (S/C = 2,000)
BIPHEMP: 92.5% ee (S/C = 2,000)

→ → levofloxacin

Scheme 17 Asymmetric hydrogenation of hydroxyacetone

section 3) that the narrower bite angle provides improved enantioselectivity. This optically active diol **19** is used for levofloxacin, a new quinolone antibacterial agent [67].

The SEGPHOS–Ru(II) catalyst is also effective for the hydrogenation of a carbonyl group in a ketoester. For example, α-ketoester **20** is converted into the corresponding alcohol **21** under 50 atm of hydrogen pressure at 50 °C in ethanol with 95.7% ee (Scheme 18) [13]. A 1500 S/C for the catalyst was used. Again, the enantiomeric excess is higher with the SEGPHOS–Ru(II) catalyst than with a BINAP–Ru(II) catalyst. The hydroxyester product **21** can be used for enalapril. Furthermore, the bulky α-ketoester **22** can also be hydrogenated to give the corresponding α-hydroxyester **23**. The ee with the SEGPHOS–Ru(II) catalyst is also higher than with a Tol-BINAP–Ru(II) catalyst.

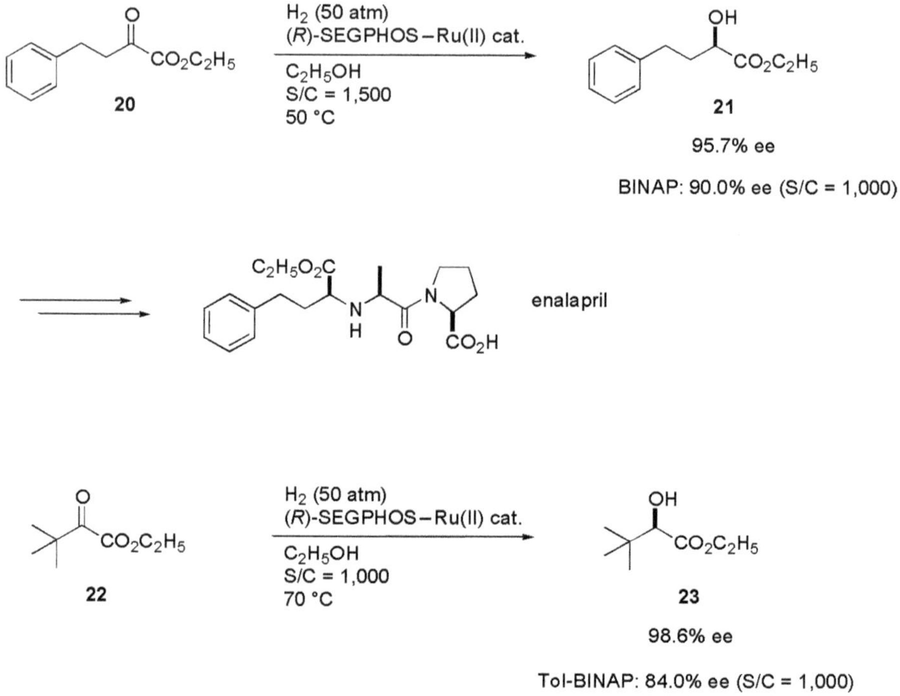

$CO_2C_2H_5$ **20** — H_2 (50 atm), (*R*)-SEGPHOS–Ru(II) cat., C_2H_5OH, S/C = 1,500, 50 °C → OH $CO_2C_2H_5$ **21**, 95.7% ee

BINAP: 90.0% ee (S/C = 1,000)

→ $C_2H_5O_2C$ N H O N CO_2H enalapril

$CO_2C_2H_5$ **22** — H_2 (50 atm), (*R*)-SEGPHOS–Ru(II) cat., C_2H_5OH, S/C = 1,000, 70 °C → OH $CO_2C_2H_5$ **23**, 98.6% ee

Tol-BINAP: 84.0% ee (S/C = 1,000)

Scheme 18 Asymmetric hydrogenation of α-ketoesters

β-Ketoesters can also be hydrogenated with high optical yields using the SEGPHOS–Ru catalyst (Scheme 19) [68]. The aromatic β-ketoester **24** is transformed into the corresponding alcohol **25** in 98.0% ee. Tol-BINAP gives 87.0% ee. The long-alkyl-chain β-hydroxyester **26** is obtained from the corresponding β-ketoester with 99.2% ee. The functionalized β-hydroxyester **27** which has an ester group in the same molecule at the ω-position is also obtained with 99.6% ee.

4-Chloroacetoacetate **28**, having a halogen atom at the C4-position, is transformed into the corresponding alcohol **29** in 98.5% ee (Scheme 20). The ob-

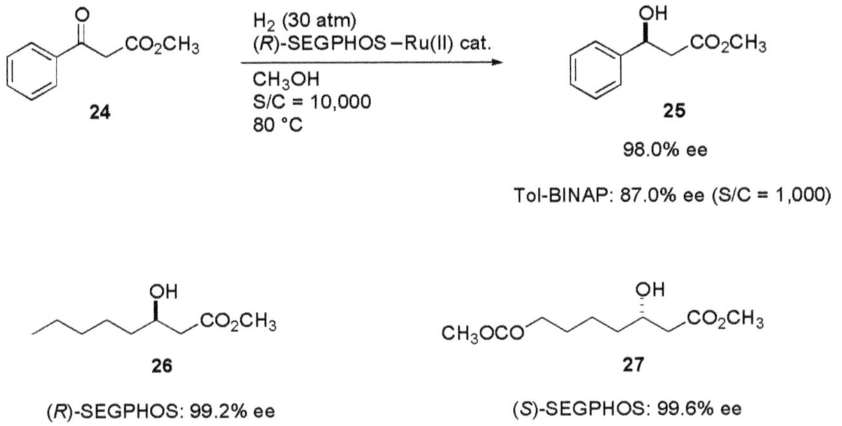

24 → 25: H_2 (30 atm), (*R*)-SEGPHOS–Ru(II) cat., CH_3OH, S/C = 10,000, 80 °C

25: 98.0% ee

Tol-BINAP: 87.0% ee (S/C = 1,000)

26: (*R*)-SEGPHOS: 99.2% ee

27: (*S*)-SEGPHOS: 99.6% ee

Scheme 19 Asymmetric hydrogenation of β-ketoesters

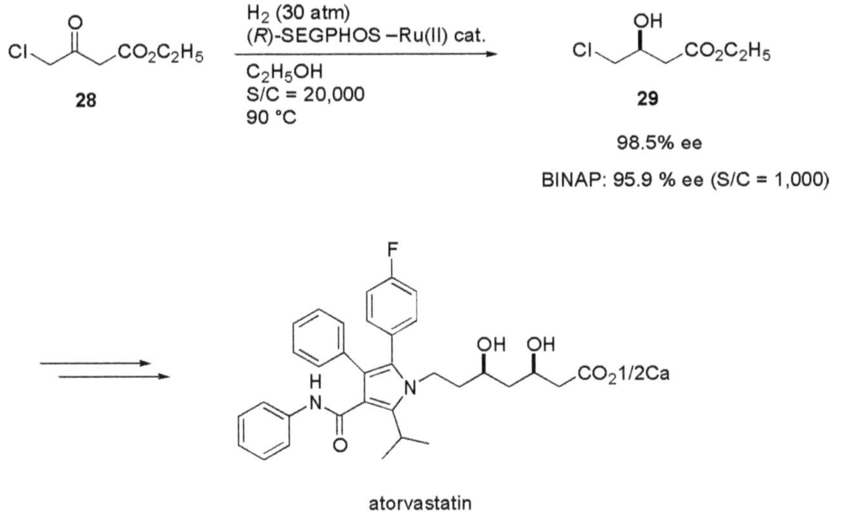

28 → 29: H_2 (30 atm), (*R*)-SEGPHOS–Ru(II) cat., C_2H_5OH, S/C = 20,000, 90 °C

29: 98.5% ee

BINAP: 95.9 % ee (S/C = 1,000)

atorvastatin

Scheme 20 Asymmetric hydrogenation of ethyl 4-chloroacetoacetate

tained 4-chloro-3-hydroxyester **29** can be transformed into the HMG-CoA reductase inhibitor [69] atorvastatin [70].

4-Benzyloxyacetoacetate **30** is an example of a β-ketoester which has a benzyloxy group at the C4-position. It can also be transformed into the corresponding alcohol **31** in 99.4% ee (Scheme 21). The obtained alcohol **31** is then transformed into β-ketoester **32** and the same asymmetric hydrogenation is repeated to give a *syn* diol **33** [71]. Acetonide formation and transformation of the methyl ester to the *tert*-butyl ester provide a common key intermediate, *tert*-butyl (3*R*,5*S*)-3,5-*O*-isopropylidene-3,5,6-trihydroxyhexanoate (**34**), to several HMG-CoA reductase inhibitors, such as simvastatin [72] and pravastatin [73].

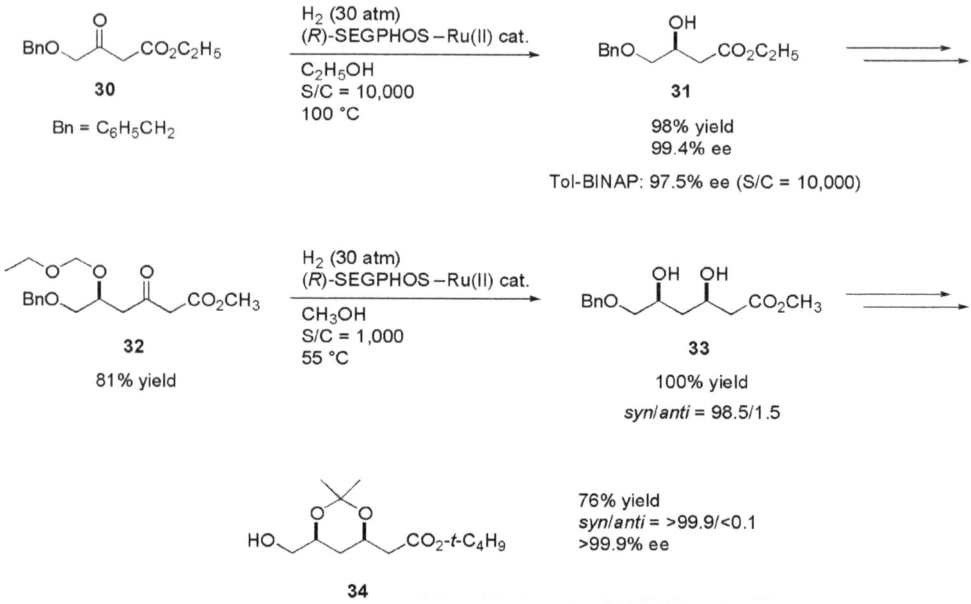

BnO–CH₂–CO–CH₂–$CO_2C_2H_5$ **30** ($Bn = C_6H_5CH_2$) → H_2 (30 atm), (*R*)-SEGPHOS–Ru(II) cat., C_2H_5OH, S/C = 10,000, 100 °C → **31** (98% yield, 99.4% ee; Tol-BINAP: 97.5% ee (S/C = 10,000))

32 (81% yield) → H_2 (30 atm), (*R*)-SEGPHOS–Ru(II) cat., CH_3OH, S/C = 1,000, 55 °C → **33** (100% yield, *syn*/*anti* = 98.5/1.5)

HO–…–CO_2-*t*-C_4H_9 **34** (76% yield, *syn*/*anti* = >99.9/<0.1, >99.9% ee)

Scheme 21 Synthesis of *tert*-butyl (3*R*, 5*S*)-3,5-*O*-isopropylidene-3,5,6-trihydroxyhexanoate

Several β-ketoesters can be transformed into oxazolidinones [74] and β-amino alcohols (Scheme 22). As mentioned above, β-ketoesters **35** can be transformed into the corresponding β-hydroxyesters **36** with the BINAP–Ru catalyst in 92–97% ee. The esters **36** are transformed into hydrazides **37**. The hydrazides **37** are then recrystallized to increase the ee's. The Curtius rearrangement generates the desired oxazolidinone **38** with more than 98% ee, which is then hydrolyzed to the amino alcohol **39** in 98–100% ee.

In this way, oxygen and nitrogen atom-containing oxazolidinones **38a** and **38b** are obtained from γ-chloroacetoacetate **28**. Other products that can be ob-

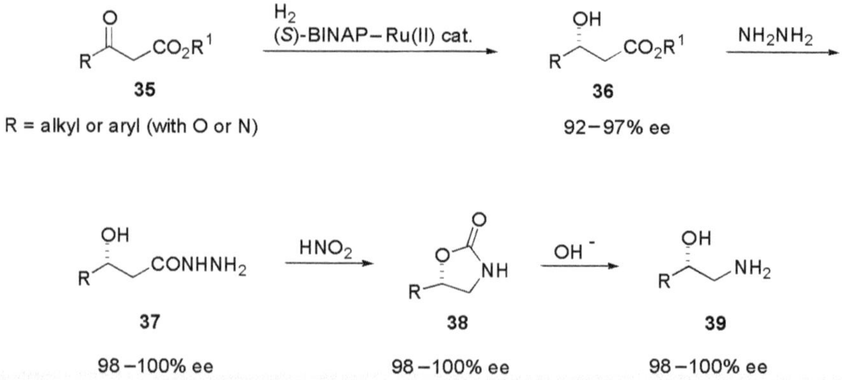

35 ($R = alkyl or aryl (with O or N)$; R–CO–CH₂–CO_2R^1) → H_2, (*S*)-BINAP–Ru(II) cat. → **36** (92–97% ee) → NH_2NH_2 → **37** (R–CH(OH)–CH₂–$CONHNH_2$; 98–100% ee) → HNO_2 → **38** (98–100% ee) → OH^- → **39** (R–CH(OH)–CH₂–NH_2; 98–100% ee)

Scheme 22 Synthesis of 5-substituted 2-oxazolidinones and β-amino alcohols

tained from the γ-chloroacetoacetate **28** include β-amino alcohols **39**, hydroxypyrrolidone **40** [75, 76], hydroxylactone **41** [77], and hydroxytetrahydrofuran **42** [77] (Scheme 23). Further, the γ-chloroacetoacetate **28** can be a starting material for the key intermediates **34** and **43** of HMG-CoA reductase inhibitors.

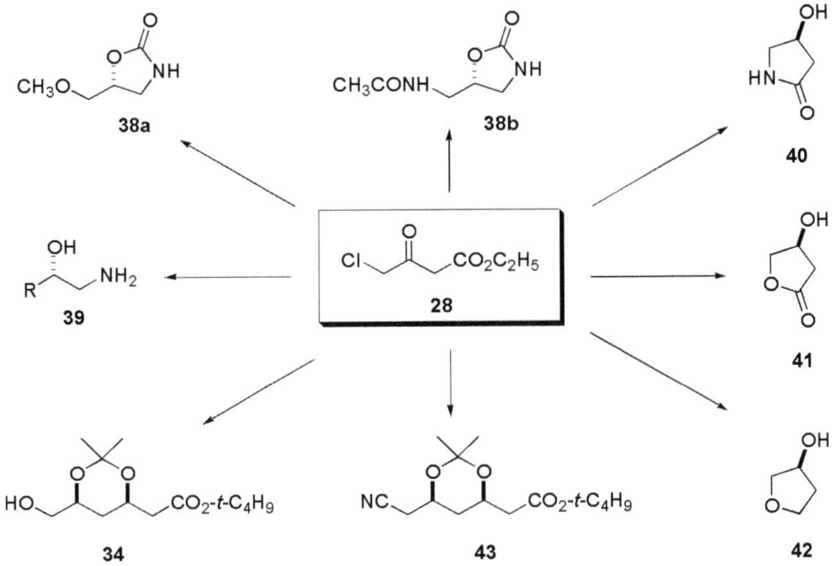

Scheme 23 The chemistry of γ-chloroacetoacetate

γ-Ketoester **44** can also be easily hydrogenated with the SEGPHOS–Ru catalyst (Scheme 24). The alcohol **45** is cyclized under acidic conditions into the lactone **46** obtained in 99.0% ee. The same process with Tol-BINAP gives a lower optical yield of 97.2%.

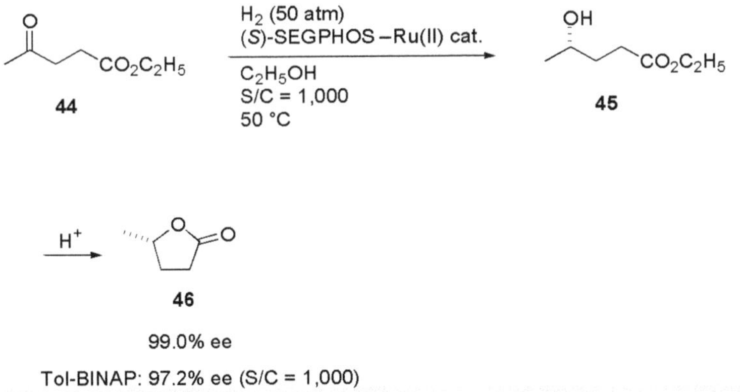

H_2 (50 atm)
(*S*)-SEGPHOS–Ru(II) cat.
C_2H_5OH
S/C = 1,000
50 °C

99.0% ee
Tol-BINAP: 97.2% ee (S/C = 1,000)

Scheme 24 Asymmetric hydrogenation of γ-ketoester

5.3.1.2
Dynamic Kinetic Resolution

The chiral lability of 2-substituted 3-oxo carboxylic esters, coupled with the high chiral recognition ability of the BINAP–Ru(II) complexes, has prompted us to investigate the possibility of stereoselective hydrogenation utilizing the dynamic kinetic resolution outlined in Scheme 25. If the racemization of enantiomers **A** and **B** is rapid enough with respect to the hydrogenation giving compound **C**, and the rates of the hydrogenation of **A** and **B** are substantially different, the hydrogenation can form one isomer predominantly out of the four possible stereoisomers. Thus, asymmetric hydrogenation of 2-substituted 3-oxo carboxylic esters provides an opportunity to obtain one enantiomer in a diastereo- and enantioselective manner [78].

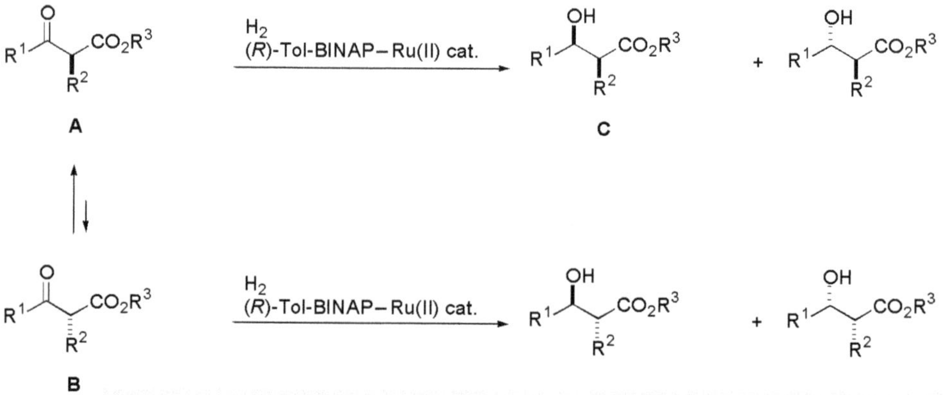

Scheme 25 Dynamic kinetic resolution

Scheme 26 shows several examples of diastereoselective hydrogenation using SEGPHOS–Ru(II) catalysts. The β-hydroxyester **47** is obtained from the corresponding β-ketoester with a DM-SEGPHOS–Ru(II) catalyst in 97.5% de and more than 98% ee. The optical yield is higher than with Tol-BINAP. This alcohol **47** is used for a moisturizer ceramide. Threonine derivative **48** is obtained with DTBM-SEGPHOS in 98% de and 99% ee, and the lactam alcohol **49** is obtained in 98% de and 98% ee. Finally, chlorohydrin **50** is obtained in 97.3% de and 99.5% ee.

The diastereoselective hydrogenation of the α-chloro-β-ketoester **51** is applied to the synthesis of several β-amino-α-hydroxyalkanoic acids **52a** and **52b** via epoxide **53** (Scheme 27). These β-amino-α-hydroxyalkanoic acids can be used for the key intermediates of aminopeptidases [79], renin [80, 81], and HIV protease inhibitors [82, 83].

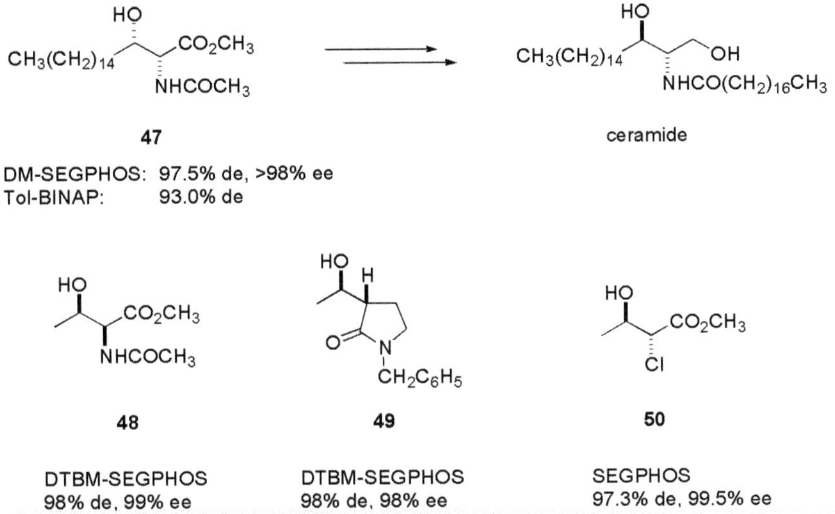

47 → ceramide

DM-SEGPHOS: 97.5% de, >98% ee
Tol-BINAP: 93.0% de

48
DTBM-SEGPHOS
98% de, 99% ee

49
DTBM-SEGPHOS
98% de, 98% ee

50
SEGPHOS
97.3% de, 99.5% ee

Scheme 26 Diastereoselective hydrogenation

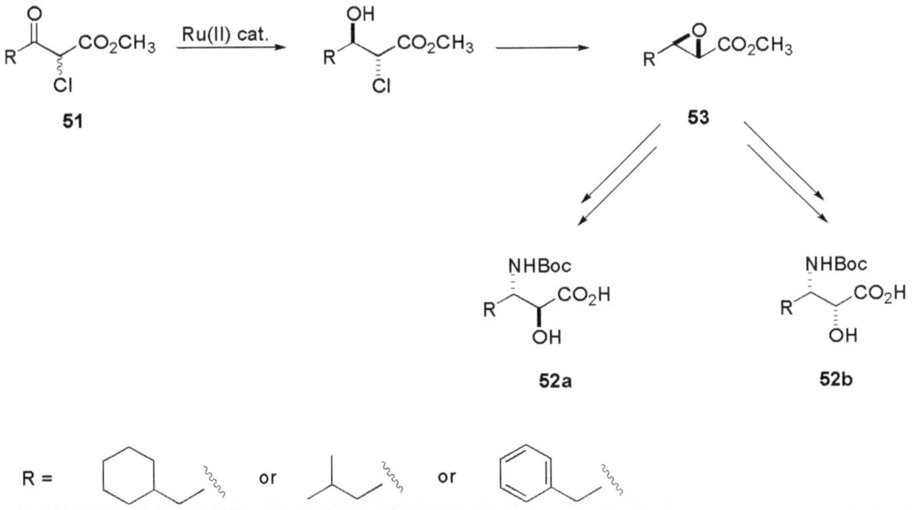

51 —Ru(II) cat.→ → 53 → 52a, 52b

Scheme 27 Synthesis of β-amino-α-hydroxyalkonoic acids

5.3.1.3
1β-Methylcarbapenem

Scheme 28 shows the TAKASAGO process for 4-acetoxy-2-azetidinone which started production in 1992. An amidomethyl group is introduced into the α-position in β-ketoester **54**. Dynamic kinetic resolution is used to control the two chiral centers simultaneously from the racemic β-ketoester **55** [84], which is

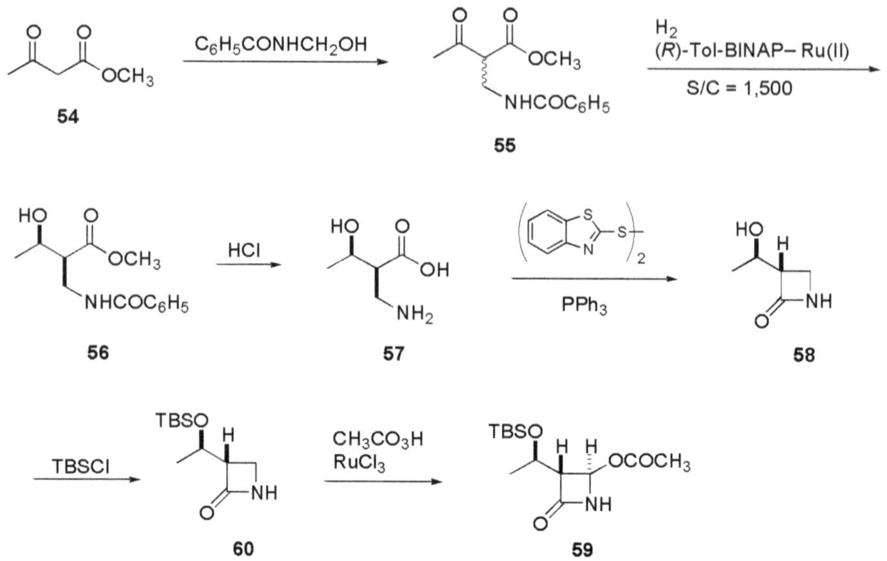

Scheme 28 Industrial process for 4-acetoxy-2-azetidinone

then hydrogenated diastereoselectively with a Tol-BINAP–Ru catalyst to *syn*-β-hydroxyester **56**. Hydrolysis of the β-hydroxyester **56** gives the free amino acid **57**, which is then cyclized into azetidinone **58**. After the hydroxy group is protected as a *tert*-butyldimethylsilyl ether, the C4-position is oxidized to make the 4-acetoxy-2-azetidinone **59**. The ruthenium-catalyzed oxidation of 2-azetidinone **60** with peroxides gave the corresponding 4-acetoxy-2-azetidinone **59** highly efficiently [85]. The combination of these two key reactions, which consist of the dynamic kinetic resolution and the ruthenium-catalyzed oxidation of 2-azetidinones, established a new industrial process for the production of a key intermediate in the synthesis of carbapenem antibiotics.

In this process, the desired β-hydroxyester **56** is obtained diastereoselectively in 86.0% de and 98.0% ee using a Tol-BINAP–Ru catalyst. To increase the enantioselectivity and the diastereoselectivity, several ligands were screened (Table 1). In the reduction of carbonyl compounds, SEGPHOS, with its narrow dihedral angle, is effective, but in this reaction, the ligand gives only 79.6% de. By changing the ligand sterically and electronically, the diastereoselectivity (de) and enantioselectivity (ee) were much improved. The resultant ligand, DTBM-SEGPHOS, gives improved optical yields of 98.6% de and 99.4% ee. With its improved properties DTBM-SEGPHOS will be used for diastereoselective hydrogenation in the industrial production of 4-acetoxy-2-azetidinone **59**.

More recently, our research has been focusing on the development of an efficient auxiliary for the construction of the 1β-methylcarbapenem nucleus (Scheme 29). We were interested in the use of 4-methyl-*N*-(1-phenylethyl)benzenesulfonamide (**61**) as the auxiliary, which can be prepared in a single step from easily available and inexpensive (*R*)-1-phenylethylamine. The magnesium eno-

Table 1 Dynamic kinetic resolution

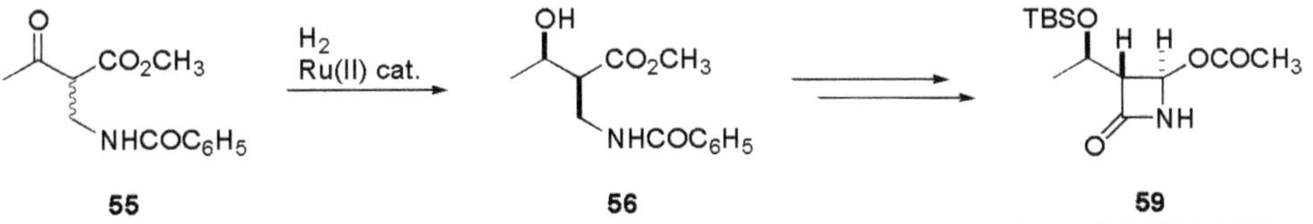

Ligand	S/C	H_2 atm	Temperature °C	% de	% ee
(*R*)-Tol-BINAP	1500	30	60	86.0	98.0
(*R*)-SEGPHOS	1000	10	70	79.6	-
(*R*)-DM-SEGPHOS	3000	30	60	93.5	98.0
(*R*)-DTBM-SEGPHOS	3000	30	80	98.6	99.4

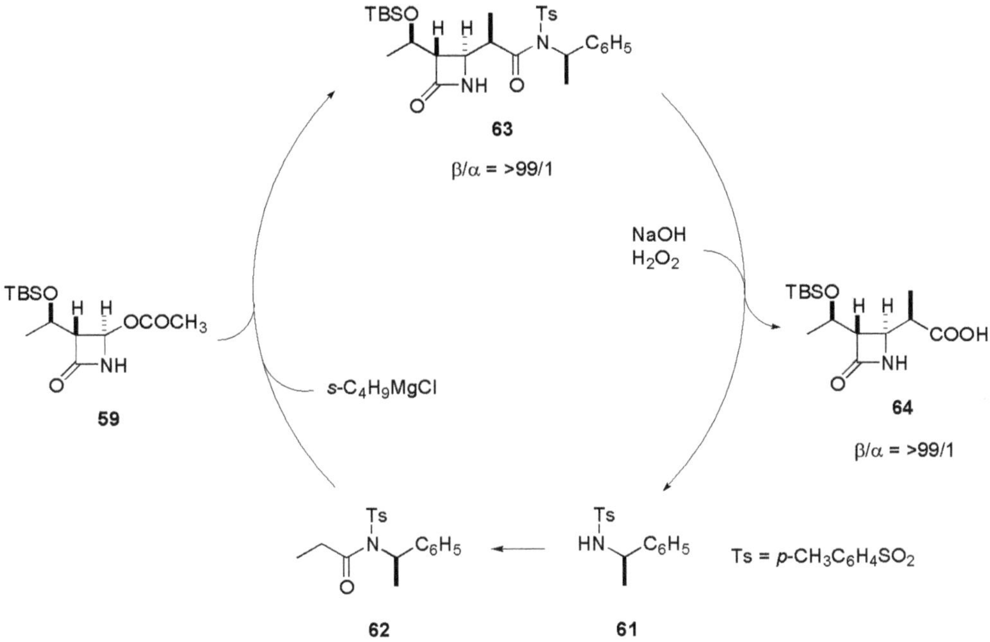

Scheme 29 A key intermediate of 1β-methylcarbapenems: β-methylcarboxylic acid

late-mediated aldol-type reaction of the chiral *N*-propionylsulfonamide **62** gave adduct **63** in high yield with high selectivity. Treatment of the product **63** with alkaline hydroperoxide led to the carboxylic acid **64**; thus, the auxiliary **61** was recovered and reusable [86].

Since we have succeeded in the synthesis of the β-methylcarboxylic acid **64**, the synthesis of enol phosphate **65** which is the key intermediate for several carbapenems was carried out via the diazo keto ester **66**, adopting Merck's process

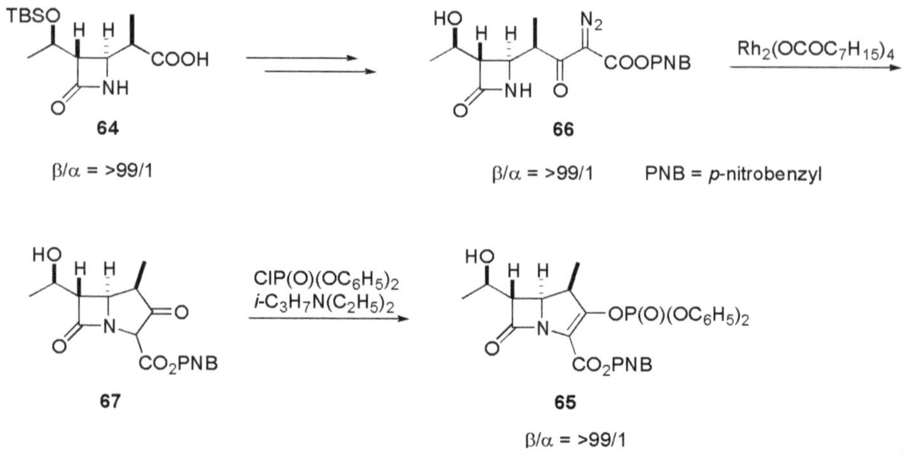

TBSO H H COOH NH O **64** β/α = >99/1

HO H H N_2 COOPNB NH O O **66** β/α = >99/1 PNB = *p*-nitrobenzyl

$Rh_2(OCOC_7H_{15})_4$

HO H H N O CO_2PNB **67**

$ClP(O)(OC_6H_5)_2$ *i*-$C_3H_7N(C_2H_5)_2$

HO H H N $OP(O)(OC_6H_5)_2$ CO_2PNB **65** β/α = >99/1

Scheme 30 A key intermediate for carbapenems: enol phosphate

[87] (Scheme 30). Under the influence of a catalytic amount of rhodium(II) octanoate, the diazo compound **66** was converted to bicyclic keto ester **67** which was then treated with diphenyl chlorophosphate to give the desired enol phosphate **65**.

5.3.1.4
Simple Ketone

Some phosphine–ruthenium complexes were known to hydrogenate simple ketones [88], but their activities were normally lower than those of the rhodium complexes. A trinuclear ruthenium complex, $[RuHCl(dppb)]_3$ [dppb=1,4-bis(diphenylphosphino)butane] as reported by James et al. [89], slowly catalyzes the hydrogenation of acetophenone under 1 atm of H_2 at 50 °C in *N,N*-dimethylacetamide.

The development of practical methods effecting the stereoselective hydrogenation of simple ketones is highly desirable, since the existing homogeneous catalysts lack reactivity, stereoselectivity, or both. A new catalyst system consisting of $RuCl_2[P(C_6H_5)_3]_3$, $NH_2(CH_2)_2NH_2$, and KOH in a 1:1:2 mole ratio effects facile hydrogenation of acetophenone in 2-propanol to give 1-phenylethanol [90]. Enantioselective hydrogenation of simple ketones is achievable when the Ru catalyst is modified by 2,2′-bis(diphenylphosphino)-1,1′-binaphthyl (BINAP) and chiral amines such as 1,2-diphenylethylenediamine (DPEN) and 1-isopropyl-2,2-bis(*p*-methoxyphenyl)-1,2-ethylenediamine (DAIPEN) [90, 91].

Noyori's asymmetric hydrogenation of simple ketones, such as acetophenone (**68**) and unsymmetrical benzophenone **69**, was also successfully industrialized (Scheme 31). In this reaction system, the combination of a BINAP–Ru catalyst and a diamine ligand is crucial for a high optical yield. The (*S*)-DM-BINAP–Ru complex and (*S*)-DAIPEN catalyze the hydrogenation of acetophenone (**68**) to give alcohol **70** in 99% ee. The catalytic activity is extremely high and the S/C

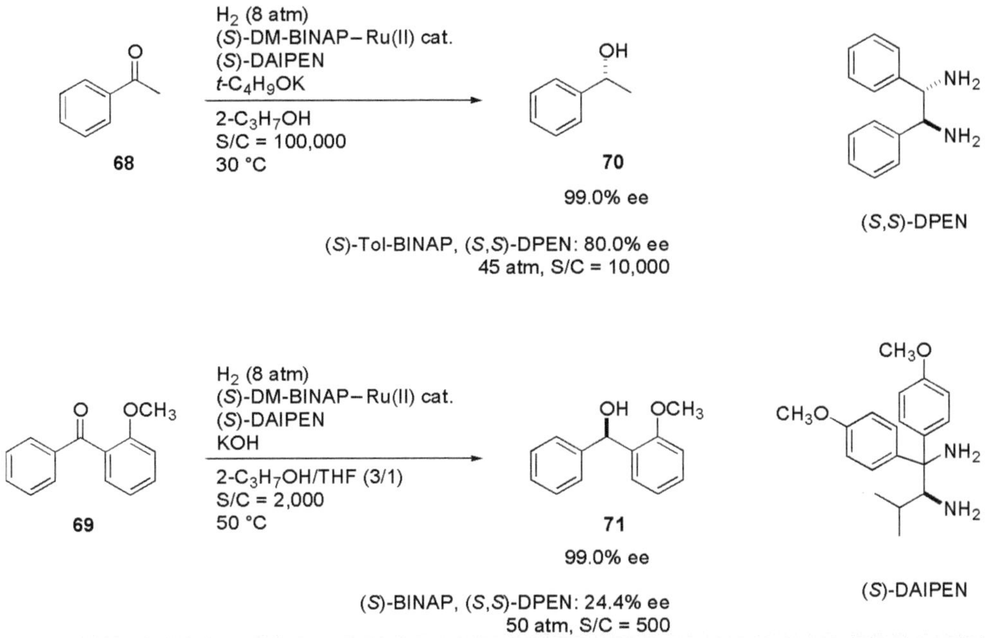

H_2 (8 atm)
(*S*)-DM-BINAP–Ru(II) cat.
(*S*)-DAIPEN
t-C_4H_9OK
2-C_3H_7OH
S/C = 100,000
30 °C

O
68
OH
70
99.0% ee

(*S*)-Tol-BINAP, (*S,S*)-DPEN: 80.0% ee
45 atm, S/C = 10,000

NH_2
NH_2
(*S,S*)-DPEN

H_2 (8 atm)
(*S*)-DM-BINAP–Ru(II) cat.
(*S*)-DAIPEN
KOH
2-C_3H_7OH/THF (3/1)
S/C = 2,000
50 °C

O OCH_3
69
OH OCH_3
71
99.0% ee

(*S*)-BINAP, (*S,S*)-DPEN: 24.4% ee
50 atm, S/C = 500

CH_3O
CH_3O
NH_2
NH_2
(*S*)-DAIPEN

Scheme 31 Asymmetric hydrogenation of simple ketones

reaches 100,000. When the (*S*)-Tol-BINAP–Ru complex and (*S,S*)-DPEN are used, the optical yield decreases. The unsymmetrical benzophenone **69** is also transformed into the corresponding alcohol **71** in this reaction system with 99.0% ee. The combination of (*S*)-BINAP and (*S,S*)-DPEN gives only 24.4% ee.

In 1996, Noyori [92] reported that the BINAP–Ru(II)–chiral diamine combined system acted as a very efficient catalyst for diastereo- and enantioselective hydrogenation of cyclic ketones (Scheme 32). Hydrogenation of racemic 2-substituted cyclohexanones in the presence of the Ru(II)–phosphine–diamine combined catalyst affords *cis* alcohols highly stereoselectively. Racemic 2-methoxycyclohexanone (**72**) gives 99.3% de and 99.2% ee with a Ru(II)–(*S*)-DM-BINAP–(*S,S*)-DPEN combined system, but 2-methylcyclohexanone (**73**) gives 95.4% de and 86.3% ee with a Ru(II)–(*S*)-DM-BINAP–(*R,R*)-DPEN combined system. The combination of the configuration of the used phosphine and diamine ligands is critical to getting high optical yields depending on the substrate. When (*S*)-BINAP instead of (*S*)-DM-BINAP is used in the combined system, the de's and ee's are decreased [93]. (1*R*,2*S*)-2-Methoxycyclohexanol (**74**) is a key chiral building block for the synthesis of the tricyclic β-lactam antibiotics, sanfetrinem [94].

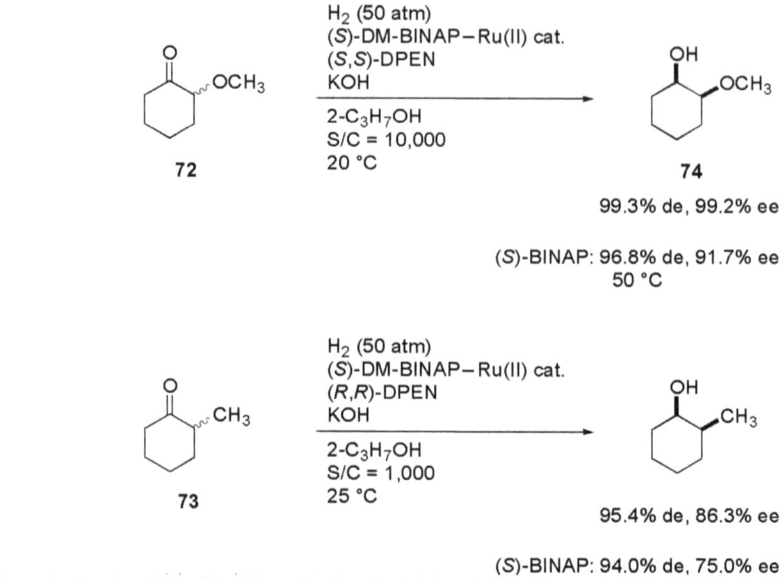

72 → 74: H_2 (50 atm), (*S*)-DM-BINAP–Ru(II) cat., (*S*,*S*)-DPEN, KOH, 2-C_3H_7OH, S/C = 10,000, 20 °C

74: 99.3% de, 99.2% ee

(*S*)-BINAP: 96.8% de, 91.7% ee, 50 °C

73 → : H_2 (50 atm), (*S*)-DM-BINAP–Ru(II) cat., (*R*,*R*)-DPEN, KOH, 2-C_3H_7OH, S/C = 1,000, 25 °C

95.4% de, 86.3% ee

(*S*)-BINAP: 94.0% de, 75.0% ee

Scheme 32 Diastereoselective hydrogenation of 2-substituted cyclohexanones

5.3.2
Olefin

Hydrogenation of olefins is one of the most significant synthetic operations. The ability to do it asymmetrically provides a versatile tool to create stereo-defined tertiary carbon centers of chiral organic molecules. Geraniol (**75**) and nerol convert quantitatively to the citronellol (**76**) enantiomers with 96–99% enantiomeric purity. Hydrogenation of tetrahydrofarnesol (**77**) in the presence of $Ru(OCOCH_3)_2$[(*S*)-binap] affords (3*R*,7*R*)-hexahydrofarnesol (**78**) with 99% diastereomeric purity. With this system, we can obtain a side chain of vitamin E [95, 96] (Scheme 33).

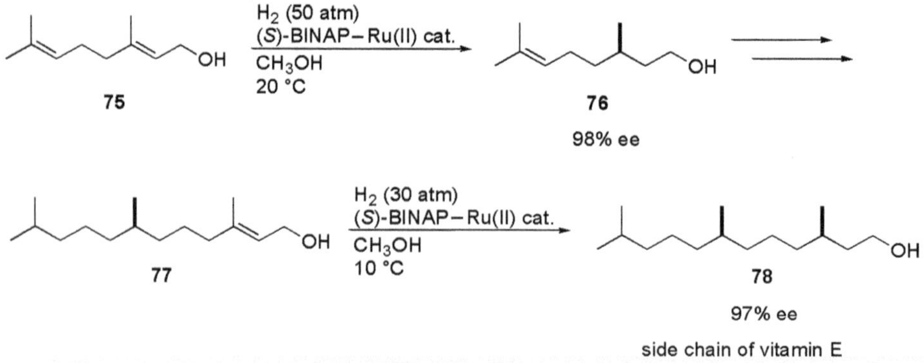

75 → 76: H_2 (50 atm), (*S*)-BINAP–Ru(II) cat., CH_3OH, 20 °C

76: 98% ee

77 → 78: H_2 (30 atm), (*S*)-BINAP–Ru(II) cat., CH_3OH, 10 °C

78: 97% ee, side chain of vitamin E

Scheme 33 Asymmetric hydrogenation of allyl alcohols

BINAP–Ru(II) dicarboxylate complexes [62] catalyze hydrogenation of α,β- and β,γ-unsaturated carboxylic acids in an enantioselective manner [97–99]. We reported that [RuI(H_8-binap)(*p*-cymene)]I served as an even more effective catalyst for the asymmetric hydrogenation of an α,β-unsaturated carboxylic acid than BINAP complexes of Ru(II) (Scheme 34) [100, 101]. An important application of this is the synthesis of naproxen [102–104], an important anti-inflammatory. In the presence of a catalytic amount of the (*S*)-H_8-BINAP–Ru catalyst, the hydrogenation of tiglic acid (**79**) proceeds smoothly at room temperature under 4 atm of hydrogen to give an optically active methylbutanoic acid (**80**), in 96% ee. A lower ee (91%) has been obtained with BINAP. (*S*)-2-Methylbutanoic acid (**80**) and its esters are important in creating artificial fruit flavors, e.g. apple, strawberry, and grape. Similarly, naproxen is prepared with the (*S*)-H_8-BINAP–Ru catalyst in 92.2% ee. Even higher ee can be obtained when the (*S*)-BINAP–Ru catalyst is used, but the reaction conditions are not suitable for industrial application [84].

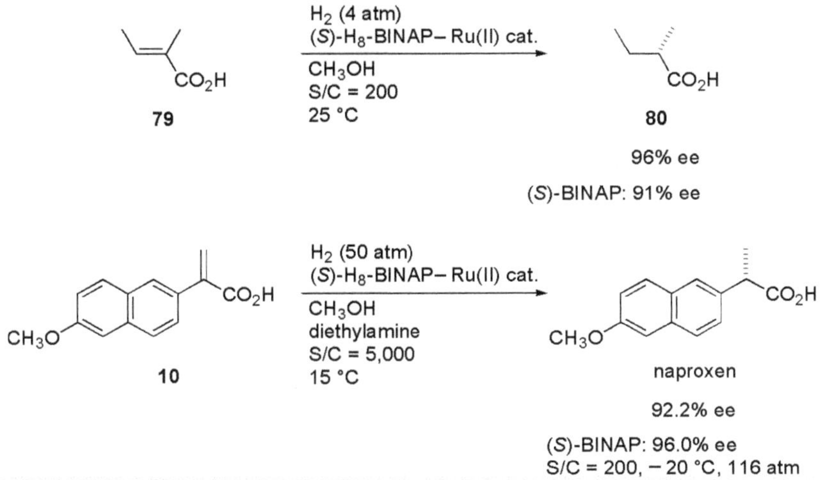

CO_2H
79
H_2 (4 atm)
(*S*)-H_8-BINAP– Ru(II) cat.
CH_3OH
S/C = 200
25 °C
CO_2H
80
96% ee
(*S*)-BINAP: 91% ee

CH_3O
CO_2H
10
H_2 (50 atm)
(*S*)-H_8-BINAP– Ru(II) cat.
CH_3OH
diethylamine
S/C = 5,000
15 °C
CH_3O
CO_2H
naproxen
92.2% ee
(*S*)-BINAP: 96.0% ee
S/C = 200, – 20 °C, 116 atm

Scheme 34 Asymmetric hydrogenation of α,β-unsaturated carboxylic acids

Because enantiomers react at different rates in chiral environments, certain racemic allylic alcohols can be resolved by the BINAP–Ru catalyzed hydrogenation [105]. Some cyclic compounds are obtained in >99% ee at ~50% conversion. This method provides a practical way to prepare (*R*)-4-hydroxy-2-cyclopentenone (**81**), a useful building block for prostaglandin synthesis [106], in 98% ee at 68% conversion (Scheme 35).

The reduction of the olefin group in diketene (**82**) is possible with a SEGPHOS–Ru catalyst (Scheme 36). The optical yield is similar to that with the Tol-BINAP–Ru catalyst [107–109], but the catalytic activities between them are very different. To hydrogenate diketene (**82**), Tol-BINAP needs an S/C of 1000. However, the SEGPHOS–Ru-catalyst's activity is much higher and the S/C is 12,270 [110]. The lactone **83** is the raw material for biodegradable polymers [111, 112].

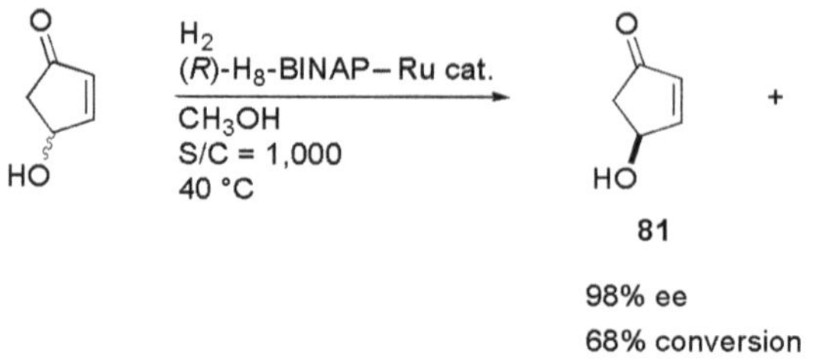

H_2
(*R*)-H_8-BINAP–Ru cat.
CH_3OH
S/C = 1,000
40 °C

81

98% ee
68% conversion

Scheme 35 Kinetic resolution of 4-hydroxy-2-cyclopentenone

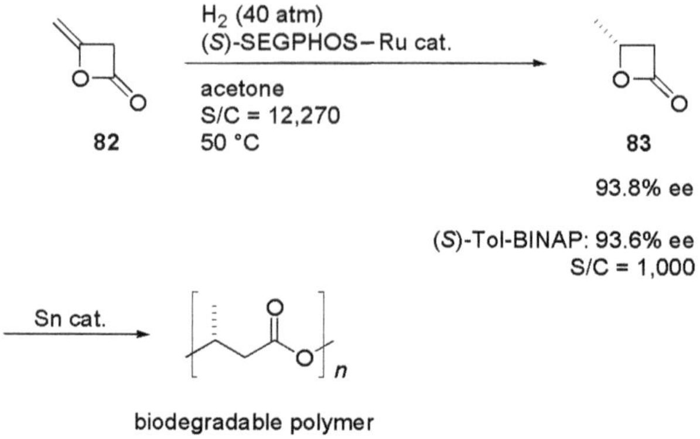

H_2 (40 atm)
(*S*)-SEGPHOS–Ru cat.
acetone
S/C = 12,270
50 °C

82

83

93.8% ee

(*S*)-Tol-BINAP: 93.6% ee
S/C = 1,000

Sn cat.

biodegradable polymer

Scheme 36 Asymmetric hydrogenation of diketene

5.3.2.1
Ritalin

Racemic *threo*-methylphenidate hydrochloride (Ritalin® hydrochloride) is a mild nervous system stimulant and is currently the most widely used drug for the treatment of children with attention-deficit hyperactivity disorder (ADHD) or hyperkinetic child syndrome [113, 114]. Racemic (±)-methylphenidate possesses side effects, e.g. anorexia, insomnia, weight loss, dizziness, dysphoria, and euphoria. To segregate the desired pharmacological activities from the side effects, there is a great interest in preparing enantiomerically pure (2*S*,2′*R*)-methylphenidate (**84**) on a large scale.

The enantioselective synthesis of (2*S*,2′*R*)-methylphenidate (**84**) involving the asymmetric hydrogenation of enamine **85** as the key step with [RuI(H_8-binap)(*p*-cymene)]I provides an approach to (2*R*,2′*R*)-methylphenidate (**86**) after epimerization (Scheme 37) [115].

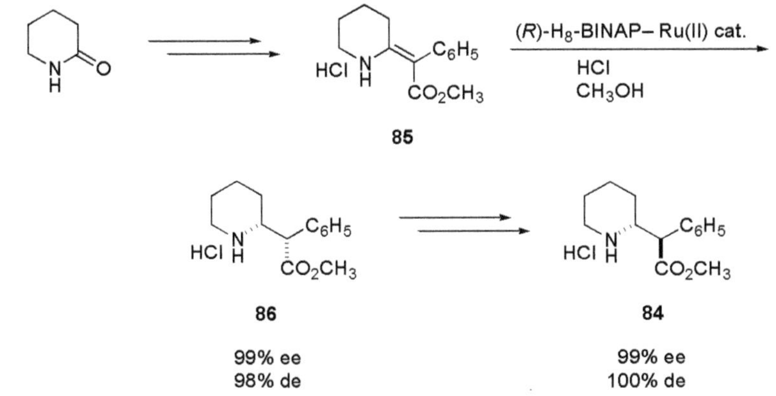

(*R*)-H_8-BINAP–Ru(II) cat.
HCl
CH_3OH

HCl C_6H_5 CO_2CH_3

85

86

99% ee
98% de

84

99% ee
100% de

Scheme 37 Synthesis of (2*R*,2′*R*)-(+)-*threo*-methylphenidate

5.4 Asymmetric Hydrogen Transfer Reaction

Chiral propargylic alcohols are useful building blocks for the synthesis of various biologically active compounds [116–127]. Although asymmetric hydrogenation would be the most straightforward approach, none of the currently available catalyst systems can convert α,β-acetylenic ketones to propargylic alcohols in a chemoselective and enantioselective manner. The $RuCl_2(phosphine)_3$/1,2-diamine/KOH system [91] cannot be used. Noyori reported the first asymmetric transfer hydrogenation of acetylenic ketones using chiral Ru(II) catalysts and 2-propanol as the hydrogen donor [126, 127]. This method allows highly selective reduction of structurally diverse acetylenic ketones to propargylic alcohols of high enantiomeric purity leaving the triple bond intact (Scheme 38). We have succeeded in industrializing this asymmetric hydrogen transfer reduction.

$(CH_3)_3Si$ — Ru(*p*-cymene)[(*R*,*R*)-Tsdpen], 2-C_3H_7OH → $(CH_3)_3Si$ … OH

NaOH → OH

98.3% ee

Ru(*p*-cymene)[(*R*,*R*)-Tsdpen]

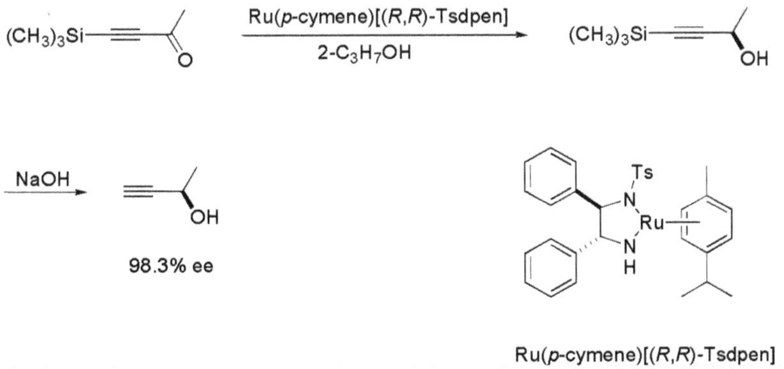

Scheme 38 Asymmetric hydrogen transfer reduction

6
Summary

The BINAP chemistry developed by Noyori and Takaya has become the cornerstone for the success of many industrial asymmetric processes at TAKASAGO. The development of new chiral ligands and catalysts resulted in the discovery and applications of SEGPHOS–Ru catalysts whose efficiency generally surpasses that of BINAP–Ru catalysts.

Asymmetric hydrogenation can offer either the *R*- or *S*-isomer in the same manner, by selecting the appropriate substrate and catalyst, and often exceeds biocatalysis in its generality. The operations of the reaction, isolation, separation, and purification involved are simple, easily performed, and well-suited for mass production. Industrial demand for asymmetric catalytic reactions will certainly continue to increase enormously.

Acknowledgement. We are grateful to all our coworkers who have so effectively participated in our work described here and whose names are shown in the list of references.

References

1. Stinson SC (2001) Chem Eng News 79(40):79
2. Dang TP, Kagan HB (1971) J Chem Soc Chem Commun 481
3. Knowles WS, Sabacky MJ, Vineyard BD (1972) J Chem Soc Chem Commun 10
4. Vineyard BD, Knowles WS, Sabacky MJ, Bachman GL, Weinkauff DJ (1977) J Am Chem Soc 99:5946
5. Brunner H, Zettlmeier W (1993) Handbook of enantioselective catalysis with transition metal compounds, vol 2. VCH, Weinheim
6. Brunner H, Zettlmeier W (1993) Handbook of enantioselective catalysis with transition metal compounds, vol 1. VCH, Weinheim
7. Miyashita A, Yasuda A, Takaya H, Toriumi K, Ito T, Souchi T, Noyori R (1980) J Am Chem Soc 102:7932
8. Takaya H, Mashima K, Koyano K, Yagi M, Kumobayashi H, Taketomi T, Akutagawa S, Noyori R (1986) J Org Chem 51:629
9. Sayo N, Zhang X, Ohmoto T, Yoshida A, Yokozawa T (1997) US Patent 5 693 868A
10. Zhang X, Sayo N (1999) US Patent 5 922 918A
11. Zhang X, Mashima K, Koyano K, Sayo N, Kumobayashi H, Akutagawa S, Takaya H (1991) Tetrahedron Lett 32:7283
12. Zhang X, Mashima K, Koyano K, Sayo N, Kumobayashi H, Akutagawa S, Takaya H (1994) J Chem Soc, Perkin Trans I 2309
13. Saito T, Yokozawa T, Ishizaki T, Moroi T, Sayo N, Miura T, Kumobayashi H (2001) Adv Synth Catal 343:264
14. Schmid R, Cereghetti M, Heiser B, Schönholzer P, Hansen H-J (1988) Helv Chem Acta 71:897
15. Schmid R, Foricher J, Cereghetti M, Schönholzer P (1991) Helv Chem Acta 74:370
16. Artz SP, Cram DJ (1984) J Am Chem Soc 106:2160
17. Sainsbury M (1980) Tetrahedron 36:3327
18. Kumobayashi H, Akutagawa S, Otsuka S (1978) J Am Chem Soc 100:3949
19. Tani K, Yamagata T, Otsuka S, Akutagawa S, Kumobayashi H, Taketomi T, Takaya H, Miyashita A, Noyori R (1982) J Chem Soc Chem Commun 600
20. Tani K, Yamagata T, Tatsuno Y, Yamagata Y, Tomita K, Akutagawa S, Kumobayashi H, Otsuka S (1985) Angew Chem Int Ed 24:217

21. Otsuka S, Tani K, Yamagata T, Akutagawa S, Kumobayashi H, Yagi M (1983) JP Patent S58-4748A
22. Tani K, Yamagata T, Akutagawa S, Kumobayashi H, Taketomi T, Takaya H, Miyashita A, Noyori R, Otsuka S (1984) J Am Chem Soc 106:5208
23. Takabe K, Uchida Y, Okisaka S, Yamada T, Katagiri T, Okazaki T, Oketa Y, Kumobayashi H, Akutagawa S (1985) Tetrahedron Lett 26:5153
24. Henrick CA, Staal GB, Siddall JB (1973) J Agri Food Chem 21:354
25. Hayashi T, Katsumura A, Konishi M, Kumada M (1979) Tetrahedron Lett 425
26. Törös S, Kollár L, Heil B, Markó L (1982) J Organomet Chem 232:C17
27. Achiwa K, Kogure T, Ojima I (1977) Tetrahedron Lett 4431
28. Sakuraba S, Achiwa K (1995) Chem Pharm Bull 43:748
29. (1996) Drugs Fut 21:83
30. James BR (1973) Homogeneous hydrogenation. Wiley, New York
31. Kieboom APG, van Rantwijk F (1977) Hydrogenation and hydrogenolysis in synthetic organic chemistry. Delft University Press, Rotterdam
32. Birch AJ, Williamson DH (1976) Org React 24:4
33. James BR (1979) In: Stone FGA, West R (eds) Advances in organometallic chemistry (catalysis and organic syntheses), vol 17. Academic Press, New York, p 319
34. Chaloner PA, Esteruelas MA, Joo F, Oro LA (1994) Homogeneous hydrogenation. Kluwer, Dordrecht
35. Ojima I, Eguchi M, Tzamarioudaki M (1995) In: Abel EW, Stone FGA, Wilkinson G, Hegedus LS (eds) Comprehensive organometallic chemistry II, vol 12. Elsevier, Oxford, p 9
36. Rylander PN (1979) Catalytic hydrogenation in organic syntheses. Academic Press, New York
37. Rylander PN (1990) Hydrogenation methods. Academic Press, London
38. Harada K, Munegumi T (1991) In: Trost BM, Fleming I (eds) Comprehensive organic synthesis, vol 8. Pergamon Press, Oxford, p 139
39. Ohkuma T, Noyori R (1998) In: Beller M, Bolm C (eds) Transition metals for organic synthesis: building blocks and fine chemicals, vol 2. Wiley-VCH, Weinheim, p 25
40. Ohkuma T, Noyori R (1999) In: Jacobsen EN, Pfaltz A, Yamamoto H (eds) Comprehensive asymmetric catalysis, vol 1. Springer, Berlin Heidelberg New York, p 199
41. Ohkuma T, Kitamura M, Noyori R (2000) In: Ojima I (ed) Catalytic asymmetric synthesis, 2nd edn. Wiley-VCH, New York, p 1
42. Schrock RR, Osborn JA (1970) J Chem Soc Chem Commun 567
43. Tani K, Suwa K, Tanigawa E, Yoshida T, Okano T, Otsuka S (1982) Chem Lett 261
44. Tani K, Tanigawa E, Tatsuno Y, Otsuka S (1985) J Organomet Chem 279:87
45. Tolman CA (1977) Chem Rev 77:313
46. Burk MJ, Harper TGP, Lee JR, Kalberg C (1994) Tetrahedron Lett 35:4963
47. Bakos J, Tóth I, Heil B, Markó L (1985) J Organomet Chem 279:23
48. Jiang Q, Jiang Y, Xiao D, Cao P, Zhang X (1998) Angew Chem Int Ed 37:1100
49. Nagel U, Roller C (1998) Z Naturforsch 53b:267
50. Zhang X, Taketomi T, Yoshizumi T, Kumobayashi H, Akutagawa S, Mashima K, Takaya H (1993) J Am Chem Soc 115:3318
51. Robin F, Mercier F, Ricard L, Mathey F, Spagnol M (1997) Chem Eur J 3:1365
52. Burk MJ, Bienewald F, Challeger S, Derrick A, Ramsden JA (1999) J Org Chem 64:3290
53. Burk MJ (2000) Acc Chem Res 33:363
54. Hewitt BD, Burk MJ, Johnson NB (2000) WO Patent 00 55150
55. Pye PJ, Rossen K, Reamer RA, Tsou NN, Volante RP, Reider PJ (1997) J Am Chem Soc 119:6207
56. Lennon I (2001) The BACS conference at ChemSpec Europe, Amsterdam, p 17
57. Manfré F, Le-Fur I, Lelièvre S, Spagnol M (2000) ChiraSource. The Catalyst Group, Lisbon
58. Sakai N, Mano S, Nozaki K, Takaya H (1993) J Am Chem Soc 115:7033

59. Nozaki K, Li W, Horiuchi T, Takaya H, Saito T, Yoshida A, Matsumura K, Kato Y, Imai T, Miura T, Kumobayashi H (1996) J Org Chem 61:7658
60. Ikariya T, Ishii Y, Kawano H, Arai T, Saburi M, Yoshikawa S, Akutagawa S (1985) J Chem Soc Chem Commun 922
61. Ohta T, Tonomura Y, Nozaki K, Takaya H, Mashima K (1996) Organometallics 15:1521
62. Ohta T, Takaya H, Noyori R (1988) Inorg Chem 27:566
63. Mashima K, Kusano K, Ohta T, Noyori R, Takaya H (1989) J Chem Soc Chem Commun 1208
64. Noyori R, Ohkuma T, Kitamura M, Takaya H, Sayo N, Kumobayashi H, Akutagawa S (1987) J Am Chem Soc 109:5856
65. Kitamura M, Ohkuma T, Inoue S, Sayo N, Kumobayashi H, Akutagawa S, Ohta T, Takaya H, Noyori R (1988) J Am Chem Soc 110:629
66. Kawano H, Ishii Y, Saburi M, Uchida Y (1988) J Chem Soc, Chem Commun 87
67. (1994) Drugs Fut 19:693
68. Sumi K (2000) ChiraSource. Lisbon, Portugal
69. Grundy SM (1988) New Engl J Med 319:24
70. Graul A, Castañer J (1997) Drugs Fut 22:956
71. Sakurai K, Mitsuhashi S, Kumobayashi H (1994) JP Patent H6-65226A
72. (1988) Drugs Fut 13:531
73. Endo A (1985) J Med Chem 28:401
74. Mitsuhashi S, Nara H, Miura T, Sumi K, Imai T (2000) EP Patent 1 008 590A
75. Mitsuhashi S, Sumi K, Moroi K (1999) JP Patent H11-286479A
76. Mitsuhashi S, Kumobayashi H (1997) JP Patent H9-208558A
77. Yuasa Y, Sano N, Konno M (1997) JP Patent H9-77759A
78. Noyori R, Ikeda T, Ohkuma T, Widhalm M, Kitamura M, Takaya H, Akutagawa S, Sayo N, Saito T, Taketomi T, Kumobayashi H (1989) J Am Chem Soc 111:9134
79. Tobe H, Morishima H, Aoyagi T, Umezawa H, Ishiki K, Nakamura K, Yoshioka T, Shimauchi Y, Inui T (1982) Agric Biol Chem 46:1865
80. Matsuda F, Matsumoto T, Ohsaki M, Ito Y, Terashima S (1990) Chem Lett 723
81. Iizuka K, Kamijo T, Harada H, Akahane K, Kubota T, Umeyama H, Ishida T, Kiso Y (1990) J Med Chem 33:2707
82. Mimoto T, Imai J, Kisanuki S, Enomoto H, Hattori N, Akaji K, Kiso Y (1992) Chem Pharm Bull 40:2251
83. Shibata N, Itoh E, Terashima S (1998) Chem Pharm Bull 46:733
84. Mashima K, Kusano K, Sato N, Matsumura Y, Nozaki K, Kumobayashi H, Sayo N, Hori Y, Ishizaki T, Akutagawa S, Takaya H (1994) J Org Chem 59:3064
85. Murahashi S-I, Naota T, Kuwabara T, Saito T, Kumobayashi H, Akutagawa S (1990) J Am Chem Soc 112:7820
86. Kumobayashi H, Miura T, Sayo N, Saito T (1999) J Synth Org Chem Jpn 57:387
87. Shih DH, Baker F, Cama L, Christensen BG (1984) Heterocycles 21:29
88. Bennett MA, Matheson TW (1982) In: Wilkinson G, Stone FGA, Abel EW (eds) Comprehensive organometallic chemisty, vol 4. Pergamon, Oxford, p 931
89. James BR, Pacheco A, Rettig SJ, Thorburn IS, Ball RG, Ibers JA (1987) J Mol Catal 41:147
90. Ohkuma T, Ooka H, Hashiguchi S, Ikariya T, Noyori R (1995) J Am Chem Soc 117:2675
91. Ohkuma T, Ooka H, Ikariya T, Noyori R (1995) J Am Chem Soc 117:10417
92. Ohkuma T, Ooka H, Yamakawa M, Ikariya T, Noyori R (1996) J Org Chem 61:4872
93. Matsumoto T, Murayama T, Mitsuhashi S, Miura T (1999) Tetrahedron Lett 40:5043
94. Di Modugno E, Erbetti I, Ferrari L, Galassi G, Hammond SM, Xerri L (1994) Antimicrob Agents Chemother 38:2362
95. Takaya H, Ohta T, Sayo N, Kumobayashi H, Akutagawa S, Inoue S, Kasahara I, Noyori R (1987) J Am Chem Soc 109:1596
96. Takaya H, Ohta T, Sayo N, Kumobayashi H, Akutagawa S, Inoue S, Kasahara I, Noyori R (1987) J Am Chem Soc 109:4129

97. Ohta T, Takaya H, Kitamura M, Nagai K, Noyori R (1987) J Org Chem 52:3174
98. Ohta T, Takaya H, Noyori R (1990) Tetrahedron Lett 31:7189
99. Ashby MT, Halpern J (1991) J Am Chem Soc 113:589
100. Zhang X, Uemura T, Matsumura K, Sayo N, Kumobayashi H, Takaya H (1994) Synlett 501
101. Uemura T, Zhang X, Matsumura K, Sayo N, Kumobayashi H, Ohta T, Nozaki K, Takaya H (1996) J Org Chem 61:5510
102. Shen TY (1972) Angew Chem Int Ed 11:460
103. Rieu J-P, Boucherle A, Cousse H, Mouzin G (1986) Tetrahedron 42:4095
104. Botteghi C, Paganelli S, Schionato A, Marchetti M (1991) Chirality 3:355
105. Kitamura M, Kasahara I, Manabe K, Noyori R, Takaya H (1988) J Org Chem 53:708
106. Noyori R (1989) Chem Br 25:883
107. Ohta T, Miyake T, Takaya H (1992) J Chem Soc, Chem Commun 1725
108. Ohta T, Miyake T, Seido N, Kumobayashi H, Takaya H (1995) J Org Chem 60:357
109. Ohta T, Miyake T, Seido N, Kumobayashi H, Akutagawa S, Takaya H (1992) Tetrahedron Lett 33:635
110. Okeda Y, Hashimoto T, Hori Y, Hagiwara T (2000) US Patent 6 043 380A
111. Hori Y, Takahashi Y, Yamaguchi A, Nishishita T (1993) Macromolecules 26:4388
112. Hori Y, Takahashi Y, Yamaguchi A, Nishishita T (1993) Macromolecules 26:5533
113. Millichap JG (1973) Ann NY Acad Sci 205:321
114. Swanson JM, Kinsbourne M (1979) In: Hale GH, Lewis M (eds) Attention and cognitive development. Plenum Press, New York, p 249
115. Seido N, Nishikawa T, Sotoguchi T, Yuasa Y, Miura T, Kumobayashi H (1999) US Patent 5 859 249A
116. Mori K, Akao H (1978) Tetrahedron Lett 4127
117. Vigneron JP, Bloy V (1980) Tetrahedron Lett 21:1735
118. Fried J, Sih JC (1973) Tetrahedron Lett 3899
119. Kluge AF, Kertesz DJ, O-Yang C, Wu HY (1987) J Org Chem 52:2860
120. Leder J, Fujioka H, Kishi Y (1983) Tetrahedron Lett 24:1463
121. Cohen N, Lopresti RJ, Neukom C, Saucy G (1980) J Org Chem 45:582
122. Roush WR, Spada AP (1982) Tetrahedron Lett 23:3773
123. Stork G, Nakamura E (1983) J Am Chem Soc 105:5510
124. Kim HY, Stein K, Toogood PL (1996) J Chem Soc Chem Commun 1683
125. Zhu G, Lu X (1995) J Org Chem 60:1087
126. Matsumura K, Hashiguchi S, Ikariya T, Noyori R (1997) J Am Chem Soc 119:8738
127. Haack K-J, Hashiguchi S, Fujii A, Ikariya T, Noyori R (1997) Angew Chem Int Ed 36:285

Topics Organomet Chem (2004) 6: 97–122
DOI 10.1007/978-3-540-36966-0

Development of Transition Metal-Mediated Cyclopropanation Reactions

Albert J. DelMonte · Eric D. Dowdy* · Daniel J. Watson

Department of Process Research and Development, Bristol-Myers Squibb, New Brunswick, NJ 08903, USA
* Department of Process Research, Gilead Sciences, Foster City, CA 94404, USA
E-mail: albert.delmonte@bms.com, edowdy@gilead.com, daniel.watson@bms.com

Abstract A brief discussion of the methods for cyclopropane formation emphasizing organometallic reagents and catalysts is followed by several examples of the application of these methods in the manufacture of functional chemicals and pharmaceuticals. The unique challenges faced by scientists in a process environment will be emphasized.

Keywords Asymmetric Catalysis · Bisoxazoline · Cyclopropanation · Diazoacetate · Pyrethroids

1 Introduction

Throughout this book you will find a plethora of examples where organometallic reactions have been used effectively for commercial synthesis. This chapter will explore methods useful for the construction of cyclopropanes on a large scale. The cyclopropane moiety has been found in numerous biologically active natural products. Despite the inherent strain and unusual bonding characteristics of cyclopropanes, their rigid and stereochemically well-defined nature make them attractive for inclusion in functional chemicals. As a result chemists have continued to devise novel and diverse approaches to the synthesis of cyclopropanes with emphasis on both diastereo- and enantioselective transformations. The most powerful and flexible of these methods rely on organometallic reagents and catalysts to effect the key cyclopropane forming step.

2 Survey of Cyclopropanation Methods

A wide variety of methods for cyclopropane generation, many involving organometallic intermediates, have been described in the literature over the past century. The sections below are intended to illustrate the breadth of methods available while giving leading references to more thorough examinations of their scope and generality.

2.1 Main Group Metal-Mediated Cyclopropanation

2.1.1 *Nucleophilic Cyclization*

Early methods for the formation of cyclopropanes relied mostly on intramolecular alkylation as the ring forming step. When possible, this chemistry was used to take advantage of inexpensive raw materials and/or advantageously positioned functional groups. In the synthesis of the reverse transcriptase inhibitor, efavirenz (**3**), 5-chloropentyne (**1**) was converted in a one-pot process to cyclopropyl acetylene by double deprotonation followed by intramolecular alkylation

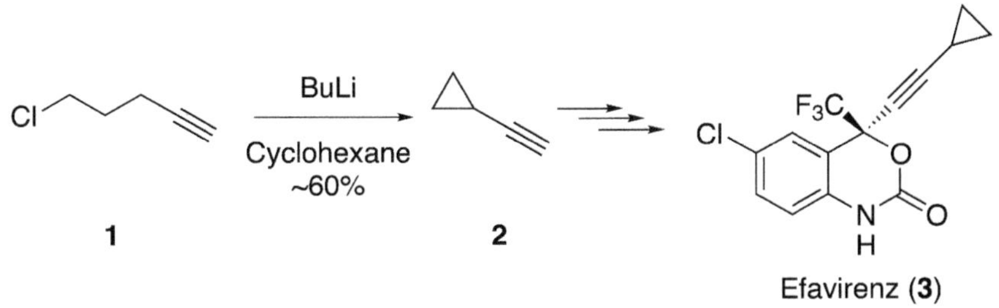

Scheme 1

(Scheme 1). This concise procedure offered operational simplicity and enhanced yields when compared with other syntheses described in the literature [1, 2].

An interesting variant of this transformation was applied in a concise synthesis of anti-ulcerative candidate, spizofurone (**7**, Scheme 2). Chloride attack on spirolactone **5** followed by decarboxylation furnished enolate **6** which was poised for attack on the pendant chloroethyl group [3–7].

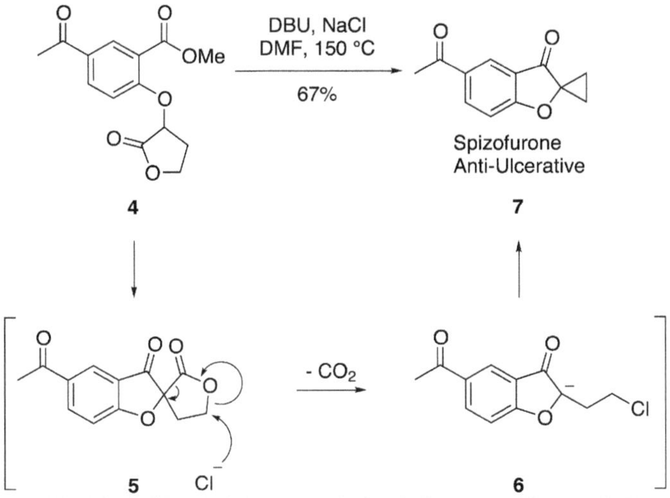

Scheme 2

2.1.2
Sulfur-Ylide Reagents

Electron-deficient double bonds can often be cyclopropanated using sulfur-ylide reagents [8]. Following an addition/elimination mechanism, the nucleophilic character of these reagents can limit their utility in substrates containing multiple electrophilic functional groups. Researchers at Eli Lilly successfully used dimethylsulfoxonium methylide to install a cyclopropane late in the synthesis of a retinoid compound (**9**) being investigated as a treatment for non-insulin-dependent diabetes (Eq. 1) [9]. Following optimization, the reaction was suitable to prepare kilogram quantities of **9**. Interestingly, this reagent was chosen in preference to an organozinc Simmons-Smith cyclopropanating reagent for scaleup because they wanted to avoid the handling of diethylzinc and diiodomethane [10]. The challenges facing the scaleup of a Simmons-Smith reaction will be discussed later in this chapter.

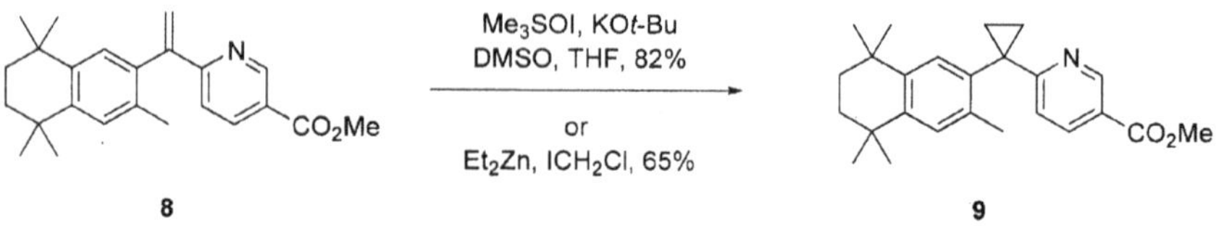

N CO_2Me **8**

Me_3SOI, KO*t*-Bu
DMSO, THF, 82%
or
Et_2Zn, ICH_2Cl, 65%

N CO_2Me **9** (1)

2.1.3 *Intramolecular Attack on an Epoxide*

Another classical approach that has proven amenable to scaleup is the conversion of an epoxide to a cyclopropane. A telescoped alkylation/cyclopropanation was reported in the patent literature for the preparation of the antidepressant Milnacipran (**14**, Scheme 3). The multistep transformation involved alkylation of benzonitrile with epichlorohydrin followed by deprotonation and intramolecular epoxide opening [11]. A comparison of this retrosynthesis and a transition metal-catalyzed olefin cyclopropanation will be presented later.

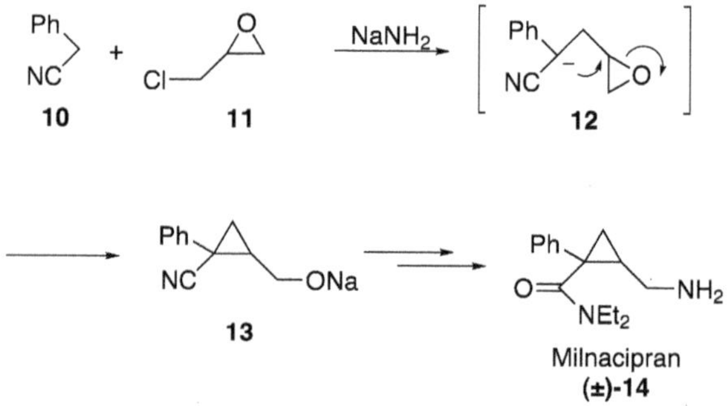

Ph NC **10** + Cl O **11** $NaNH_2$ [Ph NC O] **12**

Ph NC ONa **13** Ph O NEt$_2$ NH_2 Milnacipran (±)-**14**

Scheme 3

2.2 Transition Metal-Mediated Cyclopropanation

Process research and development is a subset of target-oriented total synthesis emphasizing commercial viability of targets defined by their biological activity. The chemistry employed is forced to fit the desired structure, sometimes requiring more powerful and selective methods than those described above. For these instances we turn to transition metal-mediated cyclopropanation.

2.2.1
Non-Catalytic Transition Metal-Mediated Cyclopropanation

2.2.1.1
Simmons-Smith Cyclopropanation

One of the now classical approaches for the generation of cyclopropanes from alkenes is the Simmons-Smith reaction [12, 13]. This reaction involves treatment of an alkene with a zinc-copper couple and diiodomethane. The reaction is stereospecific with respect to alkene geometry and product stereochemistry. Olefin substituents that have a *trans* relationship remain *trans* in the cyclopropane and, likewise, *cis* substituents maintain a *cis* relationship in the cyclopropane product. This is rationalized by invoking a concerted methylene transfer through the "butterfly" transition state, **17**, depicted in Scheme 4. Modern experimentation continues to support this hypothesis as density functional theory calculations reveal the same transition state [14].

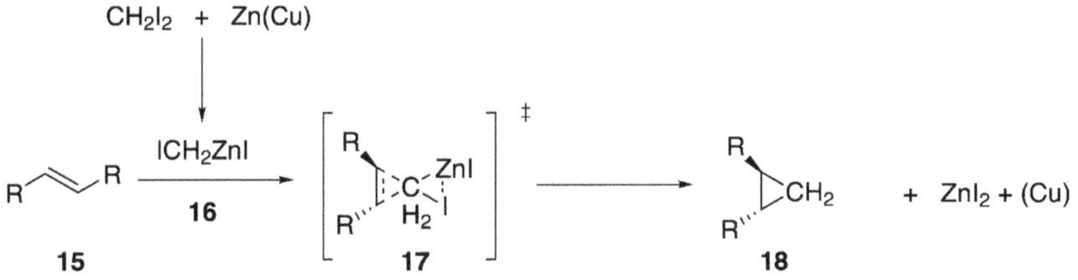

CH_2I_2 + Zn(Cu) → ICH_2ZnI (**16**); R–CH=CH–R (**15**) → [transition state **17**]‡ → cyclopropane **18** + ZnI_2 + (Cu)

Scheme 4

The mechanism, selectivity, functional group compatibility, and experimental procedures for Simmons-Smith and related cyclopropanations have been reviewed extensively [13, 15]. A number of modifications to the original Simmons-Smith cyclopropanation procedure have been reported through the years. For example, the cyclopropanation of substrates containing certain functional groups, including α,β-unsaturated aldehydes, ketones, enamines, enol ethers, and enol esters, often proceeds with low yields under the Simmons-Smith conditions. Later it was shown that the use of a zinc-silver couple led to an increase in the yield in these cases [16]. Additionally, the hydrolytic workup has been replaced with a simplified workup protocol involving the addition of an amine such as pyridine followed by filtration of the resulting zinc salts. It has also been reported that diethylzinc [17, 18] and ethylzinc iodide [19, 20] may be employed to effect cyclopropanation. Detailed spectroscopic studies have revealed that each protocol generates a structurally unique species, leading to differing selectivity and reactivity [20].

2.2.1.2
Asymmetric Simmons-Smith Methods

It has long been known that the Simmons-Smith reagent is directed by oxygen containing functional groups in the alkene bearing substrate [21]. This effect can be exploited through the use of chiral auxiliaries or protecting groups to effect enantioselective cyclopropanation [22]. Chiral modifiers have also been developed that influence the stereochemical outcome of the reaction without first having been covalently bonded to the substrate [23]. These methods often suffer from poor substrate generality and/or the expense of generating a stoichiometric amount of an enantiopure reagent. Asymmetric catalysis offers a more efficient method to transfer chirality to a prochiral substrate.

2.2.2
Transition Metal-Catalyzed Cyclopropanation

The ability of certain metals and their salts to decompose diazoacetate compounds was recognized almost a century ago [24]. Several years later, researchers learned ways to trap the reactive intermediates of this decomposition with alkenes to form cyclopropanes [25]. This method was exploited for the manufacture of early cyclopropane containing insecticides (see later). By exploiting the propensity of metal salts to bind organic ligands, scientists have improved the diastereoselectivity and added sometimes exquisite enantioselectivity to this reaction.

2.2.2.1
Salicylaldimine-Based Catalysts

The asymmetric cyclopropanation of olefins with alkyl diazoacetates enjoys a notable place in the history of asymmetric catalysis. One of the earliest examples of asymmetric organometallic catalysis was the asymmetric cyclopropanation of styrene with ethyl diazoacetate in the presence of a copper salen catalyst **19** (Eq. 2 and Table 1) [26, 27]. Although the enantiomeric excess initially reported was modest, this seminal research by Noyori and Nozaki gave birth to the field of asymmetric organometallic catalysis. This promising result was developed into an efficient catalytic method for the production of cyclopropane containing insecticide molecules by researchers at the Sumitomo Chemical Company (see below).

Ph **20** → (Ethyl diazoacetate; **19**, R = C*HPhMe) → **21** (CO_2Et) + **22** (CO_2Et) (2)

Table 1

Catalyst	Catalyst loading	Solvent	Ratio *trans:cis* **21:22**	*ee* *trans* **21**	*ee* *cis* **22**
19 Cu	0.9 mol%	Styrene	70:30	6%	6%
23 Cu	1 mol%	$ClCH_2CH_2Cl$	73:27	92%	80%
24 Cu	1 mol%	$CHCl_3$	66:34	3%	8%
25 Cu	1 mol%	$CHCl_3$	77:23	98%	93%
26 Cu	1 mol%	$CHCl_3$	73:27	99%	97%
28 Ru	2 mol%	CH_2Cl_2	91:9	89%	79%

2.2.2.2
Bisoxazoline Ligated Catalysts

Developing new Lewis basic ligands for asymmetric catalysis has been a popular research theme in both academic and industrial laboratories. Many of these catalyst systems have been demonstrated on the asymmetric cyclopropanation reaction. In 1988, Pfaltz introduced the use of C2-symmetric chiral semicorrin ligands (**23**, Fig. 1) bound to copper [28–30]. This provided a huge leap forward in the enantioselective catalytic cyclopropanation as measured by the reaction of ethyl diazoacetate with styrene (Table 1). Refinement of this idea led to bisoxazoline ligands that have proven even more selective [31–33]. These ligands are readily prepared from the corresponding β-amino alcohols, allowing for a series of catalysts with varying steric and electronic properties [34]. In parallel research reported almost simultaneously, the Evans and Masamune laboratories studied the asymmetric cyclopropanation of *mono*- and 1,1-disubstituted olefins with achiral diazoacetates using copper complexes of bisoxazoline ligands. Bisoxazoline ligands **25** and **26** were shown to be superior to **24** by forming a six-membered chelate [33]. The most effective catalyst reported was a 1:1 complex of CuOTf and bisoxazoline **26** formed in situ. This particular copper catalyst provided >99% enantiomeric excess in the cyclopropanation of isobutylene with ethyl diazoacetate in the presence of 0.1 mol% catalyst.

By exploring a series of catalysts with varying steric properties, the following trends were revealed [31, 32]. The substitution pattern of the olefin and the ster-

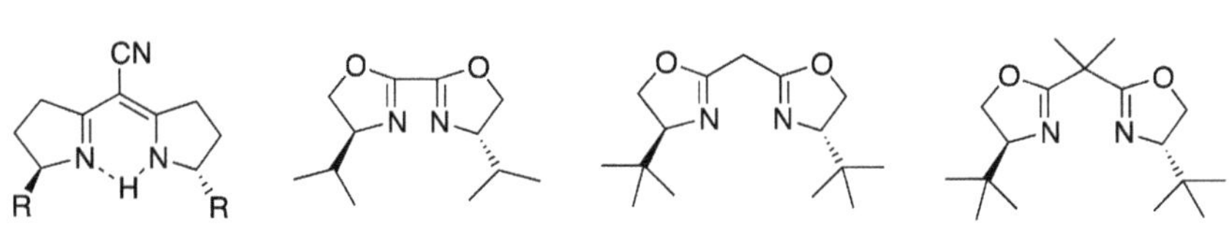

23 R = CMe_2OH **24** **25** **26**

Fig. 1

ic bulk of the diazoacetate affect the diastereoselectivity of the transformation. The enantioselectivity depends on the steric bulk of the bisoxazoline ligand. Highly stereoselective cyclopropanation can be achieved by properly tuning the catalyst, reagents and reaction conditions.

2.2.2.3
Bis-oxazolinylpyridine (Pybox) Ligated Catalysts

In 1994, Nishiyama and co-workers adapted their chiral ruthenium bis-oxazolinylpyridine (pybox) hydrosilylation catalyst to the cyclopropanation of olefins [35, 36]. By mixing commercially available dichloro(*p*-cymene) ruthenium(II) dimer with pybox ligand **27** in the presence of ethylene, complex **28** was formed (Eq. 3). This catalyst proved effective for the cyclopropanation of aromatic and aliphatic olefins in high enantiomeric excess. A procedure was developed to generate the catalyst in situ, obviating the need for gaseous ethylene. A later study demonstrated that modification of the ligand could tune the catalytic activity providing even higher enantiomeric selectivity [37]. The large scale synthesis and application of this catalyst in a pilot plant setting will be discussed later in this chapter.

$$\mathbf{27} \xrightarrow[\substack{CH_2Cl_2 \\ CH_2CH_2}]{[RuCl_2(p\text{-cymene})]_2} \mathbf{28}\ (88\%\ \text{yield}) \tag{3}$$

2.2.2.4
Rhodium Carboxamidate Catalysts

In the early 1980s rhodium (II) carboxylates were shown to be mild catalysts for the cyclopropanation of olefins [38, 39]. These dirhodium catalysts, containing four bridging carboxylate ligands, were quite active but provided low diastereo- and chemoselectivity in the cyclopropanation of various alkenes and dienes. Later it was discovered that the di-rhodium (II) complexes containing carboxamidates (**30**), commonly known as 'Doyle catalysts', were less reactive but more diastereoselective in the cyclopropanation of alkenes as well as more chemoselective in the cyclopropanation of dienes [40–42]. Catalysts of this type bearing chiral ligands have proven especially useful for the enantioselective intramolecular formation of cyclopropane-fused lactones (Eq. 4) [43].

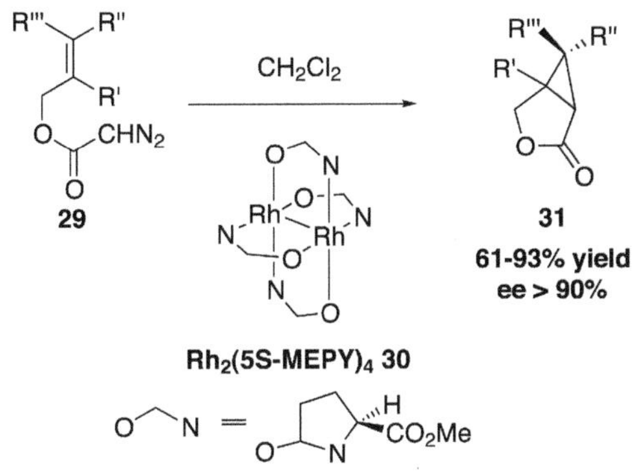

(4)

3 Application of Organometallic Cyclopropanation Chemistry and Scaleup Concerns

3.1 Simmons-Smith Reaction

In 1982, a group from Sandoz Pharmaceuticals reported a practical, scalable and safe Simmons-Smith cyclopropanation (Eq. 5) [44]. The cyclopropane product (33) was further exploited for the preparation of amides and hydrazides useful for the treatment of atherosclerosis [45, 46].

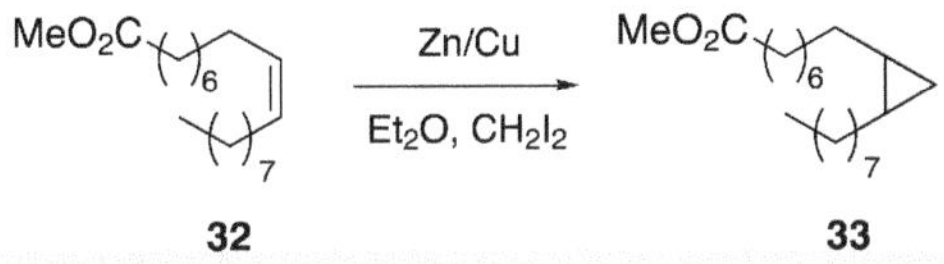

(5)

Early in the development of the cyclopropanation reaction, a delayed and violent exotherm was identified as a key safety concern. This exotherm was exacerbated on large scale and resulted in the unwanted distillation of the reaction solvent (diethyl ether). Diethyl ether is quite volatile and potentially hazardous on scale due to its low flash point (-45 °C). One solution to this problem involved replacing the diethyl ether with linear ethers having boiling points greater than or equal to 70 °C [47]. 1,2-Dimethoxyethane was the solvent of choice because its higher boiling point (85 °C) and heat capacity helped to absorb the reaction exotherm.

The high cost of utilizing zinc-copper amalgam led to the development of a cheaper process employing various forms of metallic zinc (mossy, rods or foil). The reaction was performed using ultrasonic irradiation, thus requiring no chemical activation of the zinc. This method also helped to eliminate the delayed exotherm, and the reaction time and yield were more reproducible. Initially, the reaction was performed using zinc powder, which often led to excessive foaming. This was replaced with a solid piece of zinc, which not only reduced the foaming, but also dampened the exotherm due to a decrease in the reactive surface area [48]. The zinc block could be removed at any time to slow the reaction. At the conclusion of the reaction the zinc block was conveniently removed from the mixture before workup. Interestingly, on a 1-l scale, zinc paddles attached to the glass stirrer shaft were tested. Even though they partially dissolved throughout the reaction, an excess of zinc was used in the design so that efficient stirring continued to the end of the reaction. This curious design was later replaced with zinc cones, a shape that provided a practical surface area to weight ratio. It is noteworthy that although sonication is sufficient to activate the zinc, chemical activators such as TMSCl and $TiCl_4$ have been used to reduce the induction period of the reaction [15, 49].

Of interest to scientists attempting to develop a robust Simmons-Smith cyclopropanation process is a report that trace amounts of lead present in certain zinc sources can inhibit the desired reaction (Table 2) [50]. Electrolytic zinc powder prepared by hydrometallurgy was reported to be free of lead, while zinc powder produced by pyrometallurgy (distillation) contained trace amounts of lead (0.04–0.07 mol%). These low levels of lead almost entirely inhibited the cyclopropanation of cyclooctene. Fortuitously, the negative influence of trace levels of lead could be countered by adding 2 mol% TMSCl to the reaction.

CH_2I_2, Zn
Et_2O, 40 °C, 8 h

34 **35**

(6)

Table 2

Zinc source	Lead content (mol% of Zn)	Additive	Conversion (%)	SM (%)
Electrolytic	0	None	96	2
Distilled	0.04	None	7	91
Distilled	0.04	2 mol% TMSCl	92	2

3.2 Catalytic Cyclopropanation Involving Diazo Compounds

3.2.1 *Safety Considerations in Using Ethyl Diazoacetate (EDA)*

Safety is a prime concern in all laboratory research and manufacturing scale up. Ethyl diazoacetate (EDA), a common reagent used in metal-catalyzed asymmetric cyclopropanations, is both flammable and toxic [51]. Although EDA is one of the more stable compounds in its class, in general it is recommended that diazocompounds be stored at 0 °C in the dark and precautions be taken to avoid contact with any sources of metal or strong acids. Although it is relatively easy to purchase and ship in small quantities, bulk chemical suppliers are reluctant to ship larger quantities because of the aforementioned safety issues. This leaves the process chemist with limited options: adopting an alternate synthetic route, outsourcing the chemistry to a third party with the facilities to manufacture bulk quantities of EDA on-site, or developing the required expertise and equipment to prepare and handle EDA safely.

In a series of papers, scientists from Monsanto have discussed the handling and thermal characteristics of EDA. Standardized detonation tube tests revealed that both dilute (18 wt%) and concentrated (97 wt%) toluene solutions of EDA did not detonate when exposed to shock [52]. Accelerating rate calorimetry experiments found that the onset temperature for decomposition varied from 55 °C for neat EDA to 100 °C for an 11 wt% solution in toluene [53]. The maximum self-heating rate varied dramatically from 0.35 °C/min for a dilute solution to 130 °C/min for neat EDA. The decomposition also liberated gas, with a maximum pressure rise rate of 2280 psi/min. Similar safety studies performed in other laboratories agree with the magnitude of the thermal hazards posed by EDA [54–57], and reveal that while not shock sensitive [58], EDA is sensitive to friction at 40 Newtons [54]. These studies highlight that while not benign, EDA can be synthesized and utilized safely in a manufacturing setting.

An oft-cited literature preparation of EDA involved the diazotization of ethyl glycinate hydrochloride (**36**) in a dichloromethane/water mixture with sodium nitrite and sulfuric acid (Eq. 7) [59]. Several modifications to this procedure to improve scalability were disclosed by Bristol-Myers Squibb [54–57]. Phosphoric acid was used to acidify the mixture and a borate buffer was added to provide highly reproducible reaction initiation. The organic solvent was changed from methylene chloride to toluene. In addition to being more environmentally benign, processing with toluene minimized the handling of EDA-rich layers in the workup. To avoid the storage of any unused EDA, a procedure was developed for its controlled destruction. Adding EDA to 10 mol equivalents of 50 wt% of acetic acid in water resulted in smooth decomposition. These modifications allowed for the safe and practical large scale production of 2 mol/l EDA in toluene in 83 mol% yield. Monsanto scientists mention similar improvements to the preparation of EDA resulting in 89–91 mol% yield of high quality reagent [60].

$$\underset{\mathbf{36}}{HCl\bullet H_2NCH_2C(O)OEt} \xrightarrow[\text{or}\ 1.05\ \text{eq. } NaNO_2,\ H_3PO_4 + Na_2B_4O_7,\ \text{pH } 4.5\text{-}5.0;\ \text{Toluene, } H_2O,\ -5\ \text{to}\ 20\ ^\circ C]{1.2\ \text{eq. } NaNO_2,\ \text{cat. } H_2SO_4;\ CH_2Cl_2,\ H_2O,\ -20\ ^\circ C} \underset{\mathbf{37}}{\text{Ethyl diazoacetate}} \quad (7)$$

3.2.2
Metal Toxicity

One common impediment to the application of organometallic chemistry in the preparation of pharmaceutical compounds is the toxicity of some organometallic reagents. Very low levels of metal impurities in the active pharmaceutical ingredient (API) may be required to meet purity specifications. In addition, metal impurities may have a deleterious effect on subsequent steps in the overall process. For the common reagents and catalysts employed in the cyclopropanation reactions described in this chapter, the potential for toxicity is relatively mild. Zinc [61] and copper [62] are associated with minimal toxicity as both are required for normal biochemical function in the body. Ruthenium complexes are considered "slightly toxic", warranting some care in their handling [63]. The toxicity of rhodium complexes also varies, with certain members testing positive for mutagenicity in an Ames assay [64].

To avoid any issue of metal toxicity in pharmaceutical products, metals are typically removed to very low levels. Novel techniques and reagents developed to assist in this task are described elsewhere in this volume.

3.2.3
Pyrethroids

Pyrethrum, an extract from certain chrysanthemum species, has been used for insect control for centuries [65]. The major active component of the extract was identified by Staudinger and Ruzicka in 1924 as being a *trans*-cyclopropanecarboxylate which they labeled pyrethrin I (**38**, Fig. 2) [66]. Owing to their effective pest control and low mammalian toxicity, **38** and synthetic analogs have been manufactured and marketed as insecticides.

Ethyl Chrysanthemate
(1*R*,3*R*)-39 ⟹ **40** (8)

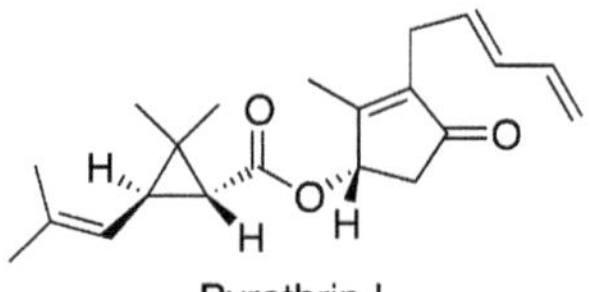

Pyrethrin I
38

Fig. 2

Ethyl chrysanthemate (**39**), a key intermediate in the preparation of synthetic pyrethroids, has long been prepared by the cyclopropanation of 2,5-dimethyl-2,4-hexadiene (**40**). Due to the commercial potential of a successful insecticide, years of research have been invested in improving the synthesis and manufacture of this compound. While early processes produced isomeric mixtures (racemic mixtures of *cis*- and *trans*- chrysanthemates), technological improvements and new approaches gradually have led to the selective synthesis of the most active isomers.

3.2.3.1
Batch Process for Ethyl Chrysanthemate

The first synthetic pyrethroid to be produced in large quantities was the isomeric mixture, allethrin (**41**, Fig. 3) [65]. The cyclopropane moiety was installed by the copper powder catalyzed reaction of ethyl diazoacetate with 2,5-dimethyl-2,4-hexadiene (Eq. 9). Ethyl chrysanthemate (**39**) was isolated in 64% yield as a 3:1 mixture favoring the *trans* isomer present in natural pyrethrin I (**38**). The most biologically active stereoisomer of allethrin comprised <20% of the final product mixture due to the lack of absolute stereocontrol and low relative stereocontrol during the synthesis.

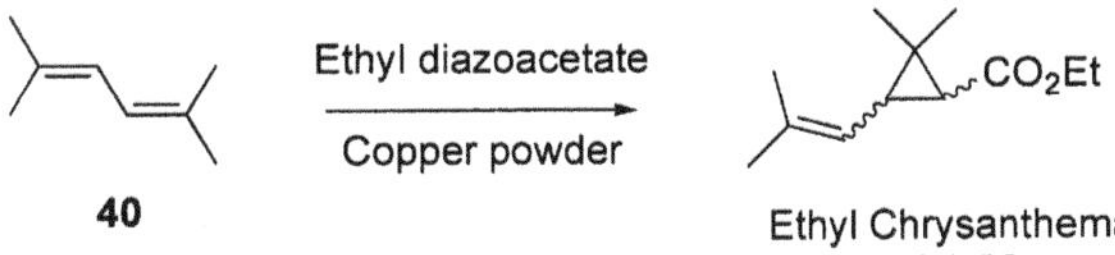

40 — Ethyl diazoacetate / Copper powder → CO_2Et Ethyl Chrysanthemate (±)-**39** ~3:1 *trans*:*cis* 65% yield (9)

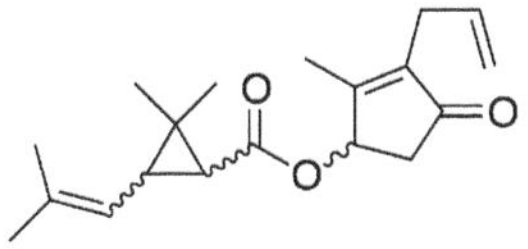

Allethrin **41**

Fig. 3

As mentioned previously the generation and handling of ethyl diazoacetate on large scale is not a trivial operation. To mitigate the risk to human operators, early manufacturers were reported to conduct the cyclopropanation reaction by remote control from behind concrete barriers [67]. Another creative processing approach to minimize the handling of and exposure to ethyl diazoacetate is embodied in a patent from ICI [68]. They used a large excess of **40** as the organic solvent to extract the EDA as it was generated. After removing the aqueous phase, copper was added as a catalyst and the mixture was heated to effect cyclopropanation. The excess diene was recycled into the process during the distillative isolation of the product.

3.2.3.2
Continuous Process for Ethyl Chrysanthemate

The ultimate reduction in handling and storage of EDA was realized in a continuous process for the production of ethyl chrysanthemate (**39**), patented by Stauffer Chemical Company in 1975 (Fig. 4) [69]. Their approach was to extract EDA into an organic solvent as it was formed by the acid catalyzed reaction of sodium nitrite and ethyl glycinate at 5 °C. This organic extract was continuously

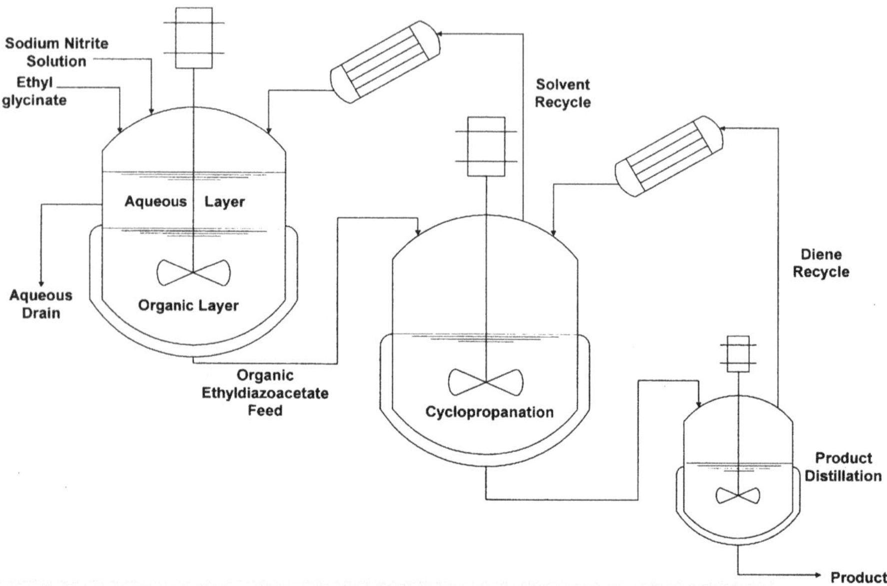

Fig. 4

fed into a reactor containing a copper catalyst and 2,5-dimethyl-2,4-hexadiene at 100–125 °C. The product solution could be collected and distilled as a single batch, or drained and distilled continuously. This allowed the excess diene to be recycled into the cyclopropanation vessel. Several modifications were made to the reaction conditions to facilitate the smooth operation of this continuous process. Dichloroethane was selected as the solvent for EDA extraction due to several advantageous properties. It did not form emulsions with the aqueous reaction mixture, thus allowing continuous removal of the EDA-rich organic phase. The solubility of water in dichloroethane was low enough that drying of the EDA solution was not required before use, and the boiling point was suitable to allow rapid reaction while distilling the solvent for recycle. To insure that the concentration of EDA would not build up before the cyclopropanation could begin, the copper/diene mixture was heated before initiating EDA feed.

While the goal of devising the continuous process was to minimize worker exposure to ethyl diazoacetate, the process also realized a higher yield (70–80%) when compared to a typical batch process.

3.2.3.3
Catalytic Asymmetric Production of Chrysanthemates

Early in the development of pyrethroid insecticides, it was discovered that certain stereoisomers possessed far greater potency than others. For this reason much research was conducted into methods for the stereoselective synthesis of cyclopropane carboxylates.

40 $\xrightarrow[\text{Cu(L*)n}]{N_2CHCO_2R}$ 39 (R = Et) (10)

The first generation of synthetic pyrethroids were most active when the cyclopropane unit possessed the same *trans* relationship found in naturally obtained pyrethrin I (**38**). Using Noyori's copper salicylaldimine based catalysts (**19**) as a starting point, scientists at Sumitomo Chemical Company optimized the conversion of 2,5-dimethyl-2,4-hexadiene to *trans*-ethyl chrysanthemate [70]. Their first approach was to optimize the catalyst structure to improve the *cis/trans* and enantioselectivity of the transformation [71–73]. They prepared an extensive series of copper complexes (**43**) using esters of amino acids as their starting point. Reaction with various Grignard reagents followed by imine formation and complexation with copper provided the desired complexes. The complexes were used to mediate the cyclopropanation of **40** with ethyl diazoacetate at fairly low catalyst loadings (0.1–1.0 mol% copper complex relative to ethyl diazoacetate).

42 → 43 (11)

1. R_2MgX
2. H_2O
3. Salicylaldehyde
4. $Cu(OAc)_2$
5. NaOH

They discovered that the catalyst structure had very little influence on the *cis/trans* selectivity of the cyclopropanation reaction. *Trans* ethyl chrysanthemate comprised 57–62% of the product mixture when catalyzed by a wide range of complexes. They found that increasing the size of R_1 from methyl to benzyl, isopropyl or isobutyl decreased the enantioselectivity of the reaction. The highest enantioselectivities were obtained when R_2 was moderately bulky (R_2=5-*tert*-butyl-2-octyloxyphenyl).

Even after considerable optimization of the catalyst structure, the ethyl chrysanthemate generated had low diastereo- and moderate enantiopurity. In the pyrethroid synthesis, the cyclopropyl ester is eventually hydrolyzed before coupling with the target alcohol. This left the Sumitomo researchers free to explore the effect of varying the diazoester on the stereochemistry of the cyclopropanation [74, 75]. On testing a series of diazoesters having increasing levels of steric encumbrance they found that larger esters gave both higher enantioselectivity and a greater preference for *trans* cyclopropanes (Fig. 5). The highest diastereo- and enantioselectivity were obtained when *l*-menthyl diazoacetate was employed.

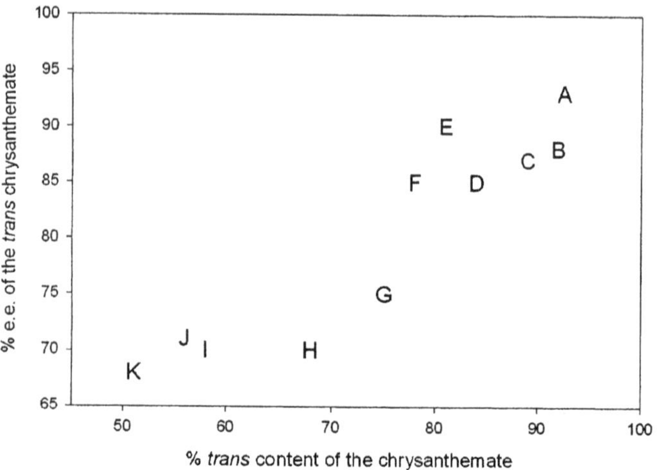

Fig. 5

Since menthyl diazoacetate is chiral, it is not surprising that matched/mismatched behavior was seen between the chiral catalyst and reagent. Interestingly, the strongest effect was in the diastereoselectivity of the cyclopropanation. Being very bulky, all the menthyl diazoacetates displayed a stronger preference for the *trans* product than ethyl diazoacetate (Table 3). The *l*-isomer was far more selective for the *trans* diastereomer than the *d*-isomer. The enantiopurity of the *trans* product was high regardless of which menthyl isomer was used. When an achiral catalyst was used (copper bronze at 123 °C), achiral ethyl chrysanthemate was produced.

Their optimized protocol was to add *l*-menthyl diazoacetate to a solution of the copper catalyst (*R*-**43**; R_1=Me; R_2=5-*tert*-butyl-2-octyloxyphenyl; 0.005 equivalents) in 2,5-dimethyl-2,4-hexadiene (10 equivalents) over 7 h. The reaction required heating to 75 °C at the beginning of diazoacetate addition in order to activate the catalyst, but could be maintained at 40 °C after cyclopropanation had begun.

Table 3

Alkyl diazoacetate	% *trans* cyclopropane	% *ee* of *cis* cyclopropane	% *ee* of *trans* cyclopropane
Ethyl	51	62	68
l-Menthyl	93	46	94
d-Menthyl	72	59	90
dl-Menthyl	81	75	90
l-Menthyl (Cu powder)	76	0.0	0.7

For the safety reasons cited previously, in processes using diazoacetates efforts are made to keep their concentration as low as possible at all times. This is usually accomplished by adding the diazoacetate to a solution of the catalyst and substrate slowly and monitoring the nitrogen off-gas to insure that it is being consumed as quickly as it is being added. This presents a challenge when using the dimeric copper salicylaldimine complexes. These complexes undergo dissociative activation in the presence of the diazoacetate to a reactive monomer before efficient catalysis begins. This is typically accomplished by adding a small portion of the diazoacetate to the catalyst mixture and heating until nitrogen evolution is observed. Once activated, the monomeric species is reactive enough for catalysis to occur at lower temperatures. This activation procedure adds to the complexity of the process, and there is a risk that the initial surge of nitrogen evolution could be uncontrolled. Chemists at Sumitomo found that this behavior could be avoided by converting the dimeric copper complexes into monomeric species by the addition of pyridine or monosubstituted hydrazines. The monomeric complexes formed were active for cyclopropanation at ambient temperature without an induction period [76].

3.2.3.4
Photo-Stable Pyrethroids

The first generation of synthetic pyrethroids (**41**) were successful products due to their effective pest control and low toxicity to mammals [65]. Their thermal and photo-lability, however, limited their use in agriculture. Structure-activity studies led to a new series of cyclopropane containing insecticides with higher activity and improved stability (**44**). Interestingly, the most active isomer of these new molecules is the *cis*- diastereomer (in contrast with natural **38**).

Roussel-Uclaf scientists reported the synthesis of this series starting from (1*R*,*cis*)-caronaldehyde (**45**) (itself prepared from *trans*-ethyl chrysanthemate **39** by an epimerization/ozonolysis sequence, Scheme 5) [65]. A more direct route would be the asymmetric cyclopropanation of an appropriate substrate. Attempts to cyclopropanate 1,1-dichloro-4-methyl-1,3-pentadiene (**46**) failed to selectively produce the desired isomer (Eq. 12) [77]. When 2-methyl-5,5,5-trichloro-2-pentene (**47**) was used as the substrate, the *cis*-cyclopropyl carboxylate was obtained in 92% yield (Scheme 6). The *cis* isomer was favored 84.6:15.4 and the *ee* of the *cis* isomer was 91%. Using *l*-menthyl diazoacetate only marginally increased the *ee* of the *cis* product. Basic hydrolysis of the es-

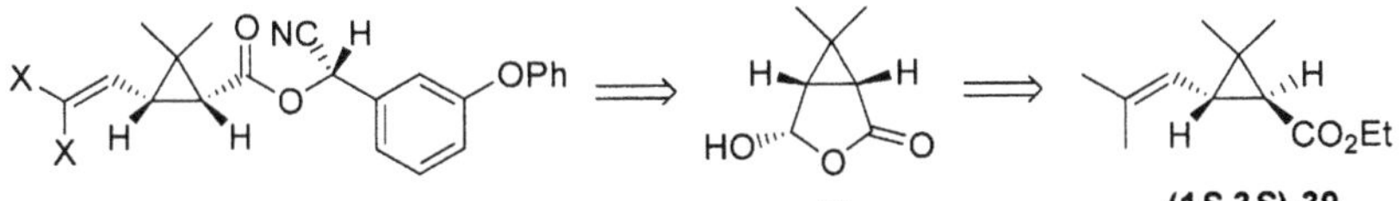

44
X = Cl, cypermethrin
X = Br, deltamethrin

45

(1*S*,3*S*)-39

Scheme 5

ter also suffices to convert the trichloroethyl group to the desired dichloroalkene (**49**).

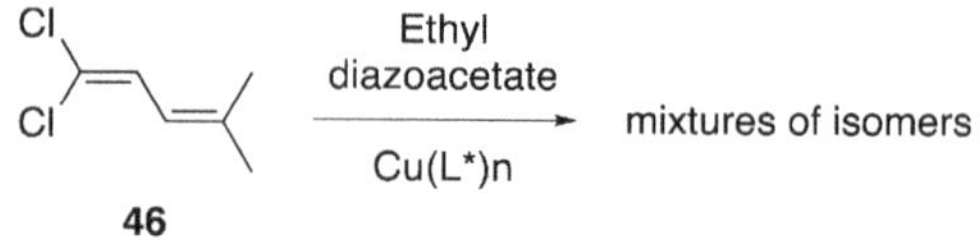

(12)

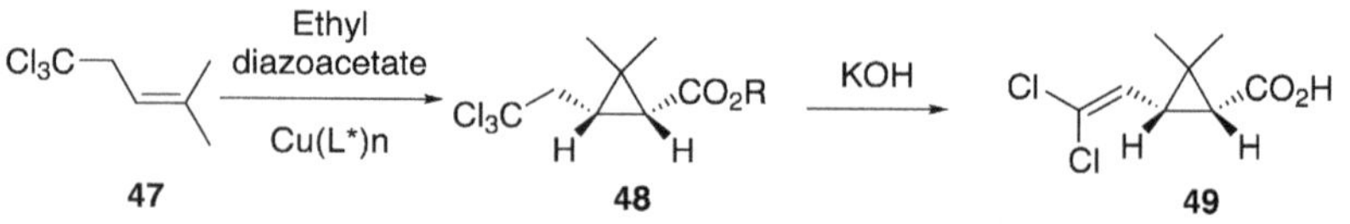

Scheme 6

3.2.3.5
Intramolecular Reaction for Novel Selectivity

Researchers at FMC envisioned another method to harness a cyclopropanation reaction for the stereoselective synthesis of **49** (Scheme 7) [78]. Racemic alcohol **50** was resolved using a chiral isocyanate. The desired enantiomer was converted in several steps to the diazoester **51**. Intramolecular cyclization was effected at low concentration in dioxane at reflux, in the presence of 5 mol% copper (II) acetylacetonate. The removal of several impurities necessitated column chromatography, and the yield of **52** was 65% following purification.

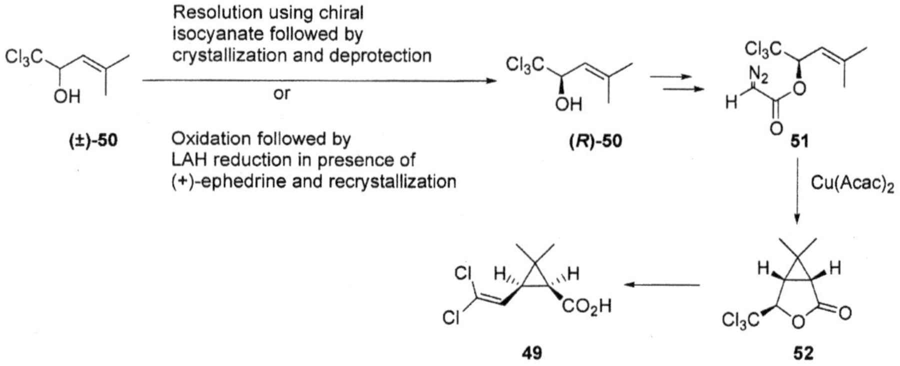

Scheme 7

Researchers at the IMI (TAMI) Institutes for Research and Development optimized a similar route for scale up (Scheme 8) [79]. An enzyme mediated hydrolysis was used to resolve the enantiomers of their starting acetate (**53**). The resolved alcohol (*R*-**50**) was transformed to the cyclopropanation substrate (**51**), with the crude solution being suitable for the cyclization reaction. This obviated

Cl_3C OAc (±)-53 → Cl_3C OH (*R*)-50 + Cl_3C OAc (*S*)-53

Cl_3C N_2 H O O 51 —Copper (II) perchlorate, CH_2Br_2, 85-90 °C→ H H Cl_3C O O 52 —Zn, MeOH→ 49

Scheme 8

the need to isolate the potentially unstable diazoester in pure form. They also found that copper (II) perchlorate performed more reliably than copper (II) acetylacetonate when crude wet solutions of the diazoester **51** were used. They found the intramolecular cyclopropanation proceeded best when run at low concentration. To accomplish this while maintaining good process throughput, they slowly added the starting material from a dilution feed tank while distilling the solvent from the reaction mixture to maintain a constant volume (Fig. 6). In

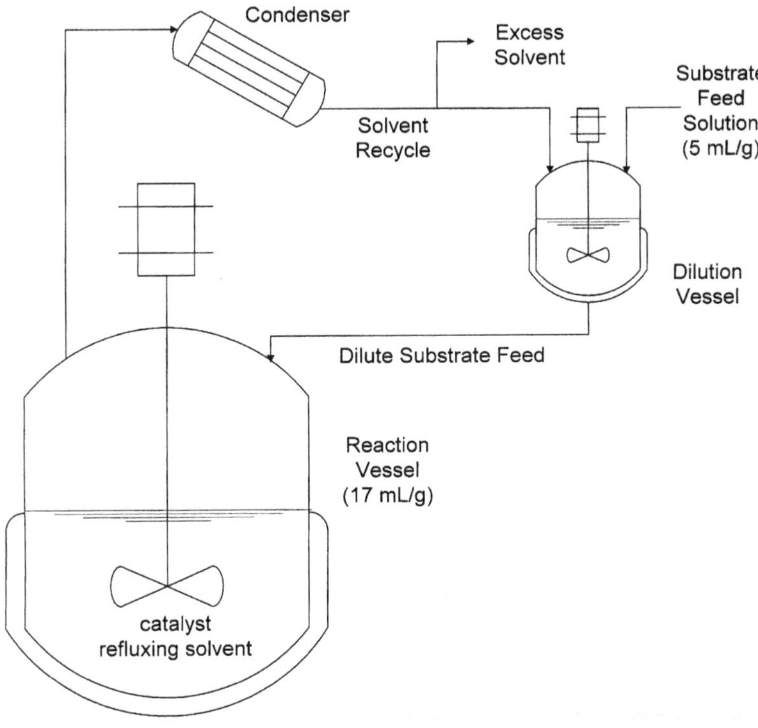

Fig. 6

this manner **51** was maintained at low concentration during the cyclization reaction while the volume yield of the process was a reasonable 60 g/l. Solvent selection for this process was a balance of several factors: reaction rate at the boiling point of the solvent, selectivity for the desired cyclopropanation over undesired C-H carbene insertion, and ease of catalyst removal on workup. Halogenated hydrocarbons gave the highest selectivity for the desired reaction, while a temperature above 80 °C was needed to affect good conversion in a reasonable time. Rarely used as a solvent, dibromomethane was selected as the optimal balance of these factors.

A similar intramolecular cyclopropanation has been demonstrated by Doyle for the synthesis of the most biologically active enantiomer of the antidepressant milnacipran (1*S*,2*R*-**14**, Scheme 9). Although the *ee* of the desired lactone (**57**) was moderate (68%), the rapid construction of the core of the molecule without relying on resolution or chiral separation shows the promise of the technique [80].

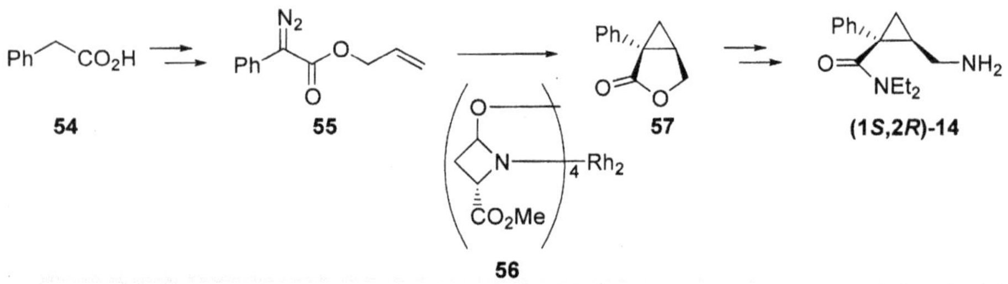

Ph CO2H 54 → Ph N2 O O 55 → (56, Rh2) → Ph 57 → Ph O NEt2 NH2 (1*S*,2*R*)-14

Scheme 9

3.2.4
Cilastatin

During Merck's development of the β-lactam antibiotic imipenem (**58**), in vivo renal metabolism was shown to limit the patient's exposure to the compound [81]. Small molecule inhibitors of dehydropeptidase-I were investigated and tested for their ability to improve the pharmacokinetic profile of **58**. Cilastatin (**61**) emerged as an effective inhibitor of renal metabolism and the combination with **58** is now marketed by Merck as Primaxin. A key intermediate in the synthesis of **61** is ethyl (+)-(1*S*)-2,2-dimethylcyclopropane carboxylate (**60**, Scheme 10) [70].

The commercial process for the preparation of **60** involved the catalytic cyclopropanation of isobutylene (**59**) with ethyl diazoacetate [82]. Minor modifications to the catalysts used for chrysanthemate production were sufficient to achieve high enantioselectivity. The reaction was amenable to scale-up and has been demonstrated on an industrial scale with an enantiomeric excess of 92%.

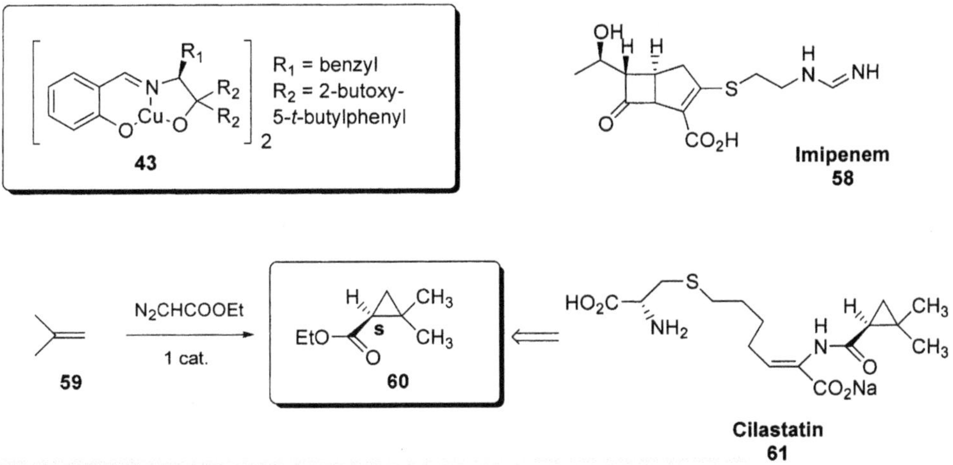

R1 = benzyl
R2 = 2-butoxy-5-*t*-butylphenyl

43

Imipenem
58

59 — $N_2CHCOOEt$, 1 cat. → 60 ⟸ Cilastatin 61

Scheme 10

3.2.5
Melatonin-Agonist BMS-214778

BMS-214778 (**64**, Scheme 11) was reported by Bristol-Myers Squibb as a melatonin agonist targeted for the treatment of sleep disorders [54–56]. Retrosynthetic analysis revealed a number of synthetic pathways that intersected at styrene **62**. Early material demands were met using conventional multi-step cyclopropanation methods. The evolution of a robust large scale synthesis from conventional multi-step methods employing main-group reagents to sophisticated asymmetric catalysis revealed many of the special challenges inherent to cyclopropanation chemistry.

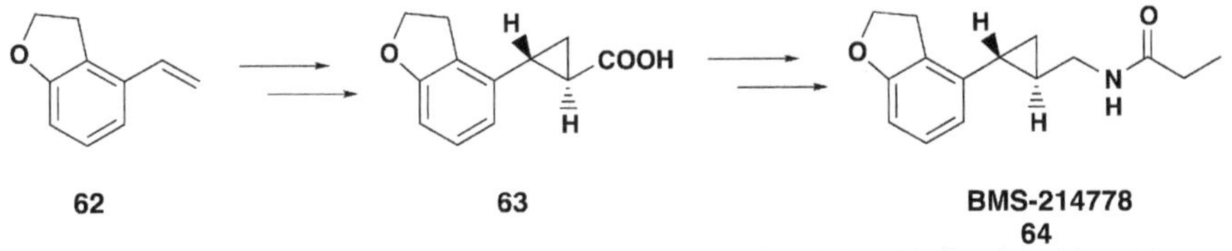

62 → 63 → BMS-214778 64

Scheme 11

The initial route investigated (Scheme 12), was completely devoid of organometallic methodology. The first step involved a non-stereoselective epoxidation of the styrene **62**. Treatment of epoxide **65** with the anion of triethylphosphonoacetate followed by hydrolysis provided the racemic acid **63** [83, 84]. Although previously reported yields for this cyclopropanation were low (20–60%), through optimization much higher yields were obtained (85–90%). Scalability and material handling were a prime concern for this route. For this reason, NaH

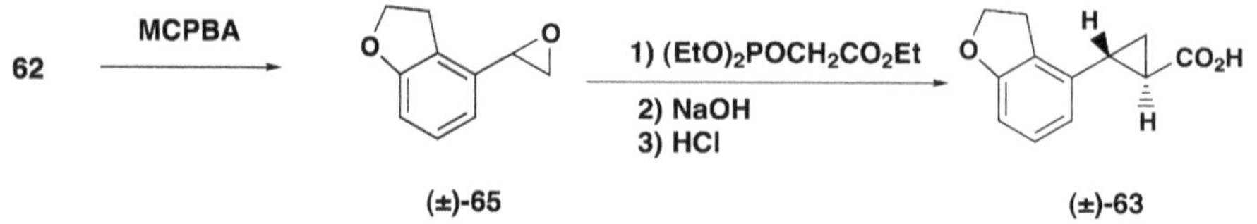

62 MCPBA → (±)-65; 1) $(EtO)_2POCH_2CO_2Et$ 2) NaOH 3) HCl → (±)-63

Scheme 12

was replaced by NaO*t*-Bu or KO*t*-Bu as the base. Enantiopure acid **63** could be isolated by a classical resolution with (+)-dehydroabietylamine ((+)-DAA). This sequence was performed on a 50-g scale with a 30% overall yield.

The second generation route employed a Sharpless asymmetric dihydroxylation (Scheme 13) to provide the epoxide **65** in excellent enantioselectivity (>99% *ee*). Transformation of the resulting diol to the epoxide using standard conditions, followed by adopting the same chemistry used previously to install the cyclopropane moiety, provided the desired acid **63** in 65% overall yield and 99% *ee* [85].

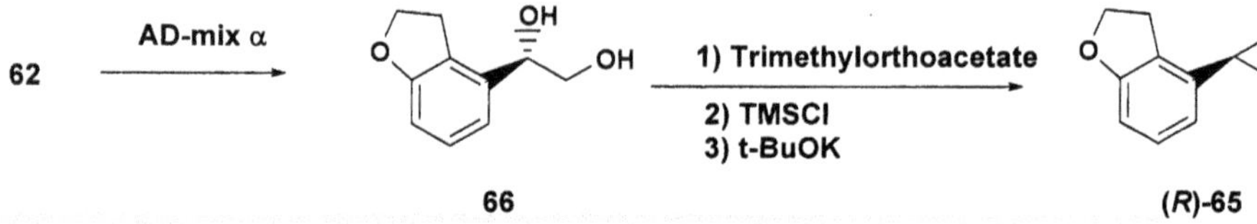

62 AD-mix α → 66; 1) Trimethylorthoacetate 2) TMSCl 3) t-BuOK → (*R*)-65

Scheme 13

Although the use of an organometallic reagent increased the overall yield from 30% to 65%, this approach required several steps and a more attractive approach employing a direct cyclopropanation of styrene **62** with EDA was explored (Eq. 13).

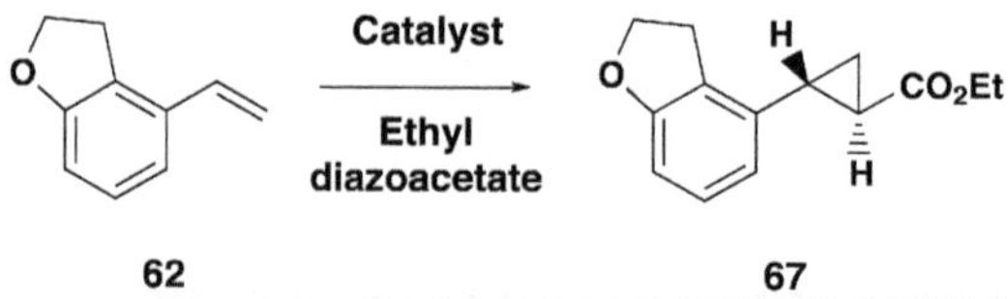

62 Catalyst, Ethyl diazoacetate → 67 (13)

Three potential catalysts (Fig. 7) were investigated for the cyclopropanation of styrene **62**. The results are summarized in Table 4. The first was Evans catalyst **68** [33] which displayed only moderate diastereoselectivity (74% *trans* cyclopropane) although the enantioselectivity was high. Reactions employing Katsuki's catalyst **69** [86, 87] never proceeded beyond 72% completion even with high catalyst loading (10 mol%). Additionally, the bromine oxidation necessary in preparation of the catalyst also made it less attractive for large scale synthesis. Nishiyama's catalyst system **28** [35, 36] functioned well with only 2 mol% catalyst and provided 90% *trans* cyclopropane in 84% *ee*. As a good balance between selectivity and catalyst loading, this system was chosen for further development.

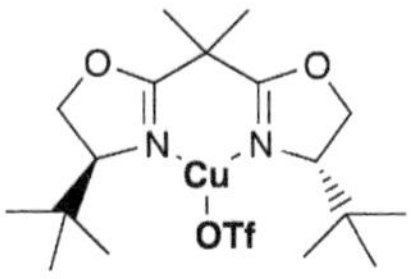

O
O
N
N
Cu
OTf

68

Ph
Ph
N
N
Co
Br
O
O
MeO
OMe

69

O
O
N
N
N
Ru
Cl
Cl

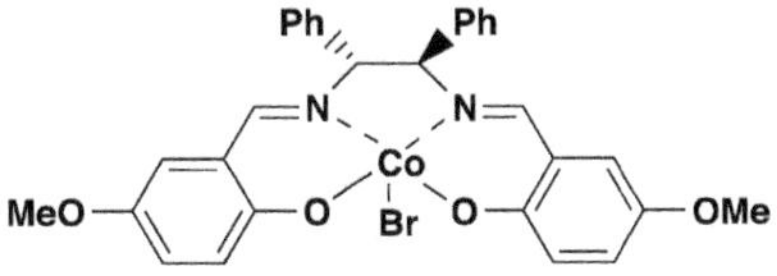

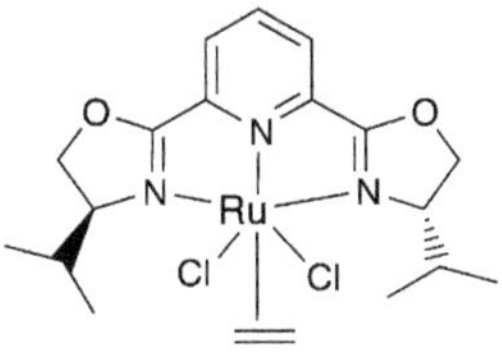

28

Fig. 7

Table 4

Catalyst	EDA addition time (h)	mol% cat.	% conv.	% *trans*	% *ee* (*trans* acid)
68	16	0.1	97	74	99
69	10	10.0	72	88	84
28	18	2.0	95	90	84

To ease the preparation of catalyst **28** on large scale, a crystallization procedure was discovered that allowed the isolation of a stable, free-flowing solid and eliminated the need for chromatography [88].

Many literature procedures for the cyclopropanation of styrenes use the olefin in large excess (5–10 equivalents). This was unacceptable for the current synthesis as the styrene was a precious synthetic intermediate. Both solvent and temperature screens showed enantio- and diastereoselectivity were relatively insensitive to these two variables. Statistical design of experiment (DOE) studies revealed a strong interaction between the amount of EDA used and the rate of addition. With a fast addition, competitive dimerization of EDA was more pronounced and larger quantities would be required to obtain complete conversion. From the temperature study it was also known that conversion was poor at lower temperatures. Based on these experiments, the optimal conditions (Eq. 14) were selected (2.5 equivalents of EDA were added over a period of 16 h to the substrate with 1 mol% catalyst at 60 °C). Under these conditions the cyclopropane carboxylate was obtained in >90% yield with >90% *trans* isomer reliably.

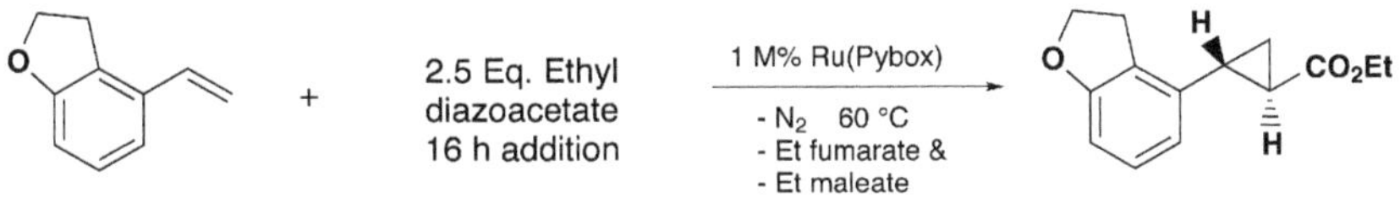

O
+
2.5 Eq. Ethyl
diazoacetate
16 h addition
1 M% Ru(Pybox)
- N2 60 °C
- Et fumarate &
- Et maleate
O
H
CO2Et
H

(14)

62
In toluene

67
in toluene

Since enantiopure cyclopropyl acid **63** was desired, a telescoped hydrolysis/resolution was investigated (Eq. 15). The optimal conditions utilized tetrabutyl ammonium hydroxide as a phase transfer catalyst. This produced the acid in 94:6 *trans*:*cis* ratio in 95 mol%. Using the resolution described for the earlier routes, the (+)-dehydroabietylamine salt **70** could be isolated in 75 mol%

and 99% *ee*. By harmonizing the reaction solvent for the EDA preparation and the cyclopropanation, **70** was the first intermediate isolated in this synthesis.

$$\mathbf{67}\ \text{In Toluene} \xrightarrow[\text{Bu}_4\text{NOH},\ 60\,^\circ\text{C},\ 4\ \text{h}]{\text{NaOH}} \xrightarrow[\text{MTBE}]{\text{H}_3\text{PO}_4} \mathbf{63}\ \text{In MTBE} \xrightarrow{\text{DAA}} \mathbf{70}\ (\text{CO}_2\text{H}\cdot\text{DAA}) \tag{15}$$

In summary, the process development of BMS-214778 (**64**) nicely demonstrated the benefits gained by increasing the sophistication of the chemistry used in its synthesis. Traditional achiral cyclopropane forming chemistry was aided by a stereoselective Sharpless dihydroxylation. This improvement was overshadowed by the direct stereoselective cyclopropanation that was ultimately scaled in the pilot plant. The research invested into the catalyst and ethyl diazoacetate preparations was repaid by a succinct final process.

4 Conclusions

Due to its unique properties, the cyclopropane group is certain to appear in new drug candidates and other functional chemicals. Researchers in both academic and industrial labs continue the search for new reagents and catalysts to form cyclopropanes with high stereoselectivity and efficiency. In the future, those groups willing to invest the effort to the application of these sophisticated methods will be rewarded with streamlined processes and corresponding improvements to their manufacturing costs.

Acknowledgements The Authors would like to thank Dr. Wendel Doubleday, Dr. Rodney Parsons, and Dr. John Scott for many helpful suggestions in the preparation of this manuscript. The authors also thank the Melatonin-agonist project team for useful discussions relating to some of the work described.

References

1. Pierce ME, Parsons RL, Radesca LA, Lo YS, Silverman S, Moore JR, Islam Q, Choudhury A, Fortunak JMD, Nguyen D, Luo C, Morgan SJ, Davis WP, Confalone PN, Chen C, Tillyer RD, Frey L, Tan L, Xu F, Zhao D, Thompson AS, Corley EG, Grabowski EJJ, Reamer R, Reider PJ (1998) J Org Chem 63:8536
2. Corley EG, Thompson AS, Huntington M (2000) Org Synth 77:231
3. Kawada M, Watanabe M, Okamoto K, Sugihara H, Hirata T, Maki Y, Imada I, Sanno Y (1984) Chem Pharm Bull 32:3532
4. Watanabe M, Kawada M, Takamoto M, Imada I, Noguchi S (1984) Chem Pharm Bull 32:3373
5. Sugihara H, Watanabe M, Kawada M, Imada I (1979) EP Patent No. 003084 A1 and B1
6. Sugihara H, Watanabe M, Kawada M, Imada I (1981) U.S. Patent No. 4,284,644
7. Budavari S, O'Neill MJ, Smith A, Heckelman PE (1989) The Merck Index, 11th edn. Merck & Co, Rahway, NJ, p 1382
8. Gololobov YG, Nesmeyanov AN, Lysenko VP, Boldeskul IE (1987) Tetrahedron 43:2609
9. Faul MM, Ratz AM, Sullivan KA, Trankle WG, Winneroski LL (2001) J Org Chem 66:5772

10. Boehm MF, Zhang L, Zhi L, McClurn R, Berger E, Wagoner M, Mais DE, Suto CM, Davies PJA, Heyman RA, Nadzan AM (1995) J Med Chem 38:3146
11. Cousse H, Mouzin G, Bonnaud B (1975) FR Patent No. 2,302,994
12. Simmons HE, Smith RD (1959) J Am Chem Soc 81:4256
13. Simmons HE, Cairns TL, Vladuchick SA, Hoiness CM (1973) Org React 20:1
14. Dargel TK, Koch W (1996) J Chem Soc Perkin Trans 2 877
15. Charette AB, Beauchemin A (2001) Org React 58:1
16. Denis JM, Girard C, Conia JM (1972) Synthesis 549
17. Furukawa J, Kawabata N, Nishimura J (1966) Tetrahedron Lett 3353
18. Furukawa J, Kawabata N, Nishimura J (1968) Tetrahedron 24:53
19. Sawada S, Inouye Y (1969) Bull Chem Soc Jpn 42:2669
20. Charette AB, Marcoux J-F (1996) J Am Chem Soc 118:4539
21. Winstein S, Sonnenberg J, de Vries L (1959) J Am Chem Soc 81:6523
22. Salaün J (1989) Chem Rev 89:1247
23. Lautens M, Klute W, Tam W (1996) Chem Rev 96:49
24. Silberrad O, Roy CS (1906) J Chem Soc 89:179
25. Dave V, Warnhoff EW (1970) Org React 18:217
26. Nozaki H, Moriuti S, Takaya H, Noyori R (1966) Tetrahedron Lett 43:5239
27. Nozaki H, Takaya H, Moriuti S, Noyori R (1968) Tetrahedron 24:3655
28. Fritschi H, Leutenegger U, Siegmann K, Pfaltz A (1988) Helv Chimi Acta 71:1541
29. Fritschi H, Leutenegger U, Pfaltz A (1988) Helv Chim Acta 71:1553
30. Pfaltz A (1993) Acc Chem Res 26:339
31. Lowenthal RE, Abiko A, Masamune S (1990) Tetrahedron Lett 31:6005
32. Lowenthal RE, Masamune S (1991) Tetrahedron Lett 32:7373
33. Evans DA, Woerpel KA, Hinman MM, Faul MM (1991) J Am Chem Soc 113:726
34. Davies IW, Gerena L, Cai D, Larsen RD, Verhoeven TR, Reider PJ (1997) Tetrahedron Lett 38:1145
35. Nishiyama H, Sakaguchi H, Nakamura T, Horihata M, Kondo M, Itoh K (1989) Organometallics 8:846
36. Nishiyama H, Itoh Y, Matsumoto H, Park S-B, Itoh K (1994) J Am Chem Soc 116:2223
37. Park S-B, Murata K, Matsumoto H, Nishiyama H (1995) Tetrahedron Asymmetry 6:2487
38. Anciaux AJ, Hubert AJ, Noels AF, Petiniot N, Teyassie P (1980) J Org Chem 45:695
39. Doyle MP, Ban Leusen D, Tamblyn WH (1981) Synthesis 787
40. Doyle MP (1986) Chem Rev 86:919
41. Doyle MP, Loh K, DeVries KM, Chinn MS (1987) Tetrahedron Lett 28:833
42. Doyle MP, Forbes DC (1998) Chem Rev 98:911
43. Timmons DJ, Doyle MP (2001) J Organomet Chem 617/618:98
44. Repic O, Vogt S (1982) Tetrahedron Lett 23:2729
45. Kathawala FG (1980) U.S. Patent No. 4,201,785
46. Kathawala FG, Heider JG (1981) U.S. Patent No. 4,248,893
47. Giger U, Repic O (1984) U.S. Patent No. 4,472,313
48. Repic O, Lee PG, Giger U (1984) Org Prep Proced Int 16:25
49. Friedrich EC, Lunetta SE, Lewis EJ (1989) J Org Chem 54:2388
50. Takai K, Kakiuchi T, Utimoto K (1994) J Org Chem 59:2671
51. Doyle MP, McKervey MA, Ye T (1998) Modern catalytic methods for organic synthesis with diazo compounds. Wiley, New York, pp 45–46
52. Clark JD, Shah AS, Peterson JC, Patelis L, Kersten RJ, Heemskerk AH (2002) Thermochim Acta 386:73
53. Clark JD, Shah AS, Peterson JC, Patelis L, Kersten RJ, Heemskerk AH, Grogan M, Camden S (2002) Thermochim Acta 386:65
54. Simpson J (2001) Development of an asymmetric cyclopropanation process using styrene as the limiting reagent. ChiraSource Conference, Philadelphia
55. Scott JW (2002) Catalytic asymetric synthesis of a melatonin agonist. ACS, Orlando

56. Kotnis AS (2002) Practical process solutions for asymmetric reactions. Chiral USA 2002, Boston
57. Kotnis AS, Simpson JH, Deshpande RP, Kacsur DJ, Hamm J, Kodersha G, Merkl W, Domina D, Wang SSY (2003) (manuscript in preparation)
58. Armstrong RK (1966) J Org Chem 31:618
59. Searle NE (1963) Organic syntheses, vol IV. Wiley, New York, Collect, p 424
60. Clark JD, Shah AS, Peterson JC, Grogan FM, Camden SK (2001) Thermochim Acta 367/368:75
61. Bertholf RL (1988) In: Seiler HG, Sigel H, Sigel A (eds) Handbook on toxicity of inorganic compounds. Marcel Dekker, New York, Chap 71
62. Sarkar B (1988) In: Seiler HG, Sigel H, Sigel A (eds) Handbook on toxicity of inorganic compounds. Marcel Dekker, New York, Chap 21
63. Léonard A (1988) In: Seiler HG, Sigel H, Sigel A (eds) Handbook on toxicity of inorganic compounds. Marcel Dekker, New York, Chap 51
64. Bradford CW, Chase BJ (1988) In: Seiler HG, Sigel H, Sigel A (eds) Handbook on toxicity of inorganic compounds. Marcel Dekker, New York, Chap 49
65. Martel J (1992) In: Collins AN, Sheldrake GN, Crosby J (eds) Chirality in industry. Wiley, New York, p 87
66. Staudinger H, Ruzieka L (1924) Helv Chim Acta 7:177
67. Anonymous (1953) Chem Week, January 24, p 38
68. Boon WR, Hall R (1955) GB Patent No. 740,014
69. Shim KS, Martin DJ (1975) US Patent No. 3,897,478
70. Aratani T (1985) Pure Appl Chem 57:1839
71. Aratani T, Yonehoshi Y, Nagase T (1975) Tetrahedron Lett 1707
72. Aratani T, Nakamura S, Nagase T, Yonehoshi Y (1977) US Patent No. 4,029,683
73. Aratani T, Nakamura S, Nagase T, Yonehoshi Y (1977) US Patent No. 4,029,690
74. Aratani T, Yoneyoshi Y, Nagase T (1977) Tetrahedron Lett 2599
75. Aratani T, Yoneyoshi Y, Fujita F, Nagase T (1980) US Patent No. 4,197,408
76. Aratani T, Yoshihara H, Susukama G (1985) US Patent No. 4,552,972
77. Aratani T, Yoneyoshi Y, Nagase T (1982) Tetrahedron Lett 23:685
78. Hatch CE III, Baum JS, Takashima T, Kondo K (1980) J Org Chem 45:3281
79. Fishman A, Kellner D, Ioffe D, Shapiro E (2000) Org Process Res Dev 4:77
80. Doyle MP, Hu W (2001) Adv Synth Catal 343:299
81. Birnbaum J, Kahan FM, Kropp H, MacDonald JS (1985) Am J Med 78:3
82. Aratani T, Yoshihara H, Susukamo G (1986) US Patent No. 4,603,218
83. Wadsworth WS Jr, Emmons WD (1961) J Am Chem Soc 83:1733
84. Denney DB, Vill JJ, Boskin MJ (1962) J Am Chem Soc 84:3944
85. (a) Singh AK, Rao MN, Simpson JH, Li W-S, Thornton JE, Kuehner DE, Kacsur DJ (2002) Org Process Res Dev 6:618; (b) Prasad JS, Vu T, Totleben MJ, Crispino GA, Kacsur DJ, Shankar S, Thorton JE, Fritz A, Singh AK (2003) Org Process Res Dev 7:821
86. Fukuda T, Katsuki T (1995) Synlett 825
87. Fukuda T, Katsuki T (1997) Tetrahedron Lett 38:3435
88. Totleben M, Prasad JS, Simpson JH, Chan SH, Vanyo DJ, Kuehner DE, Deshpande R, Kodersha GA (2001) J Org Chem 66:1057

Topics Organomet Chem (2004) 6: 123–152
DOI 10.1007/978-3-540-36966-0

Asymmetric Processes Catalyzed by Chiral (Salen)Metal Complexes

Jay F. Larrow* · Eric N. Jacobsen

*Amgen, Small Molecule Process Development, One Kendall Square, Cambridge, MA 02139, USA
Department of Chemistry and Chemical Biology, Harvard University, 12 Oxford Street, Cambridge, MA 02138, USA
E-mail: jay.larrow@amgen.com, jacobsen@chemistry.harvard.edu

Abstract A wide variety of highly selective asymmetric reactions catalyzed by chiral (salen)metal complexes have been disclosed over the past decade. Salen ligands are among the most synthetically accessible frameworks for asymmetric catalysts, and their structures are readily tuned both sterically and electronically. However, one particular chiral salen ligand (**1**) has been demonstrated to be highly effective for a wide variety of useful asymmetric transformations catalyzed by different metals. The simplicity of this ligand, the high enantioselectivities and broad substrate generality often observed, and the synthetic utility of the catalytic transformations have inspired substantial effort directed toward the commercial development of asymmetric salen chemistry. This chapter surveys the resulting progress.

Keywords (Salen)metal complexes · Asymmetric catalysis · Epoxides · Nucleophiles · Kinetic resolution

1 Introduction

The development of practical catalytic asymmetric processes for both laboratory and industrial scale applications has been enabled greatly by the discovery of a small number of ligands that exhibit high selectivity for a wide range of substrates and over a broad spectrum of reactions [1]. This group of special ligands includes BINAP and BINOL, tartaric acid derivatives, bis(oxazoline) and pybox ligands, derivatives of the cinchona alkaloids, and the Duphos bis(phosphine) ligands [2] (Fig. 1). Another member of this group of privileged ligands that has emerged over the past several years is salen ligand **1**. Metal complexes of ligand **1** have been successfully applied to a broad range of industrially important asymmetric reactions. A survey of catalytic asymmetric reactions utilizing these and related complexes and an evaluation of their utility within the pharmaceutical and fine chemical industries is the focus of this review [3].

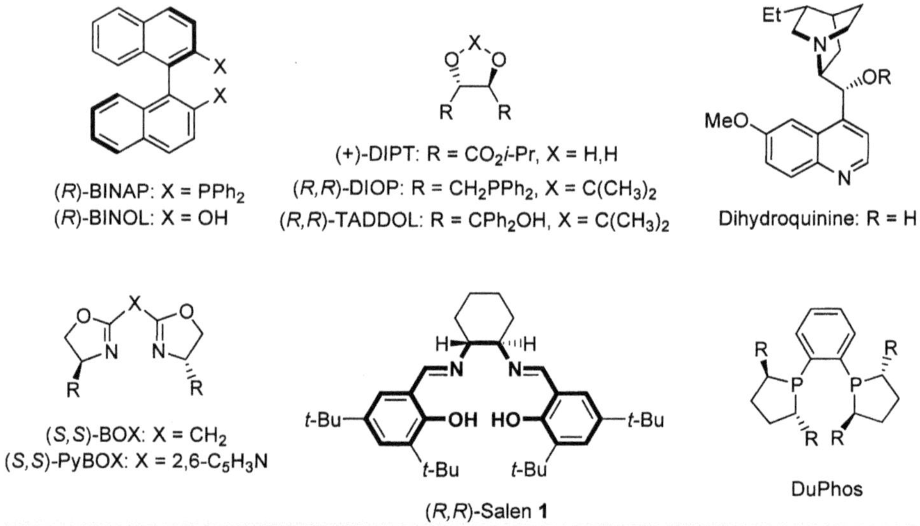

(*R*)-BINAP: X = PPh_2
(*R*)-BINOL: X = OH

(+)-DIPT: R = CO_2i-Pr, X = H,H
(*R,R*)-DIOP: R = CH_2PPh_2, X = $C(CH_3)_2$
(*R,R*)-TADDOL: R = CPh_2OH, X = $C(CH_3)_2$

Dihydroquinine: R = H

(*S,S*)-BOX: X = CH_2
(*S,S*)-PyBOX: X = 2,6-C_5H_3N

(*R,R*)-Salen **1**

DuPhos

Fig. 1 Families of privileged ligands in asymmetric catalysis which show high product selectivity and substrate generality for a number of different reactions

2 Salen Ligand Synthesis

Salen ligands are prepared by the condensation of two equivalents of a salicylaldehyde derivative with a 1,2-diamine, and the simplest, achiral version (bold structure in Fig. 1) is prepared from salicylaldehyde and ethylenediamine, abbreviations of which combine to give this ligand class its name. Chiral versions of this tetradentate bis(imine) ligand are accessed simply by using chiral 1,2-diamines, although ligands derived from other diamines (1,3-, 1,4-, etc.) are often included in this class. Chiral salen ligands have several attractive features that constitute the basis for their utility in asymmetric reactions. The salicylaldehyde and diamine components are synthetically accessible and their condensation to generate the salen ligand generally proceeds in nearly quantitative yield. Metal complexes of salen ligands are readily prepared from a variety of first row and second row transition metal salts as well as main group metals. Once the appropriate metal for the desired reactivity has been identified, the modularity of synthesis of salen ligands allows for the systematic tuning of catalyst steric and electronic properties by modification of the metal counterion, the chiral diamine or the salicylaldehyde components [4]. Although a large number of ligand structures are thus accessible, it is striking that salen ligand **1** has often been found to be the optimum ligand for a broad range of reactions catalyzed by several different metals (Fig. 2).

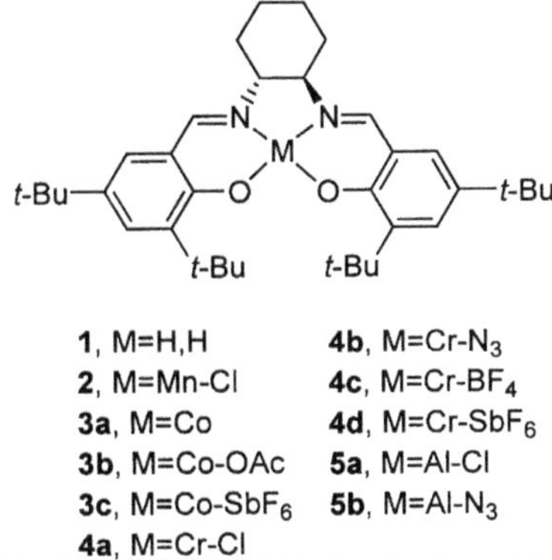

Fig. 2 Metal complexes of salen ligand **1** utilized in catalytic asymmetric processes

Salen ligand **1** was first identified in the context of Mn-catalyzed asymmetric epoxidation of unfunctionalized olefins [5]. Hundreds of salen ligand variations were investigated for this process, and several reviews are available for a discussion of the results [6]. Figure 3 shows many of the different permutations investigated, including ligands derived from chiral 1,2-, 1,3-, and 1,4-diamines, chiral tertiary1,2-diamines, chiral salicylaldehydes, and hydroxyacetophenones. Ligand **1** was selected because it showed the best balance between high selectivity for a broad range of substrates and accessibility from inexpensive raw materials. The key elements of the ligand are the bulky *tert*-butyl groups at the 3,3′- and 5,5′-positions. These groups are proposed to enforce approach of the substrate over the chiral diamine portion of the ligand where the *trans*-diaxial α-protons provide remarkably effective stereochemical communication.

The industrial synthesis of ligand **1** utilizes very inexpensive raw materials (Scheme 1) [7], and the ligand is now available commercially in both laboratory and bulk quantities. A *cis/trans* mixture of the diamine is a by-product of the conversion of adiponitrile to 1,6-hexanediamine for the Nylon industry, and is readily resolved to enantiomeric and diastereomeric purity with tartaric acid

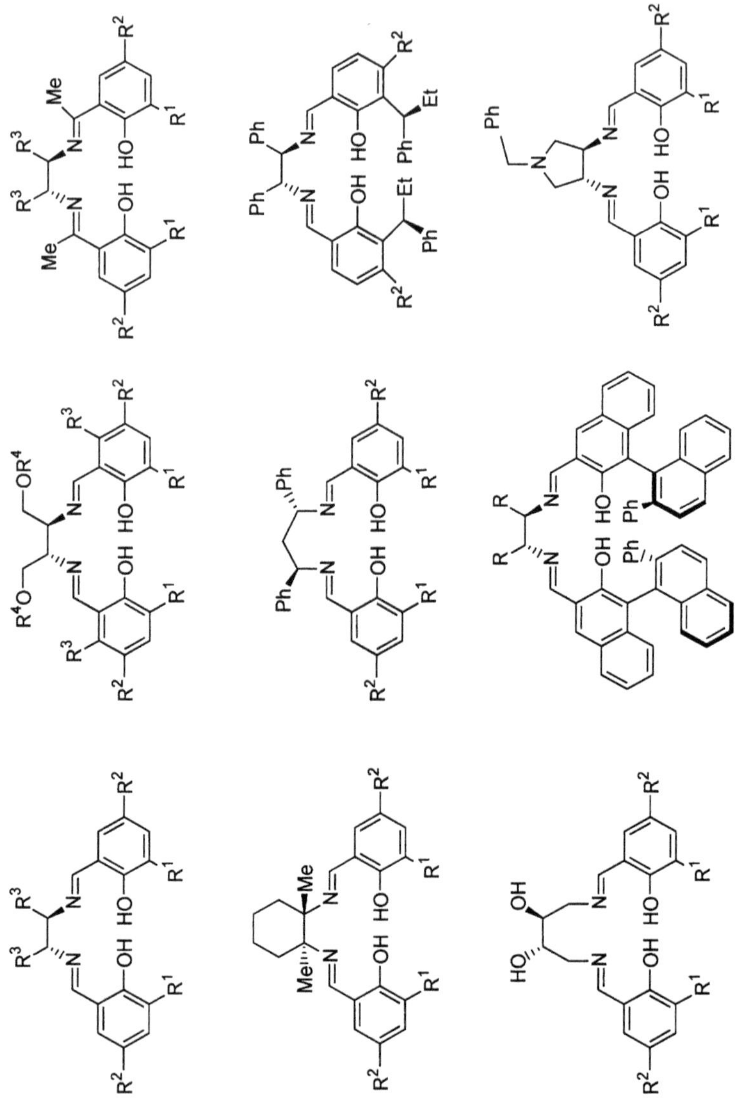

Fig. 3 Different structural variations of the salen ligand framework investigated for Mn-catalyzed asymmetric epoxidation (AE)

[7, 8]. Due to the use of unnatural (-)-tartaric acid to isolate the (*S*,*S*)-diamine, (*S*,*S*)-**1** is slightly more expensive than (*R*,*R*)-**1**. The salicylaldehyde component is prepared by formylation of 2,4-di-*tert*-butylphenol, which in turn is produced inexpensively on large scale by the double alkylation of phenol with isobutene.

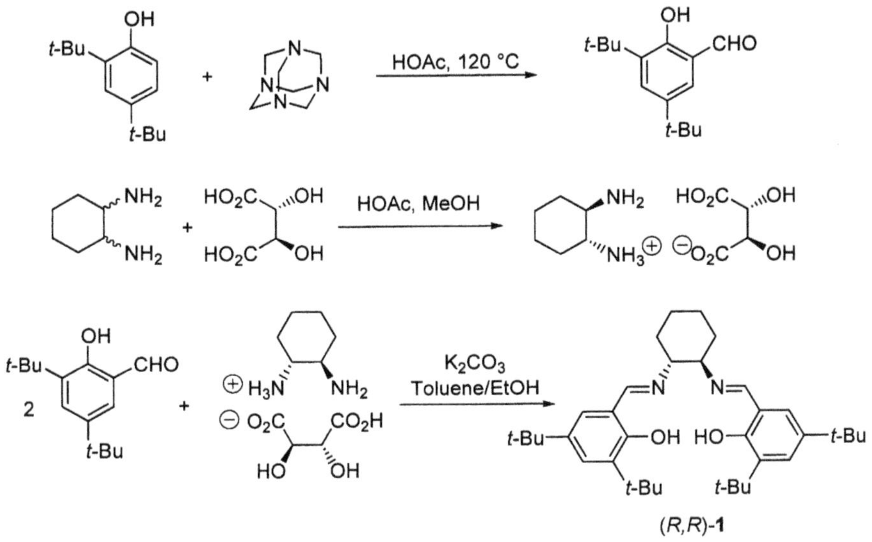

Scheme 1 Commercial synthesis of salen ligand **1**

3 Epoxidation Reactions

The first example of enantioselective epoxidation of unfunctionalized olefins catalyzed by chiral (salen)Mn(III) complexes was reported in 1990 [9], and this remains an active area of study 12 years later. One of the more practical versions of this asymmetric process utilizes Mn(III) complex **2** as the catalyst and aqueous bleach as the stoichiometric oxidant [5, 10]. Several subsequent variations of ligand structure, metal center, and terminal oxidant have been developed, but none have surpassed the utility and practicality of the process depicted in Scheme 2. The epoxidation process generally requires an unsaturated conjugat-

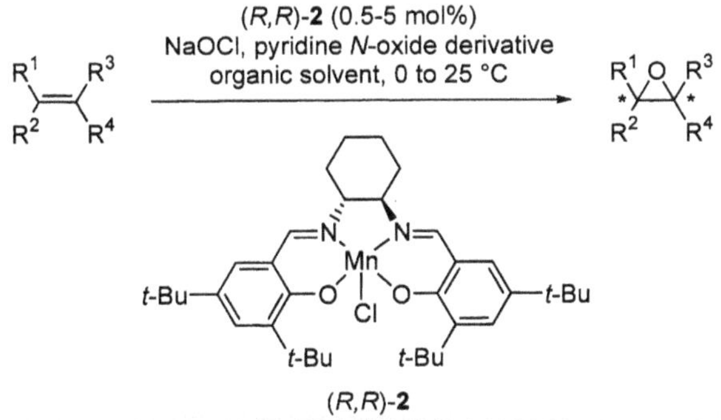

Scheme 2 General conditions for asymmetric epoxidation (AE) of conjugated olefins catalyzed by Mn complex **2**

ing group such as an arene, alkene, or alkyne for acceptable reactivity. High enantioselectivities are typically observed with most *cis*-1,2-disubstituted, as well as certain tri- and tetrasubstituted olefins [11], while low-temperature, homogeneous conditions have been developed to access terminal epoxides in up to 86% ee [12]. *trans* Olefins yield epoxides in low enantiomeric excess, but conditions have been developed that favor the production of high ee *trans* epoxides from *cis* olefins [13]. The use of amine *N*-oxide additives such as 4-phenylpyridine *N*-oxide (4-PPNO) can have a profound impact on the ratios of enantiomers and *cis/trans* epoxides produced, and serve to improve the efficiency of the catalyst [14]. These additives have been demonstrated to act as axial ligands that bind to the Mn center *trans* to the oxo ligand [15].

Diastereomeric epoxide isomers (enantiomers in the case of terminal epoxides) are often obtained as primary products from isomerically pure *cis*-olefins. This constitutes strong evidence for a mechanism of oxygen transfer from the metal center to the olefin involving non-concerted formation of the C-O bonds. Several plausible intermediates have been proposed to lie along this reaction pathway, but at this stage there is mounting support, based on both experimental and theoretical studies, for a radical pathway such as the one depicted in Fig. 4 [16].

The synthetic utility of the chiral (salen)Mn-catalyzed epoxidation reaction has been demonstrated through several important examples. The asymmetric epoxidation of *cis*-cinnamates was the key transformation in the practical synthesis of the phenylisoserine side chain of Taxol [17] and in a route to the calcium channel antagonist Diltiazem (Scheme 3) [18]. Due to the problem of *cis/trans* partitioning noted above, the asymmetric epoxidation process is most

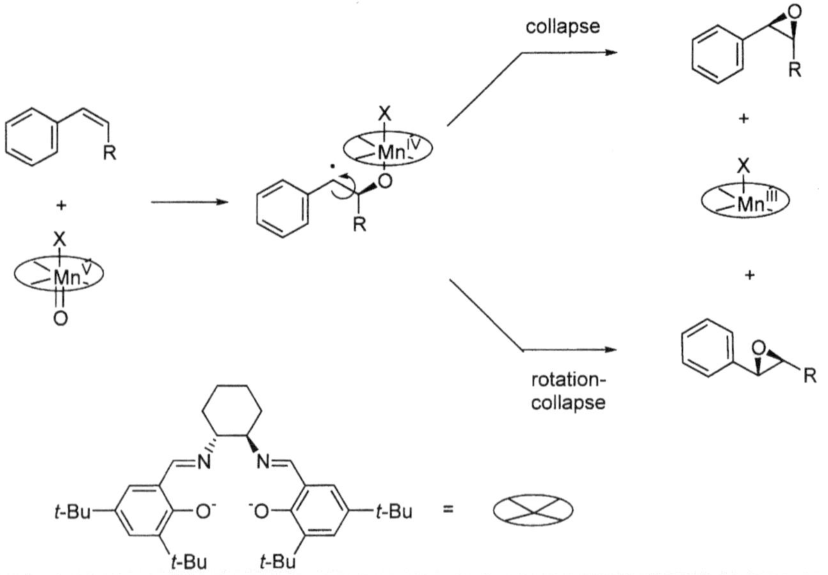

Fig. 4 Stepwise mechanism of oxygen transfer from Mn to olefin leads to diastereomeric mixtures of epoxides

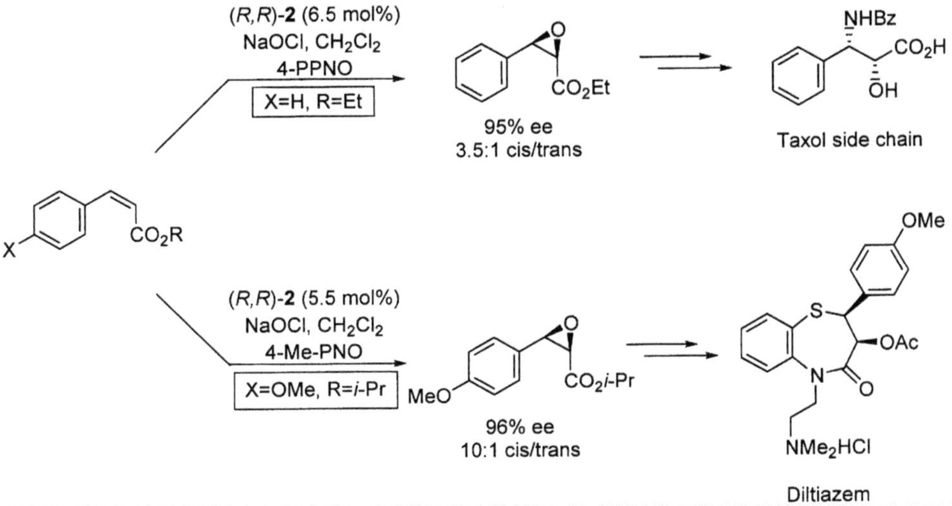

Scheme 3 AE of cinnamates

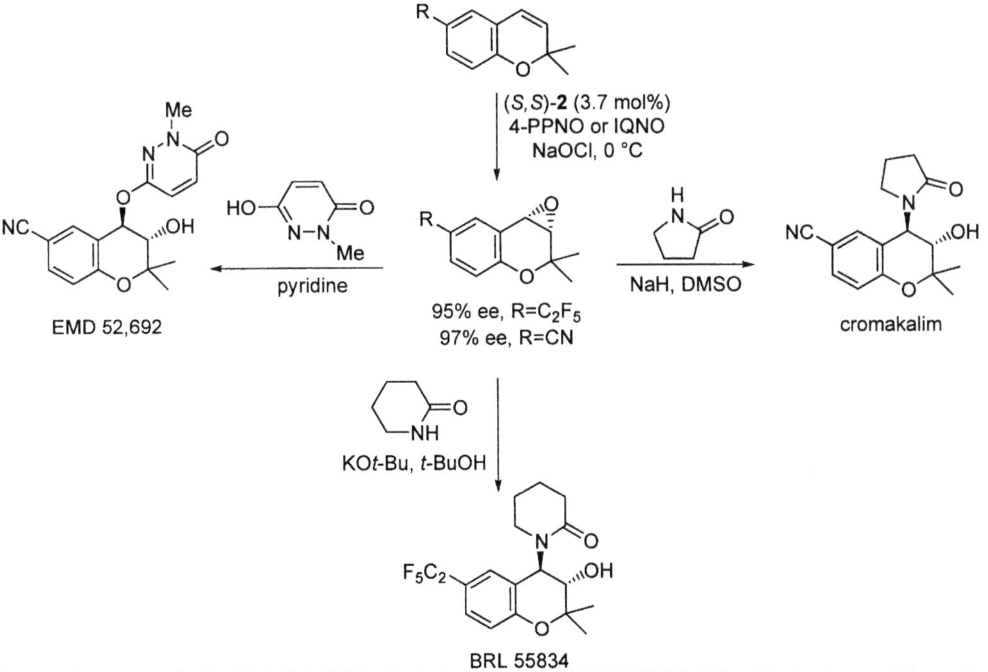

Scheme 4 AE of chromenes

practical for the epoxidation of cyclic olefins where rotation is not possible. Chromenes represent the best substrates for the AE reaction in terms of enantioselectivity, and there are several pharmaceutically important derivatives of the epoxides which have been accessed via this chemistry, including the anti-hypertensive agents cromakalim, EMD-52,692, and BRL55834 (Scheme 4) [19].

The asymmetric epoxidation of indene has also been extensively studied because the epoxide can be transformed into *cis*-1-aminoindan-2-ol [20], an effective chiral ligand and an important portion of the HIV protease inhibitor Crixivan [21]. In the optimized process for the epoxidation of indene [22], only 0.6 mol% catalyst was required using 3 mol% of additive, and the productivity of the reaction was improved by using 10–14% NaOCl solution. For the scale-up of this epoxidation, it was found that efficient mixing of the biphasic reaction mixture was critical. Thus, the reaction reaches completion in less than 15 minutes if a blender is used. The epoxide is isolated in 84–86% ee by simple distillation, and is then converted to the *cis*-amino alcohol by a Ritter reaction followed by hydrolysis. The enantiomeric excess of the product as well as the chemical purity are then upgraded by a recrystallization as the (+)-tartaric acid salt (Scheme 5). An alternative route from indene oxide to aminoindanol involves opening of the epoxide with ammonia followed by inversion at the 2-carbon via an oxazoline intermediate [23]. This process has the advantage of using the less-expensive (*R*,*R*)-**2** as the catalyst, although the overall route involves more steps.

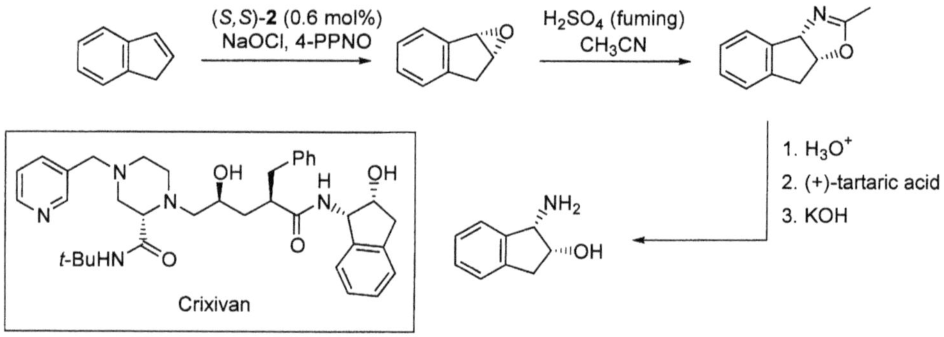

Scheme 5 AE of indene and conversion to *cis*-1-aminoindan-2-ol

In contrast to the substrate scope with (salen)Mn-catalyzed asymmetric epoxidations, (salen)Cr complexes provide high selectivity in the epoxidation of conjugated *trans* olefins [24]. Due to its enhanced stability, the Cr(V)-oxo species can be isolated and used stoichiometrically, or the corresponding (salen)Cr(III) complexes can be used catalytically with a stoichiometric oxidant such as iodosylbenzene. Although the (salen)Cr(V)-oxo complex derived from **1** epoxidizes *trans*-β-methylstyrene with good enantioselectivity (71% ee vs 24% ee with Mn complex **2**), electron-withdrawing substituents on the ligand provide optimal results [25]. The Mn and Cr systems are fairly complementary in terms of substrate scope, but the practicality of the (salen)Cr process is limited by slow reaction rates, low yields, and the necessity of iodosylarenes as the stoichiometric oxidant. In addition to his large volume of work on (salen)Mn-catalyzed asymmetric epoxidation, Katsuki has pioneered the development of photoactivated (salen)Ru complexes which catalyze the stereospecific AE of conjugated olefins [26]. Other asymmetric oxidation reactions have also been catalyzed by (salen)metal complexes including C-H oxidation [27], sulfide oxidation [28], sulfimidation [29] and the Baeyer-Villiger reaction [30], but these processes have marginal potential for broad industrial application due to substrate and selectivity limitations.

In summary, the importance of the (salen)Mn-catalyzed asymmetric epoxidation reaction lies in the fact that it was the first process to provide practical access to enantiomerically enriched epoxides without the necessity of precoordination of the substrate via a directing functional group (cf. Ti-tartrate-catalyzed asymmetric epoxidation of allylic alcohols [31]). Chiral epoxides are versatile building blocks for asymmetric synthesis due to their regio- and stereochemically predictable reactivity with a broad range of nucleophiles [32]. Thus, despite practical advantages offered by (salen)metal-catalyzed epoxidation processes, a substrate scope of conjugated olefins represents a significant limitation. Clearly, access to simple terminal epoxides is precluded, yet these are among the most important secondary building blocks of the modern chemical industry. Since worldwide efforts to develop practical asymmetric epoxidation processes for α-olefins have been unsuccessful to date [33], an alternative approach has been developed for access to these important chiral building blocks.

4
Epoxide Ring Opening Reactions

4.1
Hydrolytic Kinetic Resolution of Terminal Epoxides

The substrate limitations of the (salen)Mn-catalyzed asymmetric epoxidation were successfully overcome with the discovery of the kinetic resolution of racemic terminal epoxides with water catalyzed by chiral (salen)Co(III) complexes to produce highly enantioenriched epoxides and 1,2-diols (Scheme 6). The hydrolytic kinetic resolution (HKR) process was discovered in 1997 [34], and was quickly industrialized to provide commercial access to several important chiral building blocks such as propylene oxide, propylene glycol, epichlorohydrin, 3-chloro-1,2-propanediol, and methyl glycidate in high enantiomeric excess at large scale (>100 kg batches). Several of these chiral intermediates have already been incorporated into processes for synthesizing pharmaceutical agents.

Despite the fact that the HKR is a kinetic resolution and has a 50% maximum theoretical yield, it possesses a set of characteristics that render it nearly ideal as an industrial process [35]. First, most simple terminal epoxides are readily accessible in racemic form at very low cost, and nearly all monosubstituted epoxides examined to date have proven to be useful substrates for the HKR (nearly 100 substrates evaluated). As in any kinetic resolution, high enantiomeric excess of recovered epoxide is attainable as long as the reaction is carried out to high enough conversion. As a result of the high selectivities obtained in the HKR, re-

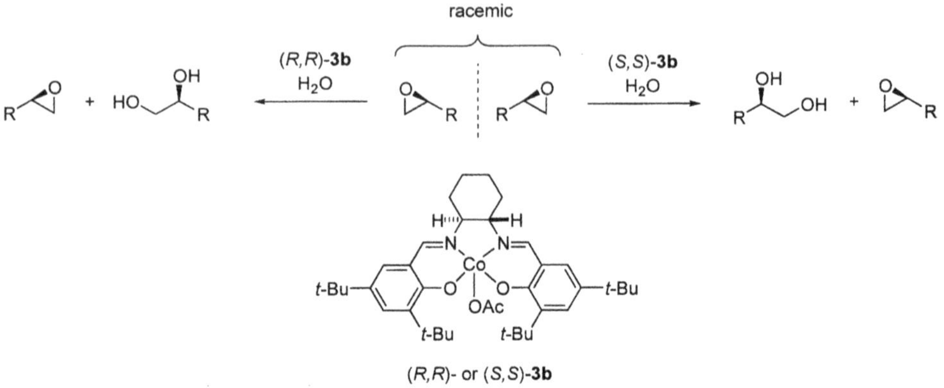

(*R,R*)- or (*S,S*)-**3b**

Scheme 6 Hydrolytic kinetic resolution (HKR) of racemic epoxides catalyzed by complex **3b**

solved epoxide can usually be recovered in >99% ee in close to the theoretical yield (40–45%) [36] (Fig. 5). Alternatively, high ee 1,2-diol products can also be obtained in a practical manner by simply carrying the resolution to slightly lower conversion using 0.45 equivalents of water. HKR reactions are typically carried out in the absence of solvent, and the resolved epoxide and diol product are often easily separated by distillation or extraction. Catalyst **3a** is available at any scale at low cost and is used at low loading levels (0.2–2.0 mol%). Furthermore, the catalyst can be recovered from HKR reactions of most epoxides and recycled without loss of activity or selectivity. The use of water as a nucleophile has tremendous practical advantages, as it is inexpensive, safe, easily handled, and of low molecular weight. The rate of water addition can also be used to modulate the rate of the reaction in order to help control the heat output from the reaction. Together, these factors combine to make the HKR process one of the most practical asymmetric transformations discovered to date.

Kinetic studies carried out on the HKR have allowed formulation of a complete rate expression for this reaction [37]. The most salient feature of the HKR as well as of other (salen)metal-catalyzed epoxide ring-opening reactions is the second-order dependence on catalyst concentration. This is consistent with a dual activation mechanism wherein epoxide is associated to one molecule of catalyst with simultaneous delivery of the hydroxide nucleophile by a second catalyst molecule. This cooperative, bimetallic mechanism also accounts for significant non-linear effects observed with this system [38]. Changes in catalyst concentration thus have an exponential impact on the rate of the reaction. In practice, this places a severe limitation on how low a catalyst loading can be employed, especially with slow-reacting epoxides.

The loss of active catalyst via reduction (**3b** to **3a**) is often observed in HKR reactions, particularly toward the final stages of the resolution process. Because of the second order dependence on catalyst concentration, this can make it very difficult to reach high enantiomeric excess (>98% ee) with slow-reacting substrates. Typically, this difficulty is overcome by the addition of excess water or by the use of higher catalyst loadings. However, the use of stronger Brønsted acids

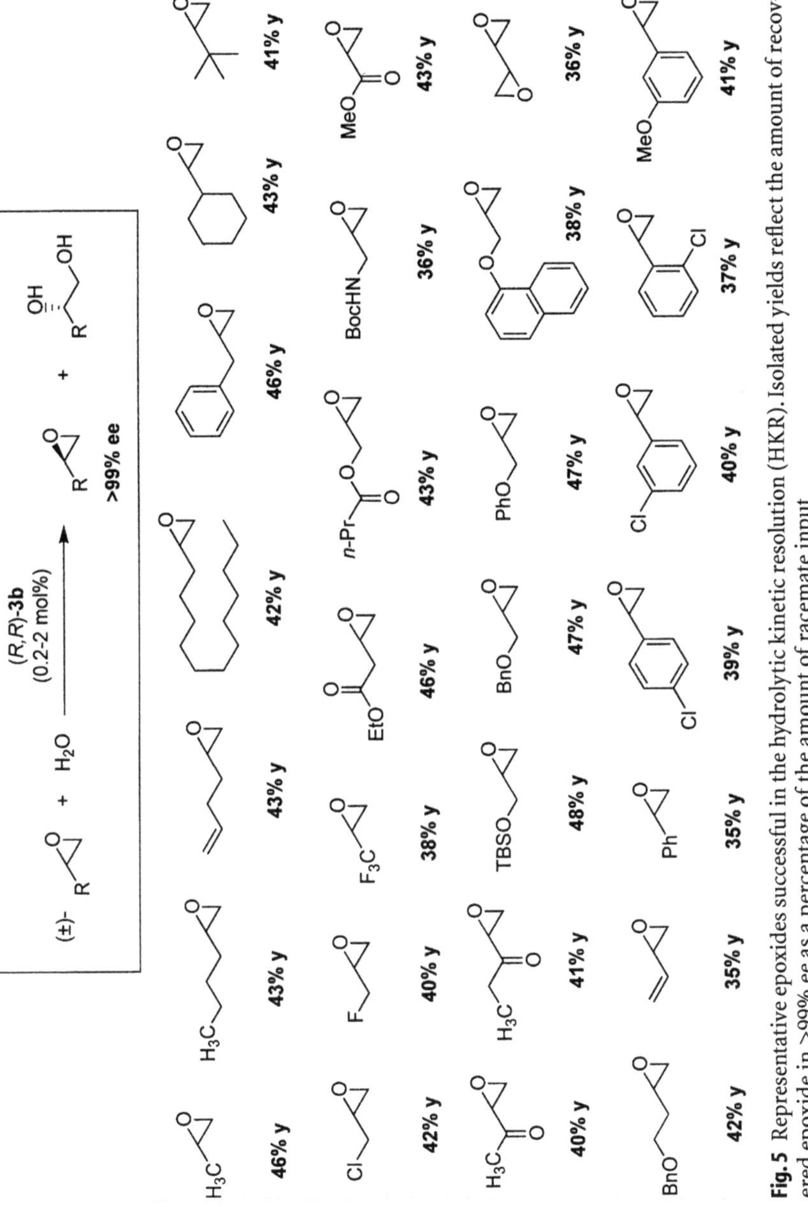

Fig. 5 Representative epoxides successful in the hydrolytic kinetic resolution (HKR). Isolated yields reflect the amount of recovered epoxide in >99% ee as a percentage of the amount of racemate input

to activate the catalyst has been developed as an alternative to overcome this problem. The choice of the acid is important because if the conjugate base is too nucleophilic, it will open the epoxide and the benefit will be lost. Electron deficient benzoic acid derivatives or sulfonic acids (e.g., 4-nitrobenzoic acid or (+)-camphorsulfonic acid) have been found to provide the optimum catalysts for slow-reacting substrates. Although the reactions are often kinetically slower with these catalysts, the reactions typically reach completion in less time due to the minimization of the catalyst reduction pathway.

In order to overcome the rate limitations on the HKR reaction imposed by the second-order dependence on catalyst concentration, attempts have been made to increase the reactivity of the catalyst by linking multiple reactive metal centers together. The practicality of such a strategy is clearly tied to the synthetic accessibility of the linked catalysts relative to the inexpensive monomeric analog **3** [39, 40]. A successful outcome was achieved by preparing C_2-symmetric (salen)Co derivatives linked together by flexible tethers at the 5,5′-positions of the salicylaldehyde units. Not only are the requisite ligands readily synthesized, but the derived cyclic oligomeric (salen)Co complexes display remarkably enhanced reactivity relative to the cobalt(III) complexes of ligand **1** [41] (Fig. 6). The first generation oligomeric catalyst (**6a**) utilized the α,α′-dichloropimelate linker as a mixture of stereoisomers for optimal electronic tuning of the metal center, but it was later found that the simple pimelate linker could be utilized with better results by varying the sulfonate counterion of the metal (**6b,c**) [42]. This change greatly simplified the synthesis of the oligomeric ligand to the extent that the commercial synthesis of the oligo(salen) catalyst is projected to cost only 2–4 times the cost of the monomeric catalyst. Also, due to the electron-rich nature of the oligo(salen) ligand, catalyst reduction is nearly eliminated as a catalyst decomposition pathway.

These novel oligomeric (salen)Co(III) complexes are more reactive than the monomeric catalyst by orders of magnitude. Not only can catalyst loadings be dramatically reduced with these complexes, but the scope of reactivity is also

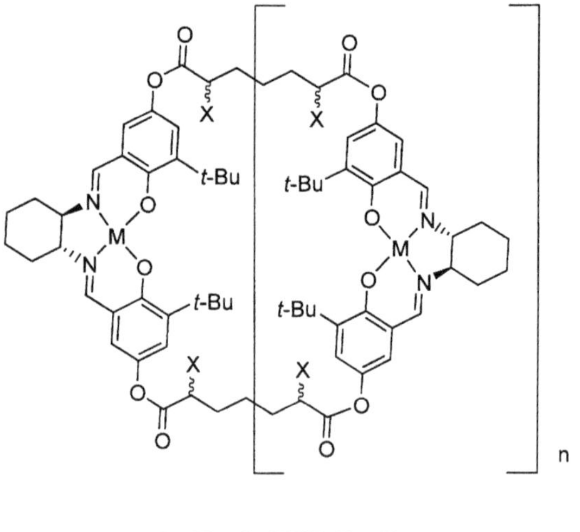

6a, M = Co·LPTS, X = Cl
6b, M = Co·CSA, X = H
6c, M = Co·3-NBS, X = H

Fig. 6 Structures of highly reactive oligomeric salen complexes

widened to include disubstituted epoxides (e.g., cyclohexene oxide) as substrates and primary alcohols as nucleophiles (see next section). The practical significance of the difference in reactivity between the oligomeric salen catalysts and the monomeric complexes is best realized by an example. In the HKR of epichlorohydrin, an optimized catalyst loading of 0.5 mol% (relative to racemate) requires 1.8 kg of catalyst **3b** per 50 kg racemate, or 3.6 wt%. However, the oligomeric catalyst currently attains the same reactivity at only 0.005 mol%, which corresponds to just 50 g of catalyst **6b** per 50 kg of racemate (0.1 wt%). At these levels, the cost contribution of the catalyst to the end product becomes virtually negligible, and product isolation is greatly facilitated.

In summary, within a remarkably short period from its initial discovery, the HKR reaction has been applied to the cost-effective industrial synthesis of several chiral building blocks. The reaction provides one-step access to enantiopure terminal epoxides that have not been practically accessible in the past. Many 1,2-diols are also accessible in high enantiomeric purity utilizing this technology. As the incorporation of these building blocks in pharmaceutical, agricultural, and fine chemical applications increases, it is anticipated that the HKR technology will be transformed from batch operation at present to a continuous mode of production utilizing a fixed-bed or similar reactor system. Efforts to immobilize both the monomeric and oligomeric salen catalysts for these applications are currently underway.

4.2 Other Nucleophilic Epoxide Opening Reactions

The first asymmetric epoxide ring opening (ARO) reaction found to be catalyzed by (salen)metal complexes was the desymmetrization of meso epoxides with $TMSN_3$ [43], discovered two years prior to the HKR. Meso epoxides undergo desymmetrization catalyzed by (salen **1**)Cr(III) complexes, with good-to-excellent enantioselectivity for epoxides derived from cyclic olefins, but lower selectivity with acyclic meso epoxides. Best results were obtained with 5-membered ring substrates (entries 2, 4–5, 7), with lower but still useful ees observed with cyclohexene oxide (entry 1). Larger ring epoxides performed poorly (entry 3), with cyclooctene oxide being completely unreactive (Table 1). The ARO reaction was extended to the kinetic resolution of racemic terminal epoxides to provide 1-azido-2-trimethylsiloxyalkane derivatives in high enantiomeric excess (Table 1, entries 9–18) [44] and of 2,2-disubstituted epoxides where the unreacted epoxides could be isolated in high ee [45]. Epoxide ring-opening with $TMSN_3$ has also been demonstrated to show a high degree of catalyst control in the regioselective opening of enantiopure dissymmetrically substituted epoxides [46].

The ARO/KR reaction is usually run in the absence of solvent, and the product can often be recovered in nearly quantitative yield by vacuum distillation. The catalyst residue (**4b**) is recyclable, and as many as ten cycles have been demonstrated. In this respect, the ARO reaction is very industrially practical since two reactants combine to form one product with no waste using a highly selective, recyclable catalyst. However, commercialization of this technology has not

Table 1 ARO and KR of epoxides via ring opening with $TMSN_3$

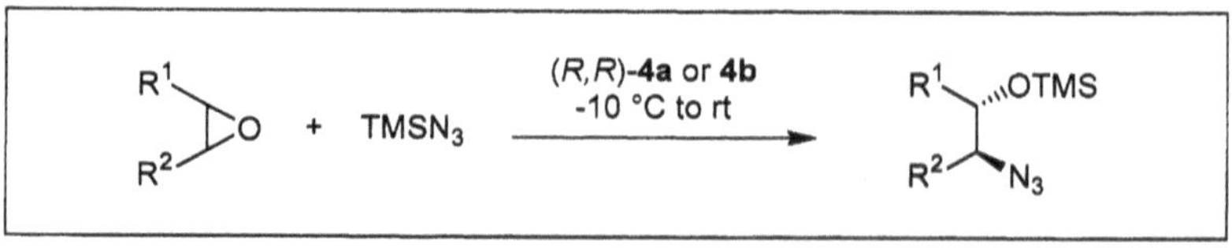

Entry	R^1	R^2	Cat. (mol%)	Yield (%)[a]	ee (%)
Meso epoxides (ARO):					
1	$-(CH_2)_4-$		**4a** (2.0)	96	85
2	$-(CH_2)_3-$		**4a** (2.0)	97	93
3	$-(CH_2)_5-$		**4a** (2.0)	99	42
4	$-CH_2OCH_2-$		**4a** (2.0)	96	97
5	$-CH_2N(C(O)CF_3)CH_2-$		**4a** (2.0)	96	97
6	$-CH_2CH{=}CHCH_2-$		**4a** (7.5)	85	92
7	$-CH_2C(O)CH_2-$		**4b** (2.0)	77	94
8	CH_3	CH_3	**4a** (2.0)	65	82
Racemic epoxides (KR):					
9	CH_3	H	**4b** (1.0)	49	97
10	CH_2CH_3	H	**4b** (2.0)	41	97
11	$(CH_2)_3CH_3$	H	**4b** (2.0)	45	97
12	CH_2Cl	H	**4b** (2.0)	47	95
13	CH_2OTBS	H	**4b** (3.0)	48	96
14	c-C_6H_{11}	H	**4b** (2.0)	42	97
15	CH_2Ph	H	**4b** (2.0)	47	93
16	$(CH_2)_2CH{=}CH_2$	H	**4b** (2.0)	47	98
17	$CH(OEt)_2$	H	**4b** (2.0)	48	89
18	CH_2CN	H	**4b** (2.0)	40	92

[a] Isolated yield of product based on epoxide input.

advanced due to the high capital costs inherent in the handling of azides on large scale.

The primary synthetic utility of the ARO technology is in the preparation of 1,2-amino alcohols in which the oxygen and nitrogen groups are differentially protected to facilitate further elaboration. Chiral amino alcohols have a rich history as asymmetric ligands and are a prominent pharmacophore in several classes of drugs [47]. Both *cis* and *trans* amino alcohol products are accessible in enantiopure form via the ARO of meso epoxides [48]. Several pharmaceutically active compounds have been synthesized in the laboratory utilizing this technology as the key asymmetric transformation (Fig. 7). The ARO desymmetrization reaction was the key transformation in the synthesis of carbocyclic nucleoside

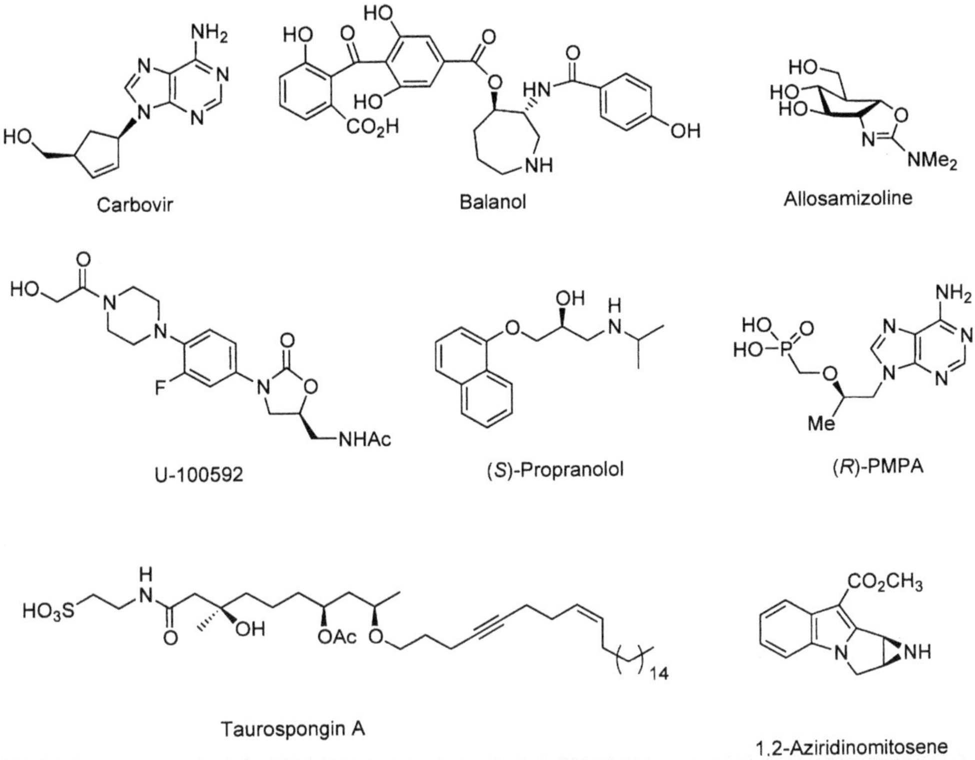

Carbovir Balanol Allosamizoline

U-100592 (*S*)-Propranolol (*R*)-PMPA

Taurospongin A 1,2-Aziridinomitosene

Fig. 7 Biologically active compounds synthesized utilizing ARO or KR as the key asymmetric transformation

analogues such as Carbovir useful in the treatment of viral infections [49]. Other applications include key intermediates in the synthesis of prostaglandins [50]; allosamizoline, a component of the allosamidin family of chitinase inhibitors [51]; Balanol, a natural product inhibitor of protein kinase C [52]; and the 1,2-aziridinomitosene ring system of the mitomycin antitumor antibiotics [53]. The kinetic resolution application was utilized in the synthesis of the β-blocker (*S*)-propranolol and the antiviral agent (*R*)-PMPA [44], as well as the natural product taurospongin A [54]. Epichlorohydrin was also found to undergo dynamic kinetic resolution, yielding the product (*S*)-3-azido-1-chloro-2-trimethylsiloxypropane in 97% ee and 76% yield based upon racemic epoxide. This product was utilized in the synthesis of the novel aryl oxazolidinone antibiotic U-100592 [55].

The ARO displays similar kinetic behavior to the HKR in that it is subject to a second-order dependence on catalyst concentration; a similar bimetallic mechanism involving simultaneous activation of nucleophile and electrophile by distinct catalyst molecules has been postulated for both reactions [56, 57]. Benzylic thiol nucleophiles work best with (salen)Cr complexes, but moderate levels of enantioselectivity limit the utility of this methodology [58]. Halides have shown excellent reactivity as nucleophiles with both catalyst systems, but enantioselectivities are generally low. Recently, complex **4a** has been reported to be a competent promoter of the enantioselective opening of epoxides with fluoride, but the efficiency of catalysis is poor and the mechanism is unclear [59]. The ARO of meso epoxides with benzoic acid catalyzed by (salen)Co(III) complexes displayed good-to-excellent levels of enantioselectivity, and the highest ees were obtained with aromatic substituted epoxides such as *cis*-stilbene oxide [60]. Although the kinetic resolution of racemic terminal epoxides with carboxylic acid derivatives was not preparatively useful, catalytic regioselective opening of resolved epoxides can be accomplished with most carboxylic acids. This strategy has been effectively utilized to prepare glycidyl butyrate by opening resolved epichlorohydrin with butyric acid followed by ring closure of the chlorohydrin to the epoxide (Scheme 7).

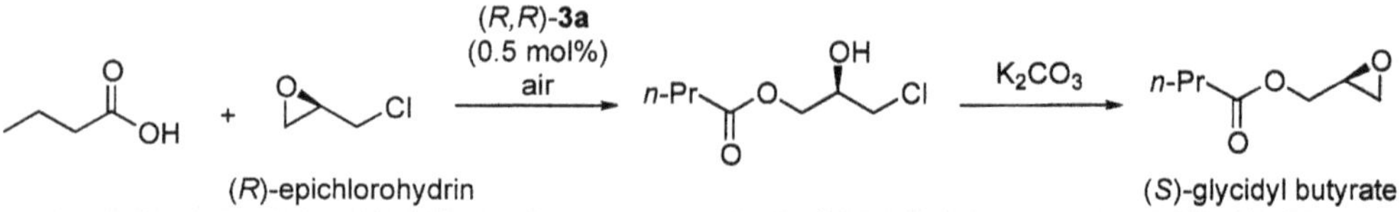

Scheme 7 Synthesis of glycidyl butyrate by regioselective ring opening of resolved epichlorohydrin with butyric acid

In another application with industrial potential, phenols were found to be effective nucleophiles for the kinetic resolution of terminal epoxides utilizing the Co(II) complex **3a** as precatalyst (Scheme 8) [61]. Initially, perfluoro-*tert*-butanol was required as the catalyst activator for good reactivity, but it has since been demonstrated that electron deficient phenols and 2,6-lutidinium *p*-toluenesulfonate are more practical catalyst activators [62]. Phenols are highly useful nucleophiles because the α-aryloxy alcohol system is a prominent pharmacophore [47]. The dynamic kinetic resolution of epibromohydrin with phenols was utilized to prepare aryl glycidyl ethers in >99% ee and 74–77% yield based on racemic epoxide [61, 63]. Aryl glycidyl ethers have been used for the manufacture of several β-blockers marketed today including (*S*)-propranolol (see Fig. 7) [47].

The ring opening of resolved propylene oxide with 2,3-difluoro-6-nitrophenol was also used in the one-step preparation of a key intermediate in the synthesis of the quinolone antibiotic Levofloxacin [64]. However, the regioselectiv-

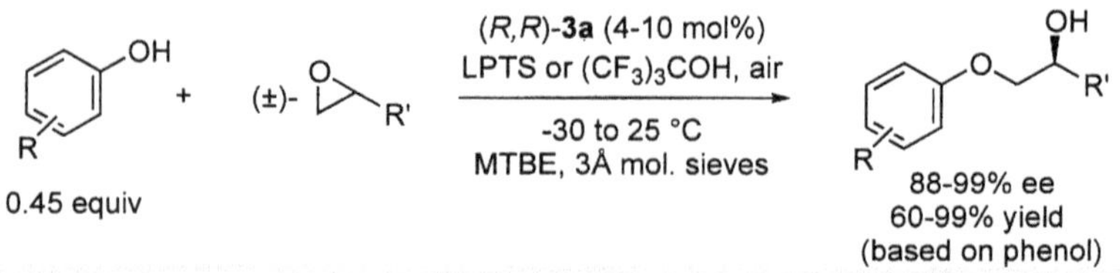

Scheme 8 Kinetic resolution (KR) of racemic epoxides with phenols

ity of opening is uncharacteristically low at 25–30:1, and the regioisomeric products are prone to equilibrate via a Smiles rearrangement. In order to prevent the erosion in regioisomeric purity, a method of silyl transfer from phenol to product was utilized to halt this rearrangement (Scheme 9) [65].

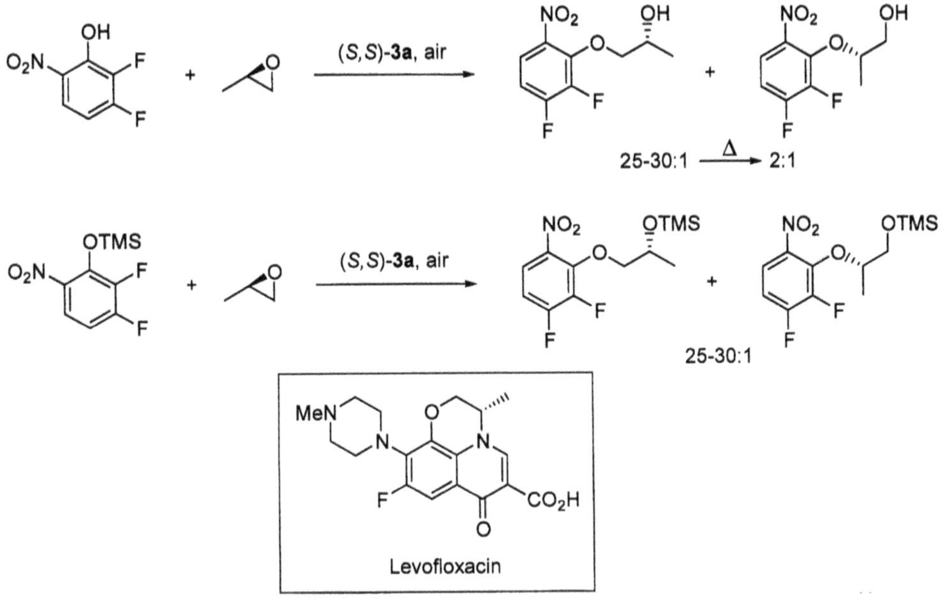

(S,S)-**3a**, air

25-30:1 —Δ→ 2:1

(S,S)-**3a**, air

25-30:1

Levofloxacin

Scheme 9 Potential industrial routes to the key intermediate in the synthesis of the quinolone antibiotic levofloxacin

As with the HKR reaction, the ring opening of epoxides with these other nucleophiles has been rendered significantly more practical from an industrial perspective by the development of the oligomeric (salen)Co catalysts. These catalysts again display enhanced reactivity and selectivity with these nucleophiles relative to the monomeric catalysts (Table 2). This is even more important with nucleophiles other than water because higher loadings of the monomeric and oligomeric catalysts are generally required for these reactions. In this context, the synthetic accessibility of the oligo(salen) catalysts becomes the paramount issue because the cost contribution of the catalyst to the product is no longer negligible as it is in the HKR example described in the previous section. In addition, the range of useful nucleophiles has been expanded to include primary alcohols (entries 4–5), which are effectively unreactive with the monomeric catalysts [66]. Oligo(salen) complexes of metals other than Co(III) have yet to be evaluated, but enhanced performance and an expansion of the pool of useful reactions can be reasonably anticipated.

In summary, chiral (salen)Co(III) and Cr(III) complexes have been found to catalyze the asymmetric ring opening of meso and racemic terminal epoxides with a high degree of selectivity using a variety of synthetically useful nucle-

Table 2 Comparison of monomeric and oligomeric catalysts in KR and ARO of epoxides with hydroxylic nucleophiles

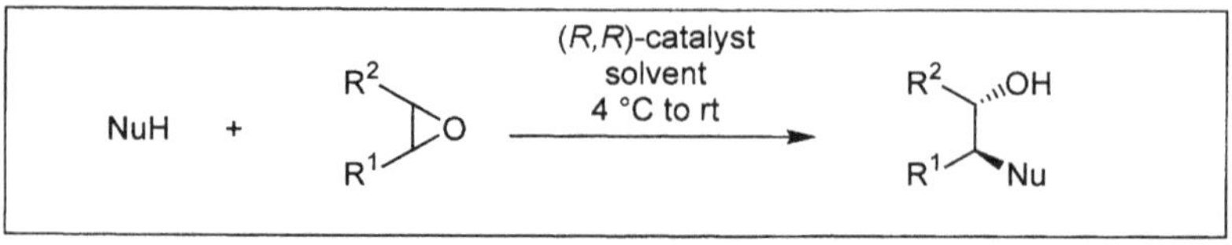

NuH + (R^2, R^1 epoxide) —(R,R)-catalyst, solvent, 4 °C to rt→ (R^2 ...OH, R^1 Nu)

Entry	NuH	Epoxide	Cat. (mol%)[a]	Time (h)	Yield (%)[b]	ee (%)
Racemic epoxides (KR):						
1	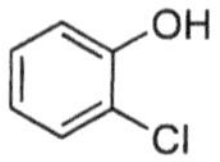 OH, Cl	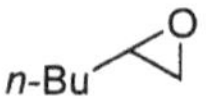 n-Bu, O	**3**·OTs (4.0)	72	40	68
			6a (0.8)	6	49	99
			6c (0.8)	10	47	>99
2	PhOH	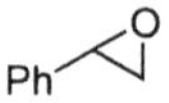 Ph, O	**3**·OTs (4.0)	240	c	nd
			6a (1.0)	8	30	97
3	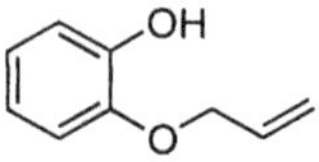 OH, O	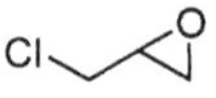 Cl, O	**3**·OTs (4.4)	24	24	84
			6a (0.25)	12	49	98
4	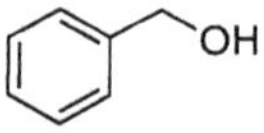 OH	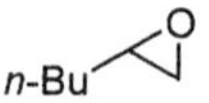 n-Bu, O	**6c** (0.25)	8	45	99
5	 Me_3Si, OH	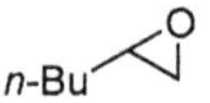 n-Bu, O	**6c** (0.1)	2	48	>99
Meso epoxides (ARO):						
6	H_2O	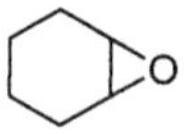 O	**3b** (5.0)	72	71	75
			6a (1.5)	3	95	86
			6b (0.5)	12	90	93
			6c (0.5)	12	92	93

[a] Catalyst loading on a per Co basis relative to nucleophile. [b] Isolated yield of product based on epoxide input. [c] After 10 d, the reaction reached 63% conversion and produced a 2:3 mixture of regioisomeric products favoring internal attack.

ophiles. Novel cyclic oligo(salen) catalysts have also shown remarkable enhancement in reactivity and selectivity relative to monomeric catalysts, leading to the prospect of additional reactions catalyzed by these complexes. Interestingly, despite a great deal of effort, cyanide and other carbon nucleophiles were completely unreactive with these catalysts [67]. However, with an overwhelming body of evidence supporting a cooperative bimetallic mechanism for nucleophilic epoxide opening reactions by simultaneous activation of the nucleophile and the epoxide by distinct catalytic centers, the question of whether or not this mechanism was general for a broader range of nucleophile-electrophile reactions catalyzed by (salen)metal complexes became paramount. With this in mind, the search began for new nucleophile-electrophile reactions catalyzed by chiral (salen)metal complexes.

5 Carbonyl Addition Processes

Because of its great synthetic significance and the fact that it utilizes a novel nucleophile-electrophile pair, the addition of hydrogen cyanide to imines (the Strecker reaction) was chosen as an ambitious target for asymmetric catalysis by chiral (salen)metal complexes. The asymmetric Strecker reaction provides facile access to optically active α-amino acids, but progress in this area was limited until very recently [68]. In the current study, a series of metal complexes of salen **1** were screened for catalysis of the reaction of *N*-allyl benzaldimine with trimethylsilylcyanide. Several (salen)metal complexes were found to catalyze the reaction with varying degrees of conversion and enantioselectivity, and the best results were observed with the Al complex **5a** [69]. The actual reactive reagent in the reaction was determined to be HCN, which could be conveniently generated in situ from TMSCN and methanol. In order to suppress the racemic background reaction, the reactions were performed at -70 °C, and the products were trifluoroacetylated to prevent racemization. A variety of alkyl and aryl *N*-allyl imines were evaluated in the reaction, and substituted aryl imines gave the highest levels of enantioselectivity (Scheme 10). Alkyl imines yielded products with more modest ees. Modification of the nitrogen substituent of the imine or of the steric or electronic properties of the salen ligand proved unsuccessful in improving the results with these synthetically useful substrates. However, a novel non-metal catalyst (7) was identified which provided excellent selectivities with all types of imine structures [70].

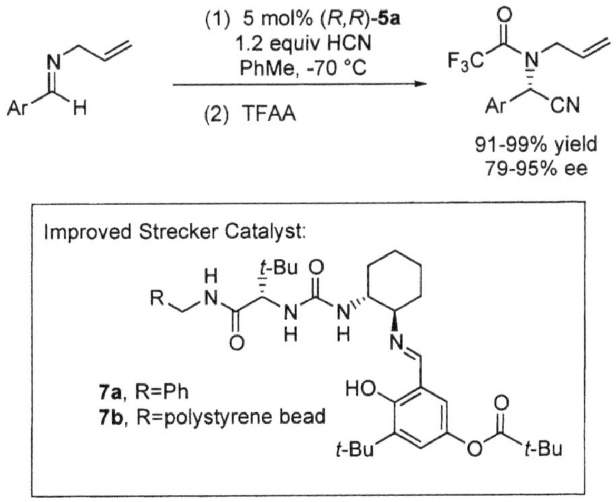

(1) 5 mol% (*R*,*R*)-**5a**
1.2 equiv HCN
PhMe, -70 °C
(2) TFAA

91-99% yield
79-95% ee

Improved Strecker Catalyst:

7a, R=Ph
7b, R=polystyrene bead

Scheme 10 Asymmetric Strecker reaction

Among the (salen)metal complexes found to be active for the addition of cyanide to imines was the Ti(IV)Cl_2 complex of ligand **1**. This complex was previously identified by North and Belokon as an asymmetric catalyst for the addition of TMSCN to aldehydes [71] and ketones [72]. In a limited screening of salen ligand substituents, ligand **1** was found to be optimum for this transformation as well. Subsequently, the Ti(IV)oxo-dimer (**8**) and the V(IV)oxo (**9**) complexes of ligand **1** were determined to be improved catalysts in terms of reactivity (**8**) or selectivity (**9**) for all substrates investigated (Scheme 11) [73]. Like the (salen)Al-catalyzed addition of HCN to imines, aromatic aldehydes were the best substrates, although a greater range of selectivities was observed.

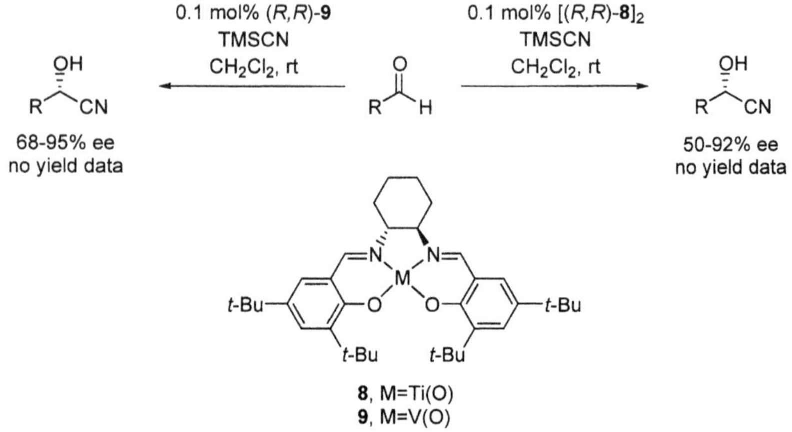

0.1 mol% (*R*,*R*)-**9**, TMSCN, CH_2Cl_2, rt

0.1 mol% [(*R*,*R*)-**8**]$_2$, TMSCN, CH_2Cl_2, rt

68-95% ee
no yield data

50-92% ee
no yield data

8, M=Ti(O)
9, M=V(O)

Scheme 11 Asymmetric hydrocyanation of carbonyls

Although incompletely elucidated, the mechanisms by which these three complexes catalyze these similar reactions appear to be distinct from the seemingly general mechanism of nucleophilic ring-opening of epoxides. The (salen)Al-catalyzed addition of HCN to imines has been determined to be first-order in catalyst, which would preclude the possibility of a cooperative bimetallic mechanism of nucleophile and electrophile activation [74]. The (salen)Ti- and (salen)V-catalyzed additions of TMSCN to carbonyls both display a non-first-order kinetic dependence on catalyst concentration [73], but this has been attributed to the formation of dimeric catalyst complexes which show greater reactivity than the monomeric species.

Another reaction found to be catalyzed by (salen **1**)Al complexes is the asymmetric conjugate addition of hydrazoic acid to α,β-unsaturated imides [75]. Using a simple protocol, excellent enantioselectivities were observed for several *N*-benzoyl imide derivatives (Scheme 12), although cinnamate derivatives suffered from poor reactivity. The β-azido imide products can be readily converted to β-amino acids.

In another interesting application of chiral (salen)metal catalysis, Bandini et al. reported an extension of the Nozaki-Hiyama-Kishi coupling to the addition

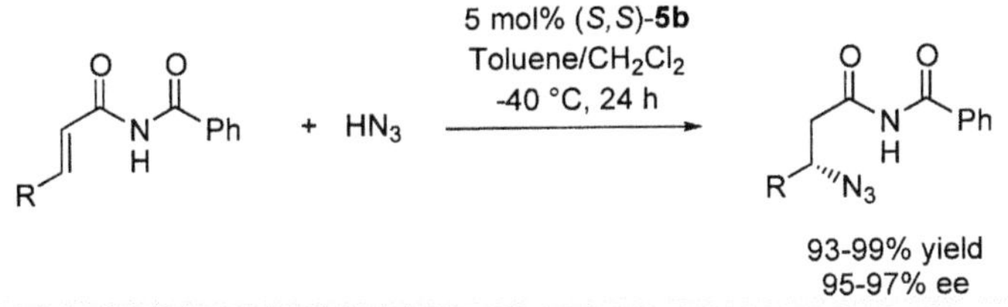
5 mol% (S,S)-5b
Toluene/CH2Cl2
-40 °C, 24 h
+ HN3
Ph
R
Ph
N3
93-99% yield
95-97% ee

Scheme 12 Asymmetric conjugate addition of hydrazoic acid

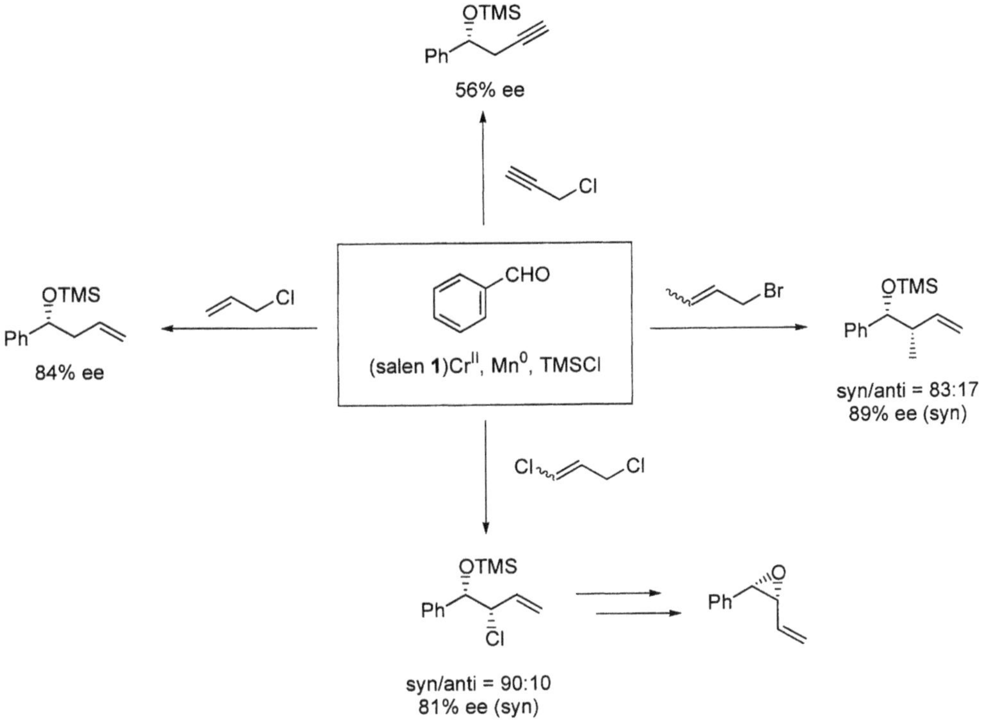
OTMS
Ph
56% ee
Cl
CHO
(salen 1)Cr^II, Mn^0, TMSCl
84% ee
Br
syn/anti = 83:17
89% ee (syn)
syn/anti = 90:10
81% ee (syn)

Scheme 13 Asymmetric addition of allylic and propargylic halides to aldehydes

of allylic and propargylic halides with aromatic aldehydes (Scheme 13) [76]. The use of 1,3-dichloropropene has also been investigated as a route to optically active vinyl epoxides [77]. Despite being much less developed than alternative tin- and silicon-based chemistry [78], this work has the advantage of directly utilizing allylic and propargylic halides from which the corresponding stannanes and silanes are typically prepared. At this stage, the practical application of these reactions is limited by the moderate-to-low yields due to competing side reactions, the long reaction times, and the requirement of high catalyst loadings (typically 10 mol% plus excess ligand).

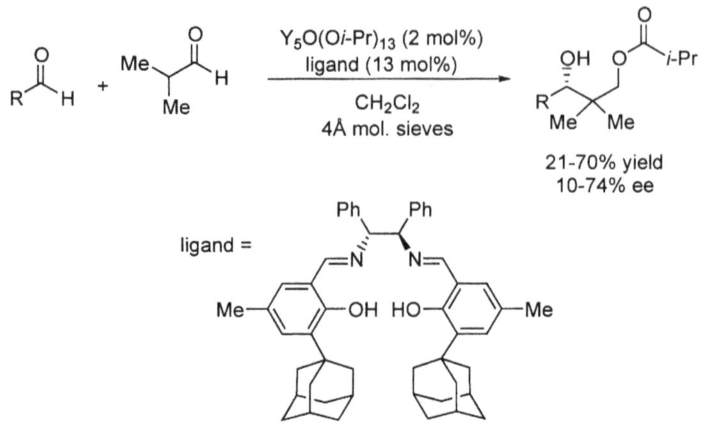

Scheme 14 Y(III)-catalyzed aldol-Tischenko reaction

Several other salen-catalyzed asymmetric transformations involving C–C bond formation have been reported recently as well. One report utilizes (salen)Y(III) complexes to catalyze the first enantioselective catalytic aldol-Tischenko reaction (Scheme 14) [79]. In this case, high-throughput screening of potential metal catalysts was conducted utilizing the commercially available salen **1**, then the ligand structure was systematically varied to optimize for enantioselectivity. Chiral (salen)Al complexes were also found to catalyze enantioselective aldol reactions of aldehydes and 5-alkoxyoxazoles (Scheme 15) [80]. These catalysts utilized 2,2′-diamino-1,1′-binaphthyl (BINAM) as the diamine component of the salen ligand, and provided the *cis* products of aromatic aldehydes with exceptionally high yield, diastereoselectivity and enantiomeric excess. The *cis* products could be thermodynamically equilibrated to the *trans* products (5:95 ratio) by treatment with base, and both diastereomers are readily

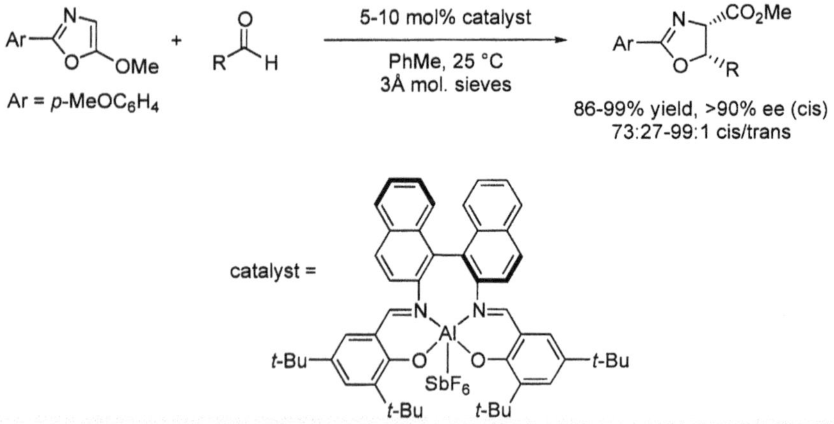

Scheme 15 Asymmetric 5-alkoxyoxazole aldol reaction

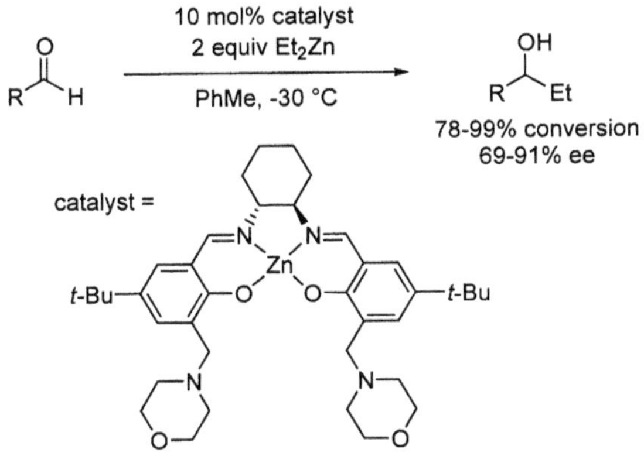

Scheme 16 Asymmetric addition of Et_2Zn to aldehydes

converted to β-hydroxy-α-amino acid derivatives. Zinc(II) complexes of salen ligands bearing secondary Lewis basic groups in the 3,3′-positions have also been found to catalyze the asymmetric addition of diethylzinc to aldehydes (Scheme 16) [81]. Rather than simultaneous activation of both nucleophile and electrophile by the metal center as in the epoxide ring opening chemistry, this example provides for electrophile activation by the metal center and nucleophile activation by the secondary basic groups. The structural features of the salen ligand allow this concept to operate without catalyst deactivation by these opposing reactive sites.

This section has summarized the extension of (salen)metal-catalyzed asymmetric reactions to include nucleophilic additions to carbonyls. The palette of reactive metals has been substantially broadened to include such diverse metals as Al, Ti, V, Y, and Zn. Most of the reactions also involve carbon-carbon bond formation, thereby expanding the pool of synthetically valuable reactions even further. The (salen)Al catalyst systems stand out due to the exceptionally high enantioselectivities observed in several reactions. The synthetic utility and high selectivity of these reactions will surely drive their application to industrial processes.

6 Cycloaddition Processes

6.1 Hetero-Diels-Alder Reaction

The search for additional reactions promoted by chiral (salen)metal complexes also led to the identification of (salen)Cr(III) complexes as competent catalysts for the asymmetric hetero-Diels-Alder (HDA) reaction of aldehydes with 1-methoxy-3-(trimethylsilyl)oxy-1,3-butadiene (Danishefsky's diene) [82]. In this reaction, salen ligand **1** was found to be the best in a limited screen of salicylidene substituents, but the identity of the counterion to chromium was found to

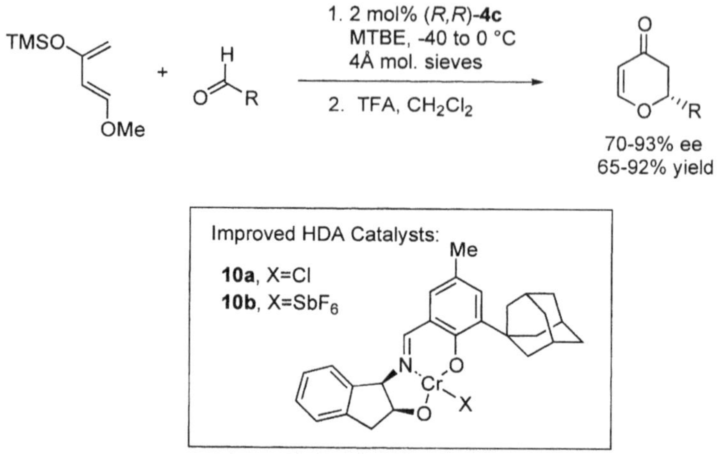

TMSO
OMe
+
H
O R
1. 2 mol% (*R*,*R*)-**4c**
MTBE, -40 to 0 °C
4Å mol. sieves
2. TFA, CH_2Cl_2
70-93% ee
65-92% yield

Improved HDA Catalysts:
10a, X=Cl
10b, X=SbF_6
Me
N
O
Cr
O
X

Scheme 17 Asymmetric hetero-Diels-Alder reaction

be particularly influential on the selectivity and reactivity of the catalysts. Catalysts possessing non-coordinating anions such as BF_4^- and PF_6^- showed greater reactivity and selectivity relative to the chloride catalyst, but only in the presence of a desiccant such as powdered molecular sieves. Using 2 mol% of the BF_4 complex **4c**, several synthetically useful aldehydes were reacted to produce dihydropyranones in 70–93% ee and 65–92% yield (Scheme 17). Control experiments revealed the HDA reaction to be a true cycloaddition process rather than a Mukaiyama aldol reaction followed by acid-catalyzed cyclization [83]. More recently, improved catalysts based upon tridentate Schiff base complexes of Cr(III) (**10**) have been discovered which dramatically improve the scope and utility of this important reaction [84].

6.2 Diels-Alder Reaction

Subsequent to the report of the asymmetric hetero-Diels-Alder reaction catalyzed by chiral (salen)Cr(III) complexes, Rawal documented the ability of these catalysts to provide enantioselection in the highly endo-selective Diels-Alder reaction between 1-amino-1,3-butadiene derivatives and substituted acroleins (Scheme 18) [85]. As with the hetero-Diels-Alder reaction, the cationic Cr complexes showed greater reactivity and product enantiomeric excess than the chloride catalysts, although the use of molecular sieves was not critical for good reactivity. The cyclohexene products of this reaction were produced in greater enantiomeric excess than the HDA reaction products (generally >90% ee). The major limitation of this transformation remained reactivity, as the reactions required several days to reach completion using 5 mol% of catalyst **4d.** This problem appears to have been overcome with a recently reported breakthrough in Diels-Alder catalysis wherein cationic (salen)Co(III) complexes display unprec-

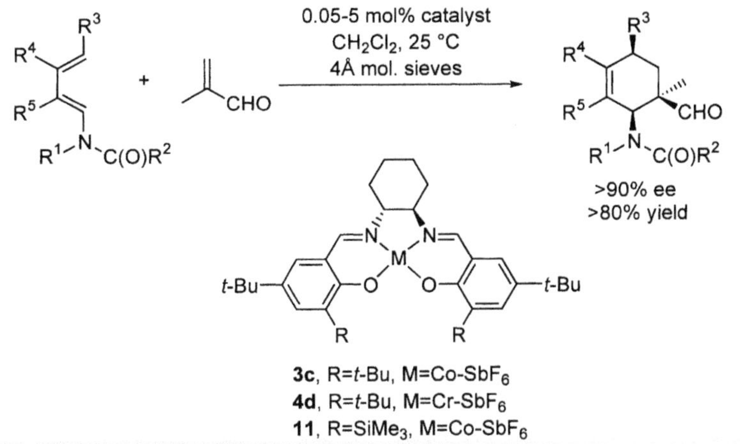

0.05-5 mol% catalyst
CH_2Cl_2, 25 °C
4Å mol. sieves

>90% ee
>80% yield

3c, R=*t*-Bu, M=Co-SbF_6
4d, R=*t*-Bu, M=Cr-SbF_6
11, R=$SiMe_3$, M=Co-SbF_6

Scheme 18 Asymmetric Diels-Alder reaction

edented catalytic activity (requiring as little as 0.05 mol% catalyst) [86]. Catalysts (**11**) possessing trialkylsilyl groups at the 3,3′-positions showed greater reactivity and selectivity than Co complex **3c.**

6.3 Cyclopropanation Reaction

Although carbene transfer from metal to olefin is mechanistically more similar to oxene transfer in epoxidation reactions, the cyclopropanation of olefins with diazoesters is formally a [2+1] cycloaddition and is thus covered in this section. If there are two unequal substituents on the carbene atom, two diastereomeric products are possible with monosubstituted olefins. Katsuki has demonstrated high *trans*- and enantioselectivity with (salen)Co(III) complexes (**12**) that have no substituent at the 3,3′-positions of the salen ligand [87]. Utilizing second generation salen ligands bearing axially chiral salicylaldehyde derivatives, Katsuki has also demonstrated high *cis* selectivity with Ru(III) (**13**) [88] and Co(II) (**14**) complexes (Scheme 19) [89]. Product yields are generally higher with the Co catalysts, while all three catalysts show exceptional enantioselectivity (>90% ee). The process to *trans* cyclopropane products seems especially practical due to the accessibility of the salen ligand.

At this stage, all three cycloaddition processes remain of academic interest, but each offers promise of future industrial utility. The utility of the asymmetric hetero-Diels-Alder reaction primarily results from the synthetic versatility of the pyranone products due to their high degree of functionalization. The unprecedented catalytic activity of readily available (salen)Co complexes in the Diels-Alder reaction offers practical advantages in the asymmetric synthesis of chiral cyclohexenes. Alternatively, the cyclopropanation processes show excellent selectivities and offer better diastereocontrol relative to traditional Cu- and Rh-catalyzed systems [90].

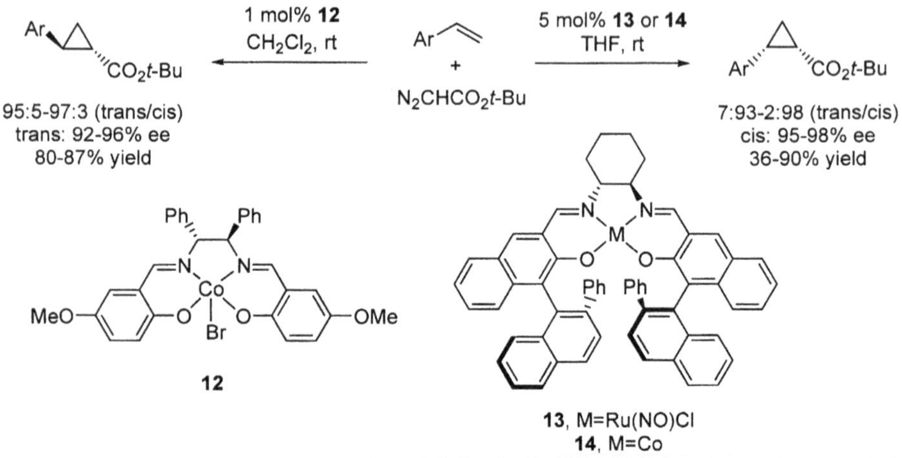

Scheme 19 Highly diastereoselective asymmetric cyclopropanation reaction

7
Conclusion

Over the past decade, chiral (salen)metal complexes have emerged as versatile catalysts for a broad range of industrially and academically interesting reactions. Since its discovery as an optimum ligand for (salen)Mn-catalyzed asymmetric epoxidation reactions in 1991 [5], metal complexes of salen ligand **1** have displayed remarkable effectiveness in a wide variety of catalytic asymmetric reactions. Ligand **1** and its metal complexes are available commercially at relatively low cost, and the Mn-catalyzed asymmetric epoxidation and Co-catalyzed hydrolytic kinetic resolution processes are currently practiced on industrial scale. In addition, new discoveries such as the extraordinarily high levels of reactivity and selectivity observed with the oligomeric (salen)Co complexes in epoxide ring opening reactions portend a continued bright future for salen ligands in asymmetric catalysis. As insight is gleaned into the factors controlling reactivity and stereoselection, rational design of improved ligands becomes likely. However, at this stage, salen ligand **1** still defines the standard by which others are evaluated.

References

1. For comprehensive coverage of the field of asymmetric catalysis see: Pfaltz A, Jacobsen EN, Yamamoto H (eds) (1999) Comprehensive asymmetric catalysis, vols 1–3. Springer, Berlin Heidelberg New York
2. For an historical review of asymmetric catalysis which discusses most of these ligand families, see: Kagan H (1999) Historical perspective. In: Pfaltz A, Jacobsen EN, Yamamoto H (eds) Comprehensive asymmetric catalysis, vol 1. Springer, Berlin Heidelberg New York, chap 2
3. Due to space limitations, salen ligands derived from chiral salicylaldehydes and/or from diamines other than 1,2-diamines will not be discussed extensively in this review.

For lead references of applications utilizing ligands of this type, see: (a) Katsuki T (1995) Coord Chem Rev 140:189; (b) Ito YN, Katsuki T (1999) Bull Chem Soc Jpn 72:603; (c) Katsuki T (2002) Adv Synth Catal 344:131
4. (a) Jacobsen EN, Zhang W, Güler ML (1991) J Am Chem Soc 113:6703; (b) Palucki M, Finney NS, Pospisil PJ, Güler ML, Ishida T, Jacobsen EN (1998) J Am Chem Soc 120:948
5. Jacobsen EN, Zhang W, Muci AR, Ecker JR, Deng L (1991) J Am Chem Soc 113:7063
6. (a) Jacobsen EN (1993) Asymmetric catalytic epoxidation of unfunctionalized olefins In: Ojima I (ed) (1993) Catalytic asymmetric synthesis, 1st edn. Wiley-VCH, New York, chap 4.2; (b) Jacobsen EN (1995) Transition metal-catalyzed oxidations: asymmetric epoxidation In: Wilkinson G, Stone FGA, Abel EW, Hegedus LS (eds) (1995) Comprehensive organometallic chemistry, 2nd edn, vol 12. Pergamon, New York, pp 1097–1135; (c) Jacobsen EN, Wu MH (1999) Epoxidation of alkenes other than allylic alcohols In: Pfaltz A, Jacobsen EN, Yamamoto H (eds) (1999) Comprehensive asymmetric catalysis, vol 2. Springer, Berlin Heidelberg New York, chap 18.2; (d) Also see [3]
7. (a) Larrow JF, Jacobsen EN, Gao Y, Hong Y, Nie X, Zepp CM (1994) J Org Chem 59:1939; (b) Larrow JF, Jacobsen EN (1997) Org Syn 75:1
8. Gasbøl F, Steenbøl P, Sørensen BS (1972) Acta Chem Scand 26:3605
9. Zhang W, Loebach JL, Wilson SR, Jacobsen EN (1990) J Am Chem Soc 112:2801
10. Zhang W, Jacobsen EN (1991) J Org Chem 56:2296
11. (a) Brandes BD, Jacobsen EN (1994) J Org Chem 59:4378; (b) Fukuda T, Irie R, Katsuki T (1995) Synlett 197; (c) Brandes BD, Jacobsen EN (1995) Tetrahedron Lett 36:5123
12. (a) Palucki M, Pospisil PJ, Zhang W, Jacobsen EN (1994) J Am Chem Soc 116:9333; (b) Palucki M, McCormick GJ, Jacobsen EN (1995) Tetrahedron Lett 36:5457
13. (a) Lee NH, Jacobsen EN (1991) Tetrahedron Lett 32:6533; (b) Chang S, Galvin JM, Jacobsen EN (1994) J Am Chem Soc 116:6937
14. (a) Irie R, Ito Y, Katsuki T (1991) Synlett 265; (b) Irie R, Noda K, Ito Y, Matsumoto N, Katsuki T (1991) Tetrahedron: Asymmetry 2:481; (c) Also see [15]
15. (a) Finney NS, Pospisil PJ, Chang S, Paluki M, Konsler RG, Hansen KB, Jacobsen EN (1997) Angew Chem Int Ed Engl 36:1720; (b) Senanayake CH, Smith GB, Ryan KM, Fredenburgh LE, Liu J, Roberts FE, Hughes DL, Larsen RD, Verhoeven TR, Reider PJ (1996) Tetrahedron Lett 37:3271
16. See [6c]. For an overview of the various mechanistic proposals, see: Linker T (1997) Angew Chem Int Ed Engl 36:2060
17. Deng L, Jacobsen EN (1992) J Org Chem 57:4320
18. Jacobsen EN, Deng L, Furukawa Y, Martínez LE (1993) Tetrahedron 50:4323
19. (a) Bell D, Davies MR, Finney FJL, Geen GR, Kincey PM, Mann IS (1996) Tetrahedron Lett 37:3895; (b) Lee NH, Muci AR, Jacobsen EN (1991) Tetrahedron Lett 32:5055
20. (a) Hughes DL, Smith GB, Liu J, Dezeny GC, Senanayake CH, Larsen RD, Verhoeven TR, Reider PJ (1997) J Org Chem 62:2222; (b) For a review of this compound, see: Senanayake CH (1998) Aldrichimica Acta 31:3
21. (a) Vacca JP, Dorsey BD, Schleif WA, Levin RB, McDaniel SL, Darke PL, Zugay J, Quintero JC, Blahy OM, Roth E, Sardana VV, Schlabach AJ, Graham PI, Condra JH, Gotlib L, Holloway MK, Lin J, Chen I-W, Vastag K, Ostovic D, Anderson PS, Emini EA, Huff JR (1994) Proc Natl Acad Sci USA 91:4096; (b) Dorsey BD, Levin RB, McDaniel SL, Vacca JP, Guare JP, Darke PL, Zugay JA, Emini EA, Schleif WA, Quintero JC, Lin JH, Chen I-W, Holloway MK, Fitzgerald PMD, Axel MG, Ostovic D, Anderson PS, Huff JR (1994) J Med Chem 37:3443
22. (a) Larrow JF, Roberts E, Verhoeven TR, Ryan KM, Senanayake CH, Reider PJ, Jacobsen EN (1999) Org Syn 76:46; (b) Also see [20a]
23. Gao Y, Hong Y, Nie X, Bakale RP, Feinberg RR, Zepp CM (1997) US Patent 559,998,518
24. (a) Bousquet C, Gilheaney DG (1995) Tetrahedron Lett 36:7739; (b) Ryan KM, Bousquet C, Gilheaney DG (1999) Tetrahedron Lett 40:3613; (c) Daly AM, Dalton CT, Renehan MF, Gilheaney DG (1999) Tetrahedron Lett 40:3617; (d) Daly AM, Renehan MF, Gilheaney DG (2001) Org Lett 3:663

25. O'Mahoney CP, McGarrigle EM, Renehan MF, Ryan KM, Kerrigan NJ, Bousquet C, Gilheaney DG (2001) Org Lett 3:3435
26. (a) Takeda T, Irie R, Shinoda Y, Katsuki T (1999) Synlett 1157; (b) Nakata K, Takeda T, Mihara J, Hamada T, Irie R, Katsuki T (2001) Chem Eur J 7:3776
27. Benzylic hydroxylation: (a) Larrow JF, Jacobsen EN (1994) J Am Chem Soc 116:12,129; (b) Hamachi K, Irie R, Katsuki T (1996) Tetrahedron Lett 37:4979; (c) Hamada T, Irie R, Mihara J, Hamachi K, Katsuki T (1998) Tetrahedron 54:10,017; (d) Komiya N, Noji S, Murahashi SI (1998) Tetrahedron Lett 39:7921. Oxidation of α-carbons in heterocycles: (e) Miyafuji A, Katsuki T (1997) Synlett 836; (f) Miyafuji A, Katsuki T (1998) Tetrahedron 54:10,339; (g) Punniyamurthy T, Miyafuji A, Katsuki T (1998) Tetrahedron Lett 39:8295; (h) Punniyamurthy T, Katsuki T (1999) Tetrahedron 55:9439. Benzylic amination: (i) Kohmura Y, Katsuki T (2001) Tetrahedron Lett 42:3339
28. Mn-catalyzed: (a) Palucki M, Hanson P, Jacobsen EN (1992) Tetrahedron Lett 33:7111; (b) Kokubo C, Katsuki T (1996) Tetrahedron 52:13,895. Ti-catalyzed: (c) Saito B, Katsuki T (2001) Tetrahedron Lett 42:3873; (d) Saito B, Katsuki T (2001) Tetrahedron Lett 42:8333; (e) Sasaki C, Nakajima K, Kojima M, Fujita J (1991) Bull Chem Soc Jpn 64:1318. V-catalyzed: (f) Nakajima K, Kojima K, Kojima M, Fujita J (1990) Bull Chem Soc Jpn 63:2620
29. (a) Nishikori H, Katsuki T (2000) Appl Catal A 194/195:475; (b) Ohta C, Katsuki T (2001) Tetrahedron Lett 42:3885; (c) Murakami M, Uchida T, Katsuki T (2001) Tetrahedron Lett 42:7071; (d) Nishikori H, Ohta C, Oberlin E, Irie R, Katsuki T (1999) Tetrahedron 55:13,937
30. Uchida T, Katsuki T (2001) Tetrahedron Lett 42:6911
31. (a) Johnson RA, Sharpless KB (2000) Catalytic asymmetric epoxidation of allylic alcohols In: Ojima I (ed) (2000) Catalytic asymmetric synthesis, 2nd edn. Wiley-VCH, New York, chap 6A; (b) Also see Bolm C (1991) Angew Chem Int Ed Engl 30:403
32. (a) Parker RE, Isaacs NS (1959) Chem Rev 59:737; (b) Behrens CH, Sharpless KB (1983) Aldrichimica Acta 16:67; (c) Hansen RM (1991) Chem Rev 91:437
33. For the most enantioselective methods developed involving synthetic catalysts, see: (a) [12a]; (b) Collman JP, Wang Z, Straumanis A, Quelquejeu M, Rose E (1999) J Am Chem Soc 121:460
34. (a) Tokunaga M, Larrow JF, Kakiuchi F, Jacobsen EN (1997) Science 277:936; (b) Furrow ME, Schaus SE, Jacobsen EN (1998) J Org Chem 63:6776
35. For a review of criteria for practical kinetic resolutions, see: Keith JM, Larrow JF, Jacobsen EN (2001) Adv Synth Catal 343:5
36. Schaus SE, Brandes BD, Larrow JF, Tokunaga M, Hansen KB, Gould AE, Furrow ME, Jacobsen EN (2002) J Am Chem Soc 124:1307
37. Hong J, Blackmond DG, Jacobsen EN (manuscript in preparation)
38. (a) Johnson DW, Singleton DA (1999) J Am Chem Soc 121:9307; (b) Blackmond DG (2001) J Am Chem Soc 123:545
39. Konsler RG, Karl J, Jacobsen EN (1998) J Am Chem Soc 120:6776
40. (a) Annis DA, Jacobsen EN (1999) J Am Chem Soc 121:4147; (b) Breinbauer R, Jacobsen EN (2000) Angew Chem Int Ed 39:3604
41. Ready JM, Jacobsen EN (2001) J Am Chem Soc 123:2687
42. Ready JM, Jacobsen EN (2002) Angew Chem Int Ed 41:1374
43. Martinéz LE, Leighton JL, Carsten DH, Jacobsen EN (1995) J Am Chem Soc 117:5897
44. Larrow JF, Schaus SE, Jacobsen EN (1996) J Am Chem Soc 118:7420
45. Lebel H, Jacobsen EN (1999) Tetrahedron Lett 40:7303
46. Brandes BD, Jacobsen EN (2001) Synlett 1013
47. For leading examples, see: Lednicer D (1998) Strategies for organic drug synthesis and design. Wiley-Interscience, New York
48. Schaus SE, Larrow JF, Jacobsen EN (1997) J Org Chem 62:4197
49. Martinéz LE, Nugent WA, Jacobsen EN (1996) J Org Chem 61:7963
50. Leighton JL, Jacobsen EN (1996) J Org Chem 61:389

51. Kassab DJ, Ganem B (1999) J Org Chem 64:1782
52. Wu MH, Jacobsen EN (1997) Tetrahedron Lett 38:1693
53. Lee S, Lee W-M, Sulikowski GA (1999) J Org Chem 64:4224
54. Lebel H, Jacobsen EN (1998) J Org Chem 63:9624
55. Schaus SE, Jacobsen EN (1996) Tetrahedron Lett 37:7937
56. Hansen KB, Leighton JL, Jacobsen EN (1996) J Am Chem Soc 118:10,924
57. For other examples supporting this mechanism, see: (a) Lida T, Yamamoto N, Matsunaga S, Woo H-G, Shibasaki M (1998) Angew Chem Int Ed 37:2223; (b) McCleland BW, Nugent WA, Finn MG (1998) J Org Chem 63:6656
58. Wu MH, Jacobsen EN (1998) J Org Chem 63:5252
59. (a) Haufe G, Bruns S (2002) Adv Synth Catal 344:165; (b) Bruns S, Haufe G (2000) J Fluorine Chem 104:247
60. (a) Jacobsen EN, Kakiuchi F, Konsler RG, Larrow JF, Tokunaga M (1997) Tetrahedron Lett 38:773. See also: (b) Kuroda T, Imashiro R, Seki M (2000) J Org Chem 65:4213; (c) Toshihiko F, Tomoyuki A (1998) Japanese Patent Application 10-072672
61. Ready JM, Jacobsen EN (1999) J Am Chem Soc 121:6086
62. Ready JM, White DE, Jacobsen EN Org Lett (in press)
63. Peukert S, Jacobsen EN (1999) Org Lett 1:1245
64. Toshihiko F, Tomoyuki A (1998) Japanese Patent Application 10-077934
65. Larrow JF (2002) US Patent 6,448,449 B2
66. For an example of intramolecular cyclizations of epoxyalcohols utilizing monomeric complex **3**, see: Wu MH, Hansen KB, Jacobsen EN (1999) Angew Chem Int Ed 38:2012
67. The reason is that cyanide binds irreversibly to the metal, thereby inactivating it to nucleophile transfer.
68. For a review of recent progress in this area, see: Yet L (2001) Angew Chem Int Ed 40:875
69. Sigman MS, Jacobsen EN (1998) J Am Chem Soc 120:5315
70. (a) Sigman MS, Jacobsen EN (1998) J Am Chem Soc 120:4901; (b) Sigman MS, Vachal P, Jacobsen EN (2000) Angew Chem Int Ed 39:1279; (c) Vachal P, Jacobsen EN (2000) Org Lett 2:867; (d) Su JT, Vachal P, Jacobsen EN (2001) Adv Synth Catal 343:197
71. (a) Tararov VI, Hibbs DE, Hursthouse MB, Ikonnikov NS, Malik KMA, North M, Orizu C, Belokon YN (1998) Chem Commun 387; (b) Belokon' Y, Flego M, Ikonnikov N, Moscalenko M, North M, Orizu C, Tararov V, Tasinazzo M (1997) J Chem Soc Perkin Trans 1 1293; (c) Belokon' YN, Caveda-Cepas S, Green B, Ikonnikov NS, Khrustalev VN, Larichev VS, Moscalenko MA, North M, Orizu C, Tararov VI, Tasinazzo M, Timofeeva GI, Yashkina LV (1999) J Am Chem Soc 121:3968. (d) For synthetic application, see: Lu S, Herbert B, Haufe G, Laue KW, Padgett WL, Oshunleti O, Daly JW, Kirk KL (2000) J Med Chem 43:1611
72. Belokon' YN, Green B, Ikonnikov N, North M, Tararov V, (1999) Tetrahedron Lett 40:6105
73. Belokon' YN, North M, Parsons T (2000) Org Lett 2:1617
74. Jacobsen EN Unpublished results
75. Myers JK, Jacobsen EN (1999) J Am Chem Soc 121:8959
76. (a) Bandini M, Cozzi PG, Melchiorre P, Umani-Ronchi A (1999) Angew Chem Int Ed 38:3357; (b) Bandini M, Cozzi PG, Umani-Ronchi A (2000) Angew Chem Int Ed (2000) 39:2327; (c) Bandini M, Cozzi PG, Melchiorre P, Tino R, Umani-Ronchi A (2001) Tetrahedron: Asymmetry 12:1063
77. Bandini M, Cozzi PG, Melchiorre P, Morganti S, Umani-Ronchi A (2001) Org Lett 3:1153
78. For a review of this area, see: Yanagisawa A (1999) Allylation of carbonyl group. In: Pfaltz A, Jacobsen EN, Yamamoto H (eds) (1999) Comprehensive asymmetric catalysis, vol 2. Springer, Berlin Heidelberg New York, chap 27
79. Mascarenhas CM, Miller SP, White PS, Morken JP (2001) Angew Chem Int Ed 40:601. Catalysts of this type have recently been utilized in the acylative kinetic resolution of secondary alcohols; see: Lin M-H, Rajanbabu TV (2002) Org Lett 4:1607

80. Evans DA, Janey JM, Magomedov N, Tedrow JS (2001) Angew Chem Int Ed 40:1884
81. DiMauro EF, Kozlowski MC (2001) Org Lett 3:3053. See also: Cozzi PG, Papa A, Umani-Ronchi A (1996) Tetrahedron Lett 37:4613
82. Schaus SE, Branalt J, Jacobsen EN (1998) J Org Chem 63:403. See also: (a) Schaus SE, Branalt J, Jacobsen EN (1998) J Org Chem 63:4876; (b) Aikawa K, Irie R, Katsuki T (2001) Tetrahedron 57:845; (c) Mihara J, Aikawa K, Uchida T, Irie R, Katsuki T (2001) Heterocycles 54:395
83. (a) Keck GE, Li XY, Krishnamurthy D (1995) J Org Chem 60:5998; (b) Corey EJ, Cywin CL, Roper TD (1992) Tetrahedron Lett 33:6907
84. Dossetter AG, Jamison TF, Jacobsen EN (1999) Angew Chem Int Ed 38:2398. For synthetic applications, see: (a) Thompson CF, Jamison TF, Jacobsen EN (2000) J Am Chem Soc 122:10,482; (b) Chavez DE, Jacobsen EN (2001) Angew Chem Int Ed 40:3667
85. (a) Huang Y, Iwama T, Rawal VH (2000) J Am Chem Soc 122:7843; (b) Huang Y, Iwama T, Rawal VH (2002) Org Lett 4:1163. For synthetic application, see: (c) Kozmin SA, Iwama T, Huang Y, Rawal VH (2001) J Am Chem Soc 124:4628
86. Huang Y, Iwama T, Rawal VH (2002) J Am Chem Soc 124:5950
87. (a) Fukuda T, Katsuki T (1995) Synlett 3:825; (b) Fukuda T, Katsuki T (1997) Tetrahedron 53:7201
88. (a) Uchida T, Irie R, Katsuki T (1999) Synlett 7:1163; (b) Uchida T, Irie R, Katsuki T (1999) Synlett 7:1793; (c) Uchida T, Irie R, Katsuki T (2000) Tetrahedron 56:3501; (d) Yao X, Qiu M, Lu W, Chen H, Zheng Z (2001) Tetrahedron: Asymmetry 12:197
89. (a) Niimi T, Uchida T, Irie R, Katsuki T (2000) Tetrahedron Lett 41:3647; (b) Niimi T, Uchida T, Irie R, Katsuki T (2001) Adv Synth Catal 343:79; (c) Uchida T, Saha B, Katsuki T (2001) Tetrahedron Lett 42:2521
90. (a) Pfaltz A (1999) Cyclopropanation and CH insertion with Cu In: Pfaltz A, Jacobsen EN, Yamamoto H (eds) (1999) Comprehensive asymmetric catalysis, vol 2. Springer, Berlin Heidelberg New York, chap 16.1; (b) Lydon KM, McKervey MA (1999) Cyclopropanation and CH insertion with Rh In: Pfaltz A, Jacobsen EN, Yamamoto H (eds) (1999) Comprehensive asymmetric catalysis, vol 2. Springer, Berlin Heidelberg New York, chap 16.2; (c) Charette AB, Lebel H (1999) Cyclopropanation and CH insertion with metals other than Cu and Rh. In: Pfaltz A, Jacobsen EN, Yamamoto H (eds) (1999) Comprehensive asymmetric catalysis, vol 2. Springer, Berlin Heidelberg New York, chap 16.3

Topics Organomet Chem (2004) 6: 153–180
DOI 10.1007/978-3-540-36966-0

Non-Salen Metal-Catalyzed Asymmetric Dihydroxylation and Asymmetric Aminohydroxylation of Alkenes. Practical Applications and Recent Advances

Steven J. Mehrman[1] · Ahmed F. Abdel-Magid[1] · Cynthia A. Maryanoff[1] · Bart P. Medaer[2]

[1] Drug Evaluation-Chemical Development, Johnson & Johnson Pharmaceutical Research and Development, Welsh & McKean Roads, Spring House, PA 19466, USA
[2] Drug Evaluation-Chemical Development, Johnson & Johnson Pharmaceutical Research and Development, Turnhoutseweg 30, 2340 Beerse, Belgium
E-mail: smehrman@prdus.jnj.com, amagid@prdus.jnj.com, cmaryanoff@prdus.jnj.com, bmedaer@prdbe.jnj.com

Abstract Over the past two decades the asymmetric dihydroxylation (AD) and asymmetric aminohydroxylation (AA) of alkenes has attracted a great deal of attention. Many literature examples have demonstrated that good to excellent ees and yields can be obtained with a diverse range of alkenes. The ability to perform reactions on small to moderate scale has led to increased utilization in larger scale applications. This review highlights recent advances in the AD and AA pertaining to practical, large-scale and industrial applications.

Keywords Asymmetric dihydroxylation · Asymmetric aminohydroxylation · Cinchona alkaloids · Process development · Scale up

List of Abbreviations

AA Asymmetric aminohydroxylation
ABS Acrylonitrile-butadiene-styrene
AD Asymmetric dihydroxylation
AQN Anthraquinone
CMR Chemzyme membrane reactor
DHQ Dihydroquinine
DHQD Dihydroquinidine
MC Microencapsulation
NMO 4-Methylmorpholine *N*-oxide
PEG Poly(ethylene glycol)
PEM Phenoxyethoxymethyl-polystyrene
PHAL Phthalazine
PHN Phenanthryl

1
Introduction

The dihydroxylation of an alkene is a chemical transformation of great utility to synthetic organic chemists. The asymmetric dihydroxylation (AD) and asymmetric aminohydroxylation (AA) of alkenes are two recent modifications of this transformation. A number of excellent reviews have been written describing the vast research conducted to determine the mechanism, as well as the scope and limitations of these transformations [1–10]. The utility of these reactions has advanced the discovery and asymmetric synthesis of new medicines and specialty chemicals [11–27]. This review will focus on the practical and large-scale applications of the AD and AA in particular, industrial applications.

A variety of metallic oxidizing agents have been used in the direct dihydroxylation and aminohydroxylation of alkenes to 1,2-diols and 2-aminoalcohols respectively. Examples of these strong oxidants include osmium(VIII), ruthenium(VIII), manganese(VII), chromium(VI), and vanadium(V) compounds [28–44]. Among the oxidants available to perform this transformation, osmium tetroxide has proven to be the reagent of choice (Eq. 1). Other oxidants often suffer from poor selectivity and overoxidation leading to byproducts. The osmium-mediated dihydroxylation of an alkene has been shown to occur in a stereospecific manner, which results from addition of the two oxygens from the same face of the alkene [37]. As most alkenes are prochiral, new stereocenters are often created by this transformation.

OH
O O
Os
O O
OH
Syn

(1)

2
Dihydroxylation and Aminohydroxylation Beginning/Ligand Acceleration

Since its discovery by Makowka in 1908 [45], the osmium-mediated dihydroxylation of alkenes has proven to be of great utility in organic synthesis. This transformation has been successfully applied to a wide range of carbon-carbon double bond substrates under a variety of reaction conditions [1–5, 9]. The first mechanistic studies examining the osmium-mediated dihydroxylation were performed in the 1930s by Criegee [46–48]. In these studies, stoichiometric amounts of osmium tetroxide were reacted with simple alkenes. The initial product was believed to be the dimeric osmium glycolate **3** (Scheme 1). This species could be cleaved to give the diol **5** and an Os(VI) species. A key observation noted by Criegee was that addition of pyridine greatly accelerated the reaction. Analysis of the pyridine containing reaction mixture showed the intermediate osmium species to be the ligated monomeric glycolate **4**, which upon workup gave the diol product **5** and an osmium(VI) species. The dramatic increase in the reaction rate was explained by the ligand acceleration model [49] shown in Scheme 1. In this model, the unligated osmium tetroxide is in equilibrium with the ligated species **1**. Each species is capable of performing the oxidation of the alkene **2**. In this simplified mechanism pyridine causes acceleration leading to shorter reaction times, $k_1>k_0$. As a result of this acceleration a majority of the alkene is subjected to oxidation by the ligated species **1**. Another important advancement in the dihydroxylation of alkenes was the development of a catalytic process by Upjohn chemists that utilized OsO_4 as a catalyst and *N*-methylmorpholine oxide (NMO) as a co-oxidant [50–52]. The NMO is used to regenerate OsO_4 from the Os(VI) byproduct produced at the end of the oxidation cycle.

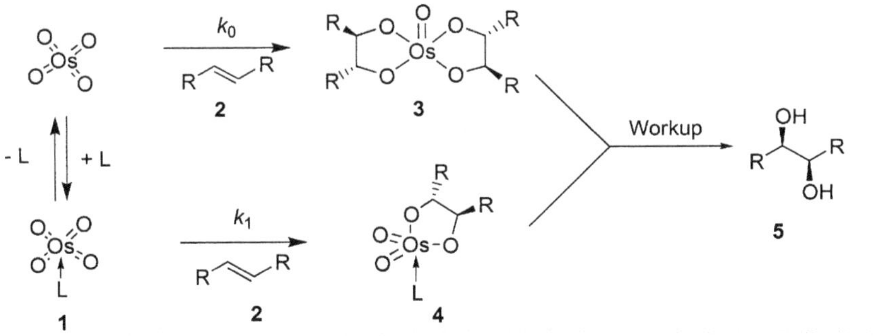

Scheme 1 Proposed dihydroxylation pathway and a ligand-accelerated model

While the dihydroxylation of alkenes was well known for a long time, the first osmium-promoted aminohydroxylation reaction was discovered in the early 1970s by Sharpless [53]. Oxidants such as trioxoimido-osmium(VIII) complex **6**, (formed in situ from OsO_4 and the Li or Na salt of an *N*-halogenated sulfonamide, alkyl carbamate, or amide) (Eq. 2) were shown to afford vicinal amino alcohols from alkenes in good yields. Even before the discovery of the asymmetric version, the ability to convert an alkene directly into an amino alcohol was a very significant advance in organic chemistry. Examples from this early study are listed in Table 1 [53]. As shown in Table 1, dihydroxylation of the alkene is usually a competing reaction and may become the major reaction depending on the conditions employed.

6

R' = SO_2R, COOR, or COR (2)

Table 1 Early examples of aminohydroxylations

entry	Alkene	Amino Alcohol	% Yield Amino Alcohol	% Yield Diol	Solvent
1	C_8H_{11}	C_8H_{11}, NH*t*Bu, OH	62	6	DCM
			89	<1	Pyridine
2	C_4H_9, C_4H_9	NH*t*Bu, C_4H_9, C_4H_9, OH	20	50 (threo)	DCM
			>95	<3	Pyridine
3	C_4H_9, C_4H_9	NH*t*Bu, C_4H_9, C_4H_9, OH	0	54	DCM
			25	42	Pyridine
4	Ph	Ph, NH*t*Bu, OH	93	<1	DCM
5	OCH_3	*t*BuHN, OCH_3, HO	0	78	DCM
			38	45	Pyridine
6		OH, NH*t*Bu	85	-	Pyridine

3 Asymmetric Dihydroxylation (AD) and Aminohydroxylation (AA)

Based upon the ligand acceleration observed by Criegee, Sharpless and Hentees developed the first AD procedure in 1980 [54, 55]. In the initial study pyridine was replaced by chiral non-racemic tertiary amines [56], the best results were obtained from the cinchona alkaloid derivatives: dihydroquinine (DHQ) acetate (**7a**) and dihydroquinidine (DHQD) acetate (**8a**) (Fig 1). These ligands served to accelerate the reaction and transfer chirality to the product. The AD of several

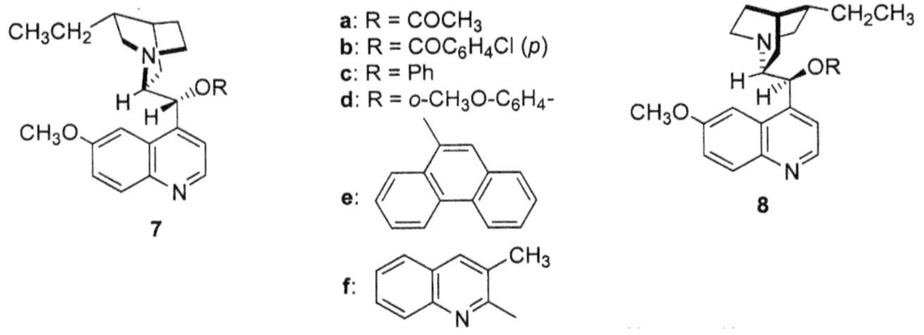

a: R = $COCH_3$
b: R = COC_6H_4Cl (*p*)
c: R = Ph
d: R = *o*-CH_3O-C_6H_4-
e:
f:

Fig. 1 Cinchona alkaloid derivatives

alkenes with stoichiometric quantities of the ligand and OsO_4 gave the corresponding enantiomerically enriched 1,2-dihydroxy compounds in fair to excellent enantioselectivities, particularly at lower reaction temperatures. The study also provided the first report of the AA. A footnote "...we have found that the presence of **8a** during the reaction of *t*-$BuNOsO_3$ with styrene results in the production of an optically active vicinal amino alcohol..." described what would be developed into the AA. It is noteworthy to point out that "the two readily available diastereomeric cinchona alkaloids (dihydroquinine and dihydroquinidine) essentially fulfill the functions of enantiomers" in the asymmetric oxidation reaction [57] as they exhibit opposite enantiofacial selectivities.

3.1
Catalytic AD and AA

Due to the known toxicity of osmium and the expense of the ligands it was necessary to develop a substoichiometric catalyst to perform the same transformation. The first catalytic process to successfully perform the AD was achieved by combining the use of the cinchona alkaloid derivatives, particularly the *p*-chlorobenzoates [57–59] (**7b** and **8b**), utilizing Upjohn's *N*-oxide-based catalytic OsO_4 dihydroxylation procedure [50–52]. The catalytic process proved very successful and produced 1,2-diols in good to excellent ees. This outstanding development made the dihydroxylation a practical and highly effective catalytic process applicable to variety of alkenes, particularly those with aromatic substituents [60]. The catalytic AA was introduced shortly after [61] and was applied successfully to a variety of alkenes. This seminal contribution provided a direct method to synthesize β-hydroxyamino functionality, a key structural unit in many biologically important molecules. Since these discoveries a continuous flow of publications have appeared resulting in better understanding the regioselectivity and enantioselectivity of this reaction [4, 6, 62–65].

To explain the results of the catalytic process, Sharpless proposed a reaction mechanism with (at least) two catalytic cycles [58, 64]. As the homogeneous catalytic cycles for the AD and AA have many similarities they are included together in Scheme 2 where X=O (OsO_4) for the AD or X=NR′ (trioxoimido-osmium(VIII) complex) for the AA [58, 66]. The primary catalytic cycle for both

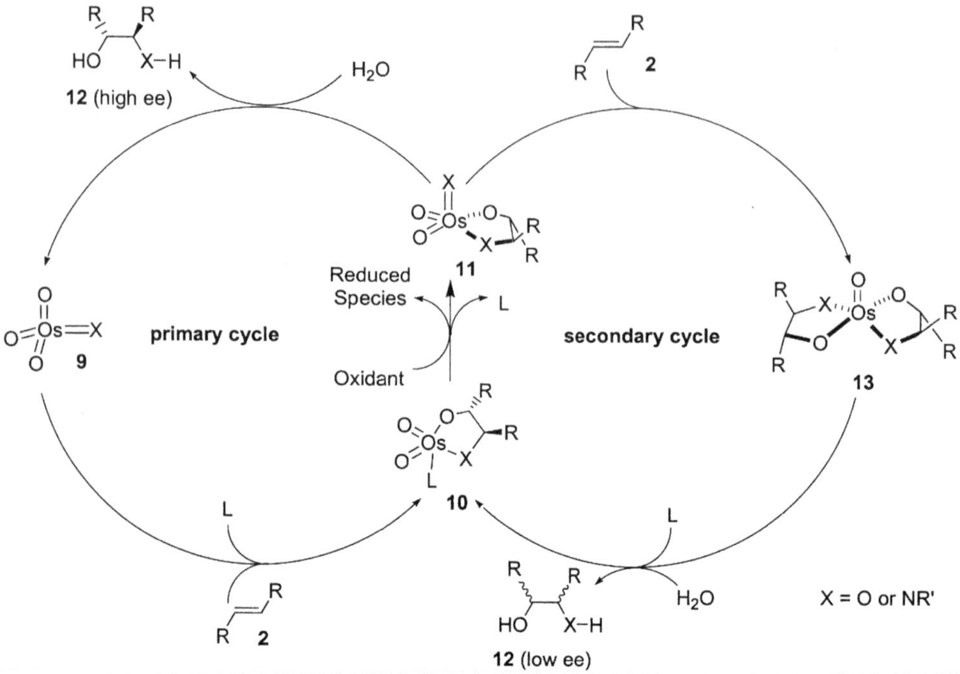

Scheme 2 Catalytic cycles of AD and AA

transformations begins with the unligated osmium species **9**, which upon coordination to the ligand **L** reacts with the alkene **2** giving osmium(VI) glycolate (AD) or azaglycolate (AA) intermediate **10**. Intermediate **11** is then formed after oxidation of the osmium(VI) species **10** by the stoichiometric co-oxidant and loss of ligand. This species is then hydrolyzed to release the product **12** and regenerate the reactive unligated osmium species **9**. Alternatively, if intermediate **11** is not quickly hydrolyzed, it may proceed to a secondary cycle where it reacts with another alkene molecule to form species **13**. Hydrolysis of **13** and ligand coordination leads back to intermediate **10** and product **12**. It has been noted the secondary cycle generally leads to lower enantioselectivity or formation of the opposite enantiomer, thereby degrading the ee of the primary cycle. The nature of the solvent, oxidant, ligand and osmium species play large roles in determining selectivity for the catalytic processes. The consequences of these two cycles and additives will be discussed in detail in their specific sections.

The realization of the existence of two catalytic cycles in the reaction mechanism with only one of the cycles giving high enantioselectivity led to further improvements to expand the scope of the reaction and include simpler alkenes with no aromatic substituents [57, 58]. For example, slow addition of the alkene to the catalytic mixture allows intermediate **11** to hydrolyze thereby suppressing the undesired secondary catalytic cycle and in general leading to higher ees than those obtained by fast addition [58]. Use of *t*-BuOH/H_2O as a solvent system has been shown to produce results similar to slow addition, which appears to suppress the secondary cycle [67].

The use of aryl ethers of DHQ and DHQD such as **7c**, **7d**, **8c**, and **8d** (instead of the *p*-chlorobenzoate derivatives **7b**) offered another improvement in the scope of the reaction to include *trans*-dialkyl-substituted alkenes (**16**) [59] and terminal alkenes (**14** and **15**) [68]. The extension of the reaction to include the highly abundant terminal alkenes greatly expanded the scope of the reaction. Thus, introduction of the aryl ether derivatives allowed a useful and practical application of the AD to four (**14–16** and **18**) of possible six classes of alkenes (**14–19**, Fig. 2) based on their substitution patterns. Monosubstituted (**14**), *gem*-disubstituted (**15**), *trans*-disubstituted (**16**) and trisubstituted (**18**) alkenes could successfully be used in these reactions to produce the corresponding 1,2-diols in high chemical yields and high enantioselectivity ranging from 79 to 99% ee [68]. The use of the solid, non-volatile potassium osmate(VI) dihydrate $[K_2OsO_2(OH)_4]$ or $[K_2OsO_2 \cdot (H_2O)_2]$ (**20**), which is completely equivalent to OsO_4 in these applications was introduced primarily for safety reasons to avoid the risk of exposure to any volatile osmium species [68].

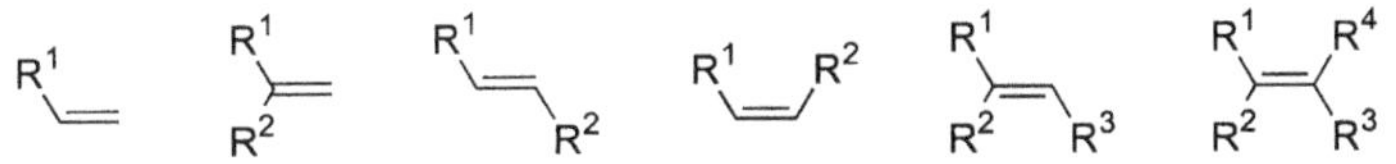

mono- **14** | *gem*-di- **15** | *trans*-di- **16** | *cis*-di- **17** | tri- **18** | tetra- **19**

Fig. 2 Substitution patterns of alkenes

Another modification that addressed the intrinsic problems of the secondary cycle in the AD was the introduction of a biphasic reaction system. Under these conditions, the osmate ester intermediate **10** is subjected to hydrolysis before oxidation thereby avoiding the secondary cycle [67, 69]. A simple model for the biphasic catalytic cycle is shown in Scheme 3. The solvent system typically consists of *t*-BuOH/H_2O, which under reaction conditions forms two heterogeneous layers. Use of $K_3Fe(CN)_6$ as the co-oxidant gave better levels of enantioselectivity. The cycle begins with OsO_4 in the organic layer which associates a ligand (L) then performs the oxidation of the alkene **2** resulting in a osmium(VI) glycolate species **10**. Due to the insolubility of the inorganic oxidant ($K_3Fe(CN)_6$) in the organic layer, intermediate **10** cannot be oxidized. The osmium must pass into the aqueous layer to be oxidized. Hydrolysis of **10** occurs leading to product **12** (1,2 diol) and a water soluble inorganic osmium(VI) species (**20**). The osmium(VI) species (**20**) then passes into the aqueous layer and is oxidized to regenerate osmium tetroxide which can return to the organic phase and complete the cycle.

The Sharpless group continued to examine other derivatives of the cinchona alkaloids as ligands for the AD of alkenes. In 1992, Sharpless reported [70] two major advances that led to the establishment of one general procedure applicable to a wide range of alkenes. These two advances are (i) the use of phthalazine ligands **21** and **22** (Fig. 3) and (ii) the use of organic sulfonamides to accelerate

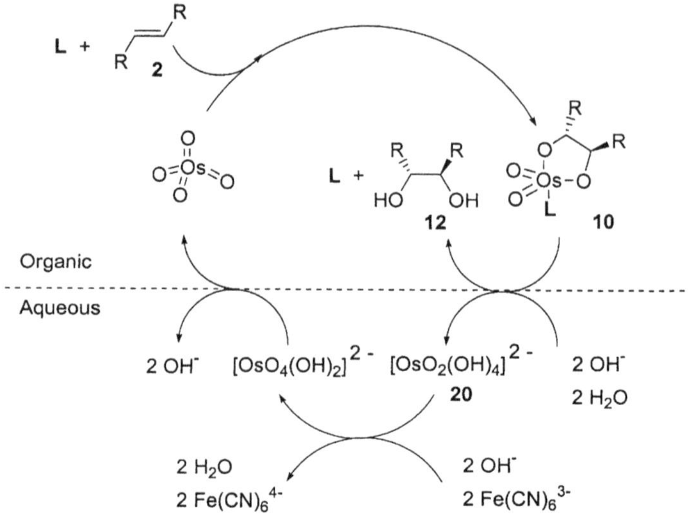

Organic

Aqueous

2 OH^- $[OsO_4(OH)_2]^{2-}$ $[OsO_2(OH)_4]^{2-}$ 2 OH^-

20 2 H_2O

2 H_2O 2 OH^-

2 $Fe(CN)_6^{4-}$ 2 $Fe(CN)_6^{3-}$

Scheme 3 Catalytic biphasic system

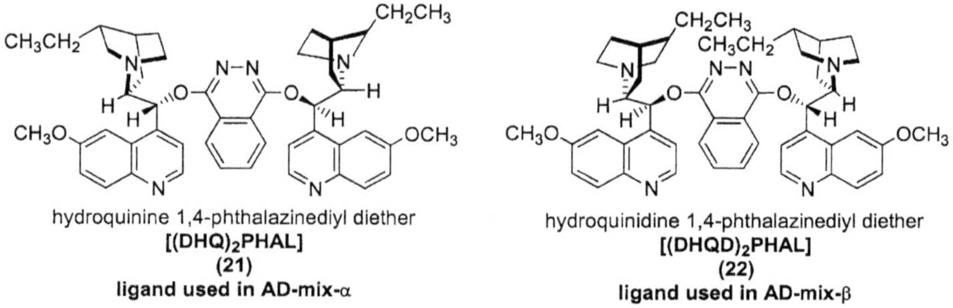

hydroquinine 1,4-phthalazinediyl diether
[(DHQ)$_2$PHAL]
(21)
ligand used in AD-mix-α

hydroquinidine 1,4-phthalazinediyl diether
[(DHQD)$_2$PHAL]
(22)
ligand used in AD-mix-β

Fig. 3 Phthalazine derivatives of cinchona alkaloids

the intermediate osmate ester hydrolysis. The four classes of alkenes (**14–16** and **18**) were conveniently dihydroxylated in high enantioselectivity under practical conditions that made large-scale applications possible. Furthermore, the experimental conditions were greatly simplified by the development of two AD-mix formulations (AD-mixes) labeled α and β. The mixes contain trace amounts of the ligand **21** or **22** and potassium osmate (total 0.6% by weight) blended with potassium ferricyanide and potassium carbonate (totaling 99.4%). The mixes are yellow powders that are stable for months when protected from prolonged exposure to moisture and are convenient to use. Normally, 1 mol% of ligands **21** or **22** is used in the catalytic process, but this amount could be further reduced without much effect on the % ee by lowering the reaction temperature [70]. Adjustment of pH [71] or addition of methanesulfonamide to the reaction mixture is recommended for non-terminal alkenes to enhance the rate of hydrolysis of

Table 2 AD examples of alkenes using AD-mixes

		AD-mix-α		AD-mix-β	
entry	Alkene	% ee	confign.	% ee	confign.
1	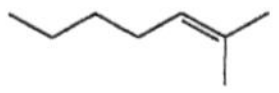	95	*S*	98	*R*
2	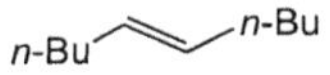 *n*-Bu, *n*-Bu	93	*S, S*	97	*R, R*
3	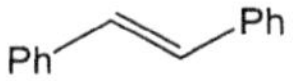 Ph, Ph	>99.5	*S, S*	>99.5	*R, R*
4	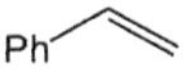 Ph	97	*S*	97	*R*
5	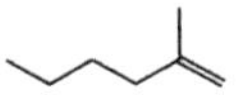	76	*S*	78	*R*
6	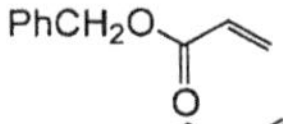 $PhCH_2O$, O	76	*S*	78	*R*
7	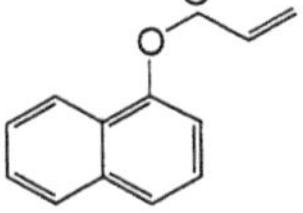 O	88	*S*	91	*R*

Reaction times ranged between 6 and 24 h. First three entries required addition of $CH_3SO_2NH_2$

the intermediate osmate(VI) ester, which leads to shorter reaction times. An example is the dihydroxylation of *trans*-5-decene which was only 70% complete after three days at 0 °C, required only 10 h in the presence of methanesulfonamide to give 97% yield and 97% ee [70]. Terminal alkenes usually react slower in the presence of methanesulfonamide and do not need this reagent. Representative examples are shown in Table 2.

The above-mentioned improvements were all designed to avoid the secondary non-selective catalytic cycle and maintain the high selectivity of the primary cycle. More recently [72, 73], the secondary catalytic cycle, which normally gives less enantioselectivity for both the dihydroxylation and aminohydroxylation of alkenes using cinchona alkaloid ligands was utilized in a very clever way to induce high enantioselectivity. The new advancement was developed based on an observation that some "special" alkenes (such as unsaturated carboxylic acids) undergo rapid and nearly quantitative aminohydroxylation or dihydroxylation with very low catalyst loading in the absence of the alkaloid ligands [74–76]. It was also observed that these "special" alkenes produce only racemic products even with large excess of the chiral alkaloid ligands. All evidence suggested that the osmium-catalyzed AA and AD of these alkenes occurred almost exclusively in the secondary non-selective catalytic cycle. While early attempts to obtain enantioselectivities failed, some new ligands designed to induce asymmetry in the osmium-catalyzed dihydroxylation and aminohydroxylation of alkenes in the secondary catalytic cycle were successful. The best ligands were *N*-toluenesulfonyl derivatives of α,β-hydroxyaminoacids such as the *N*-(*p*-toluenesulfonyl)phenylisoserine (**23**) and *N*-(*p*-toluenesulfonyl)threonine (**24**) (Fig. 4). The stereochemistry at the α-positions of these catalysts seems to determine the absolute stereochemistry of the product while the free carboxy group is essential

Fig. 4 Ligands for the secondary catalytic cycle

to maintain the catalytic activity. The enantioselectivity obtained thus far is moderate, however, this new approach has a great potential to further expand the scope of these reactions.

In the following section we'll outline many of the practical applications of the catalytic osmium mediated AD and AA of alkenes with a focus on industrial process development chemistry.

4 Examples of AD

The AD of *trans*-stilbene (**25**) to the enantiopure *threo* hydrobenzoin (**26**) (Eq. 3) is one of the first examples [77] that demonstrated the utility of the AD process. The reaction was performed on a 1 mole scale using dihydroquinidine 4-chlorobenzoate (**8b**) (0.05 mol; 5 mol%), osmium tetroxide (4 mmol) and NMO (1.5 mol) as the oxidant. The reaction was carried out at 0 °C in acetone/H_2O solvent mixture. After work up, the crude product was isolated quantitatively in 90% ee. Recrystallization of the crude product gave 72–75% yield of enantiomerically pure (*R*,*R*)-1,2-diphenyl-1,2-ethanediol (**26**).

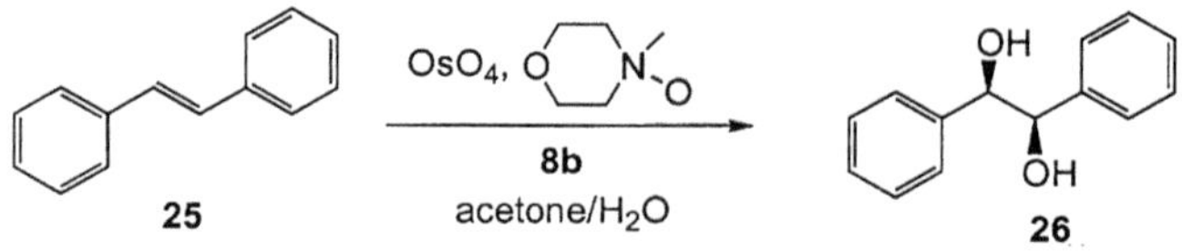

(3)

With the development of better ligands, the AD of *trans*-stilbene was greatly improved and was carried out on 1 kg scale [78]. Using only 0.25 mol% of **22** and 0.2 mol% of potassium osmate, the product **26** was obtained in 76% yield and 99% ee. Another improvement in the reaction is replacing acetone with *t*-BuOH as the reaction solvent. The poor solubility of stilbene in *t*-BuOH approximates the slow addition required for high ee. The product **26**, is also less soluble in *t*-BuOH making isolation easier.

Another early example is the AD of 3-vinyl-1,5-dihydro-3*H*-2,4-benzodioxepine (**27**) reported by Sharpless (Eq. 4) [79]. Reaction of **27** with the catalyst combination of DHQD-PHN (**8e**) (1 mol%) and potassium osmate (0.2 mol%) with potassium ferricyanide as a co-oxidant in *t*-BuOH/H_2O at 0 °C afforded the dihydroxy product **28** after separation and crystallization in 50–55% yield and 97% ee.

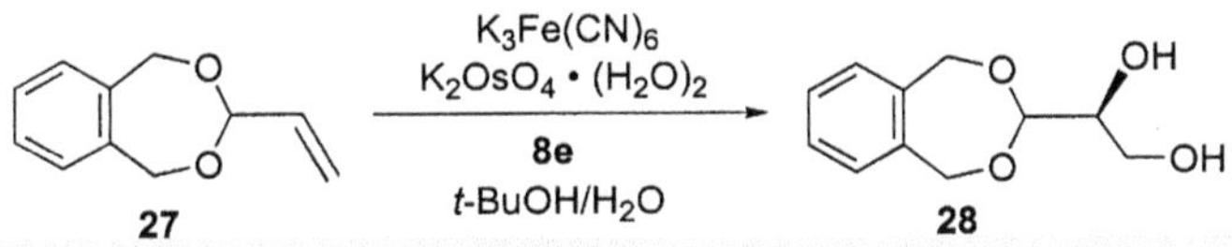

27 → 28 (4)
$K_3Fe(CN)_6$
$K_2OsO_4 \cdot (H_2O)_2$
8e
t-BuOH/H_2O

Pharmacia/Upjohn chemists have accomplished the large-scale (2.5 kg) AD of *o*-isopropoxy-*m*-methoxystyrene (**29**) [11], using **21** (0.77 mol%), $K_2OsO_4 \cdot (H_2O)_2$ (0.7 mol%) and *N*-methylmorpholine *N*-oxide (1.34 equivalents) as the reoxidant in *t*-BuOH/H_2O to obtain a 94% yield of (*S*)-1-[(2-isopropoxy-3-methoxy)phenyl]-1,2-ethanediol (**30**) in 90% ee (Eq. 5). The use of *N*-methylmorpholine *N*-oxide in place of potassium ferricyanide/potassium carbonate is practical on large-scale and leads to good selectivity. Both the ligand and the osmium catalyst can be recovered.

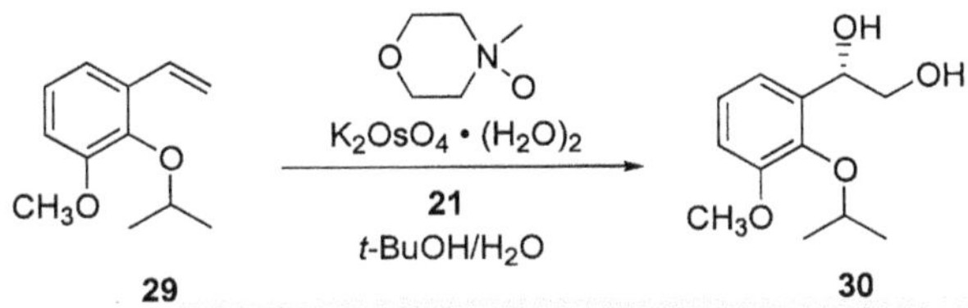

29 → 30 (5)
$K_2OsO_4 \cdot (H_2O)_2$
21
t-BuOH/H_2O

In the synthesis of the COX-2 inhibitor, **L-784512** (**33**), Merck chemists have employed the AD of trisubstituted alkene **31** (Scheme 4) as a key step [12]. Trisubstituted alkenes do not normally perform favorably under typical reaction conditions and usually result in low ee for the dihydroxy product. The initial AD using AD-mix-β and alkene **31b**, gave the dihydroxy product **32b** in only 63% ee. However, using more [$K_2OsO_4 \cdot (H_2O)_2$] (1 mol%) and **22** (5 mol%) lead to an increase of the products ee to 79%. AD of the sulfide **31a** and subsequent oxidation with H_2O_2/Na_2WO_4 in MeOH gave **32b** in 88% yield and 82% ee. The ee was further improved to >98% by a single recrystallization from *i*-PrOAc/hexane. The dihydroxy derivative **32b** was effectively converted to **L-784512** (**33**). The introduction of the quaternary stereogenic center in high stereochemical control via the AD reaction is a very effective process that is otherwise hard to accomplish.

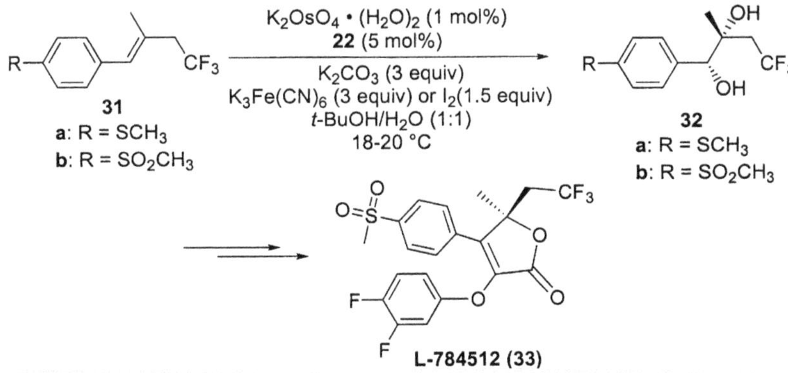

$K_2OsO_4 \cdot (H_2O)_2$ (1 mol%)
22 (5 mol%)
K_2CO_3 (3 equiv)
$K_3Fe(CN)_6$ (3 equiv) or I_2(1.5 equiv)
t-BuOH/H_2O (1:1)
18-20 °C

31
a: R = SCH_3
b: R = SO_2CH_3

32
a: R = SCH_3
b: R = SO_2CH_3

L-784512 (33)

Scheme 4 Synthesis of L-784512

Research chemists at Aventis have utilized the AD to introduce two key stereocenters during their synthesis of Flavopiridol carbocyclic analogs [13]. Alkene **34** was subjected to AD using $K_2OsO_2(OH)_4$, $K_3Fe_2(CN)_6$, K_2CO_3, $CH_3SO_2NH_2$ in *t*-BuOH/H_2O and ligand **22** at room temperature (Eq. 6). The resulting diol **35** was isolated in 94% yield, 85% ee. This route was used to provide kilograms of an advanced intermediate.

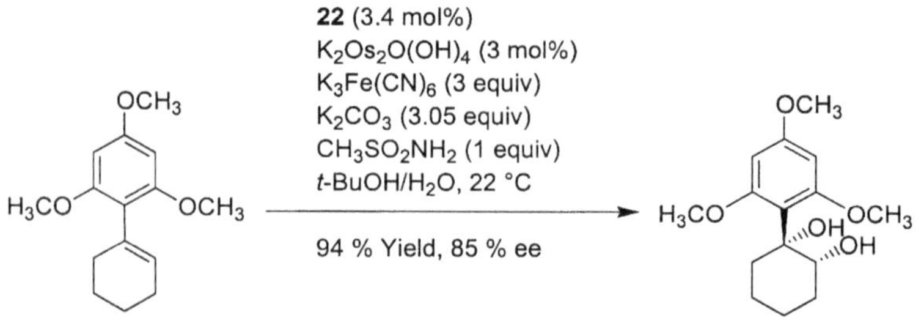

22 (3.4 mol%)
$K_2Os_2O(OH)_4$ (3 mol%)
$K_3Fe(CN)_6$ (3 equiv)
K_2CO_3 (3.05 equiv)
$CH_3SO_2NH_2$ (1 equiv)
t-BuOH/H_2O, 22 °C

94 % Yield, 85 % ee

34 → **35** (6)

The AD of *trans*-methyl cinnamate (**36**) has been used by different groups [14–16] to prepare the side chain of taxol. The initially formed dihydroxy product (**37**) is converted into the β-amino-α-hydroxy acid (**38**), which was used to introduce the taxol side chain. For example, Sharpless applied his AD procedure to convert **36** into **37** on a mole scale in 72% yield and 99% ee after recrystallization (Scheme 5). The use of NMO (instead of $K_3Fe(CN)_6$) made it possible to run the reaction at high (2 mol/l) concentration. The reaction work up included treatment with aqueous Na_2SO_3. A group from Zhong Shan University and S&P chiral-Tech in China ran into some problems on larger (50 kg) scale reaction [17]. A major drawback on that scale was an exothermic reaction that led to over-oxidation and increased impurities. The major impurities identified were the β-ketoester **39** and NMO (Scheme 5). The group found that by maintaining the reaction temperature between 35 and 40 °C, the reaction was complete in 2–3 h without loss of enantioselectivity. The use of Na_2SO_3 in the work up was proved to be important to the quality of the isolated product. Using HPLC monitoring, it was found that treatment of the reaction mixture with aqueous Na_2SO_3 at 45 °C for about 35 min eliminated any transitional Os(VIII) complex with the

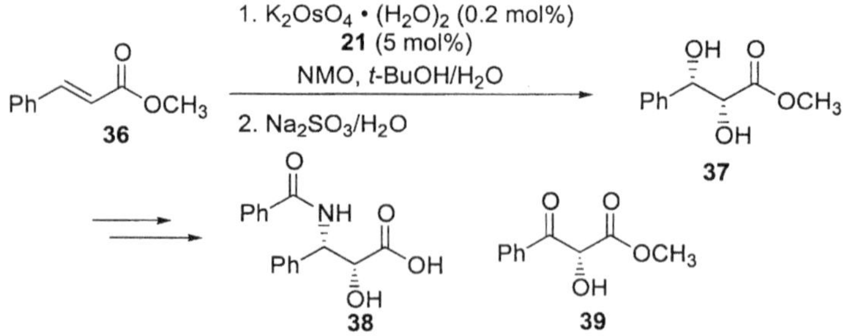

1. $K_2OsO_4 \cdot (H_2O)_2$ (0.2 mol%)
21 (5 mol%)
NMO, *t*-BuOH/H_2O
2. Na_2SO_3/H_2O

36 → **37** → **38**; **39**

Scheme 5 AD of *trans*-methyl cinnamate

diol, made isolation of the product easier, and minimized the formation of the over-oxidation ketone **39**.

In a recent report, a group from Hanyang University in South Korea applied high pressure to increase the catalyst turnover in the catalytic AD of *trans*-cinnamates. However, this was applied only on very small scale reactions [80].

An interesting modification, developed by Sepracor, for this "redox reaction" is to use of electrochemical oxidation (in place of chemical oxidation) to regenerate OsO_4 from lower valent osmium species [81]. This was accomplished directly or via a secondary oxidant that was electrolytically regenerated and in turn regenerated OsO_4 chemically from lower valent species. The process was conducted in an alkaline protic medium, usually a mixture of *t*-BuOH/H_2O (pH 8–13). In most reactions, potassium osmate (0.01–5% molar equivalent vs alkene) was used as the Os(VIII)-precursor and potassium ferricyanide as a secondary oxidant. The AD was conducted in the presence of one of the cinchona alkaloid derivatives as a chiral ligand (L*) in catalytic amounts (0.5–10% molar equivalent vs alkene). The overall reaction involved consumption of water and electricity and conversion of the alkene into the corresponding diol. The enantioselectivity was comparable to the chemical process. The electrochemical processes are outlined in Schemes 6 and 7 for the direct and indirect Os(VIII) regeneration protocols.

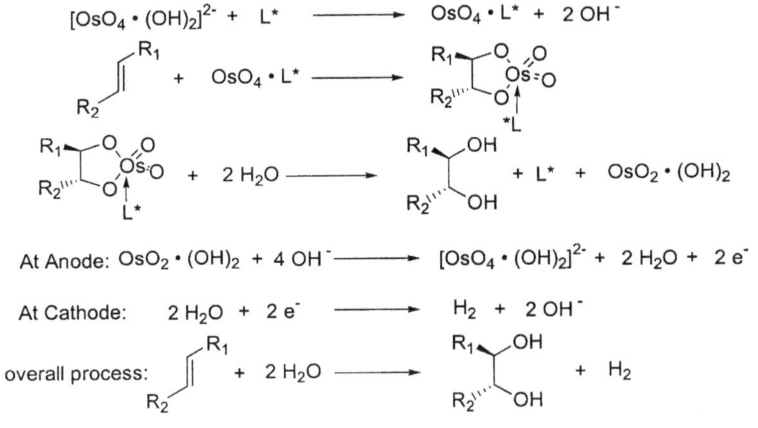

$$[OsO_4 \cdot (OH)_2]^{2-} + L^* \longrightarrow OsO_4 \cdot L^* + 2\,OH^-$$

$$R_1CH{=}CHR_2 + OsO_4 \cdot L^* \longrightarrow \text{osmate ester} \cdot L^*$$

$$\text{osmate ester} \cdot L^* + 2\,H_2O \longrightarrow R_1CH(OH)CH(OH)R_2 + L^* + OsO_2 \cdot (OH)_2$$

At Anode: $OsO_2 \cdot (OH)_2 + 4\,OH^- \longrightarrow [OsO_4 \cdot (OH)_2]^{2-} + 2\,H_2O + 2\,e^-$

At Cathode: $2\,H_2O + 2\,e^- \longrightarrow H_2 + 2\,OH^-$

overall process: $R_1CH{=}CHR_2 + 2\,H_2O \longrightarrow R_1CH(OH)CH(OH)R_2 + H_2$

Scheme 6 Direct electrochemical OsO_4 regeneration process

$$OsO_2 \cdot (OH)_2 + 2\,\text{oxidant}_{(ox)} \longrightarrow [OsO_4 \cdot (OH)_2]^{2-} + 2\,\text{oxidant}_{(red)}$$

At Anode: $2\,\text{oxidant}_{(red)} \longrightarrow 2\,\text{oxidant}_{(ox)} + 2\,e^-$

At Cathode: $2\,H_2O + 2\,e^- \longrightarrow H_2 + 2\,OH^-$

overall process: $R_1CH{=}CHR_2 + 2\,H_2O \longrightarrow R_1CH(OH)CH(OH)R_2 + H_2$

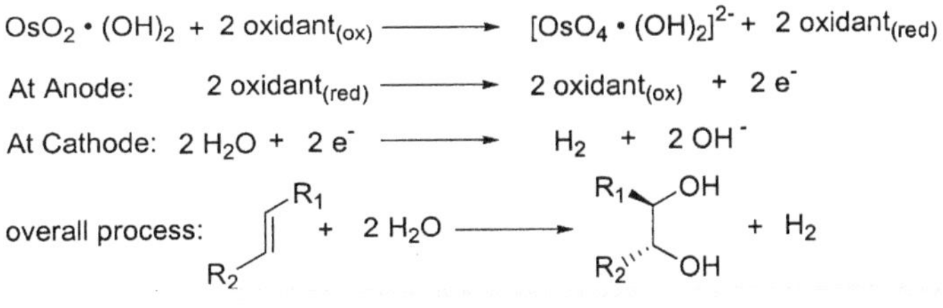

Scheme 7 Indirect electrochemical OsO_4 regeneration process

Table 3 Electrochemical AD of different alkenes

entry	Alkene	Rxn Conditions	Product	Yield	ee
1	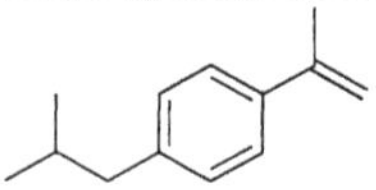 (10 mmol)	$K_4Fe(CN)_6$(4 mmol), OsO_4(0.01 mmol) **22** (0.15 mmol) K_2CO_3, *t*-BuOH/H_2O	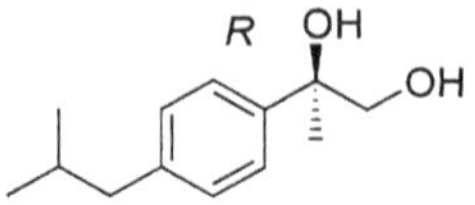 *R* OH, OH	100%	91%
2	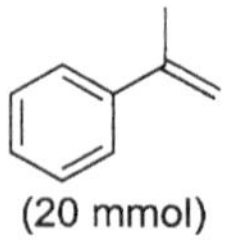 (20 mmol)	$K_4Fe(CN)_6$(8 mmol), OsO_4(0.01 mmol) **22** (0.1 mmol) K_2CO_3, *t*-BuOH/H_2O	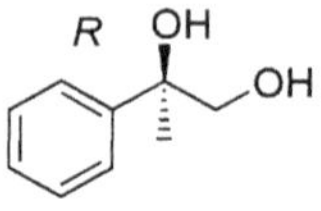 *R* OH, OH	95%	93%
3	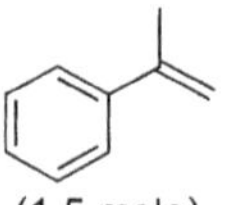 (1.5 mole)	$K_3Fe(CN)_6$(0.24 mol), OsO_4(0.0008 mol) **22** (0.012 mol) K_2CO_3, *t*-BuOH/H_2O	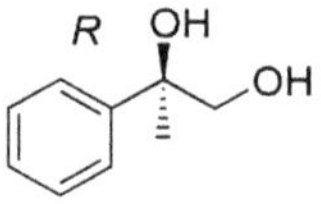 *R* OH, OH	95%	90%
4	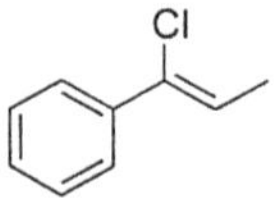 Cl (30 mmol)	$K_4Fe(CN)_6$(4.5 mmol), OsO_4(0.015 mmol) **21** (0.225 mmol) K_2CO_3, *t*-BuOH/H_2O	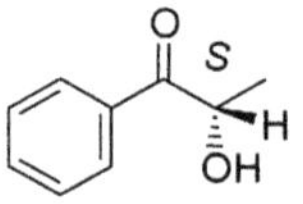 O, *S*, H, OH	--	95%
5	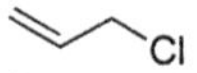 Cl (50 mmol)	$K_4Fe(CN)_6$(10 mmol), OsO_4(0.025 mmol) **21** (0.325 mmol) K_2CO_3, *t*-BuOH/H_2O	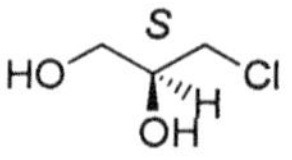 HO, *S*, Cl, H, OH	66%	55%

The initial three steps in the indirect regeneration process are the same as in the direct process. The Os(VI) species produced from the AD was chemically oxidized back to OsO_4 with the secondary oxidant (oxidant$_{(ox)}$). The reduced form of the secondary oxidant (oxidant$_{(red)}$) was re-oxidized electrochemically to the oxidizing form (oxidant$_{(ox)}$). Some of the examples reported in this study are listed in Table 3.

In both procedures the overall reaction is an electrochemical AD of alkenes with catalytic amounts of chiral ligand (L*) and OsO_4.

Another notable study was directed towards the use of oxygen [82, 83] or air [84] as oxidant to regenerate the osmium(VIII) species. The authors noted that the use of oxygen (or air) would eliminate the waste resulting from employing $K_3[Fe(CN)_6]$ or NMO as co-oxidants in large-scale reactions. The initial application utilized 1 bar pressure of pure oxygen at 50 °C and pH 10.4 in the AD of α-methylstyrene with $K_2OsO_4 \cdot (H_2O)_2$ and ligand **22** to give the dihydroxy product in 96% yield and 80% ee. This study was extended to the use of air at a pressure of 20 bar to produce a similar result. Several alkenes were dihydroxylated in good yields (48–89%) and variable enantioselectivities (53–98%) depending on the substrate under these conditions.

5
Examples of AA

Direct conversion of alkenes into enantiopure vicinal aminoalcohols (AA) is a major modification of the older and more established AD. The addition of two different heteroatoms (O and N) to the alkene adds another degree of complexity (with the regiochemical selectivity) to the enantioselectivity issues associated with AD. In many of the reported cases, dihydroxylation was a competing reaction. The aminohydroxylation catalyst is a trioxoimido-osmium(VIII) reagent ($O_3Os{=}NR$), prepared in situ by the reaction of OsO_4 with the Li or Na salt of *N*-chlorosulfonamides, *N*-chlorocarbamates or *N*-chlorocarboxamides. *N*-chlorosulfonamides were originally used as the nitrogen source in AA reactions [18, 61, 66, 85, 86]; however, their synthetic utility is limited because of their high stability, requiring forcing conditions for the removal of the sulfonyl groups [87]. While some sulfonamides such as nosylamides can be cleaved under relatively mild conditions [88] they give inferior results in the AA (compared to toluene- or methanesulfonamides). It has been shown that use of β-hydroxy-2-trimethylsilylethanesulfonamide (which gives comparable results to methyl sulfonyl), can allow for cleavage to occur by treatment with fluoride [89]. Replacing sulfonamides by carbamates [90] or carboxamides [91], has greatly increased the scope and selectivity of AA reaction [19].

The sodium salts of *N*-chloromethanesulfonamide and *N*-chlorotoluenesulfonamide are prepared by the reaction of the corresponding sulfonamides with NaOH and *tert*-butyl hypochlorite in water [85, 92, 93]. The *N*-halocarbamates are formed in situ by reaction of the carbamate with freshly prepared *tert*-butyl hypochlorite or Clorox (4–6% aqueous sodium hypochlorite solution).

Alkali metal salts of *N*-chlorocarboxamides are well known for their proclivity to undergo Hofmann rearrangement [94]. This competing reaction can however be suppressed by operating at 4 °C and use of the more stable commercially available *N*-bromo derivatives [91].

The regioselectivity of the AA has been found to be dependent on the nature of the ligand, the solvent and the *N*-protecting group introduced (Eq. 7). It's interesting to note that formation of the new carbon-nitrogen bond takes place preferably at the least substituted carbon atom of the olefin [95, 96], although some exceptions have been found [97, 98]. Furthermore, preliminary results suggested that the solvent coordination properties would play an important role in the formation of the diol as a side product [62]. In general higher yields and better enantioselectivities are obtained with sterically less demanding substituents on the nitrogen atom [66].

$$R^1\text{CH}{=}\text{CHR}^2 \xrightarrow[\text{Cinchona Ligand}]{\text{XNClM},\ K_2OsO_2(OH)_4} R^1\text{CH(NHX)CH(OH)}R^2 + R^1\text{CH(OH)CH(NHX)}R^2 \quad (7)$$

X = Ts, Ms, Cbz, Boc, TeoC, Ac; M = Na or Li

Earlier AA reactions were carried out mostly in CH_3CN/H_2O as solvent, but *t*-$BuOH/H_2O$ proved to be a perfect alternative since most *N*-sulfonyl amino alcohols are often insoluble in *t*-$BuOH/H_2O$ and can therefore be filtered directly from the reaction mixture [18, 61, 90]. The use of methanesulfonamide is usually advantageous compared to toluenesulfonamide. Excess methanesulfonamide can be readily removed by aqueous base extraction for simpler product isolation. By contrast, removal of excess toluenesulfonamide often requires a tedious chromatographic separation [66]. Additionally, most methanesulfonamides tend to crystallize more readily than toluenesulfonamides therefore enantiomeric purity near 100% can often be reached by recrystallization: e.g., the tosyl amido alcohol (ee: 45%) obtained after the AA with cyclohexene gave an ee of 99% after recrystallization from methanol [61].

Many of the cinchona alkaloid ligands including **7**, **8**, **21** and **22** (Figs. 1 and 3) in addition to the anthraquinone derivatives, $(DHQ)_2AQN$ (**40**) and $(DHQD)_2AQN$ (**41**) (Fig. 5) were used in the AA reaction.

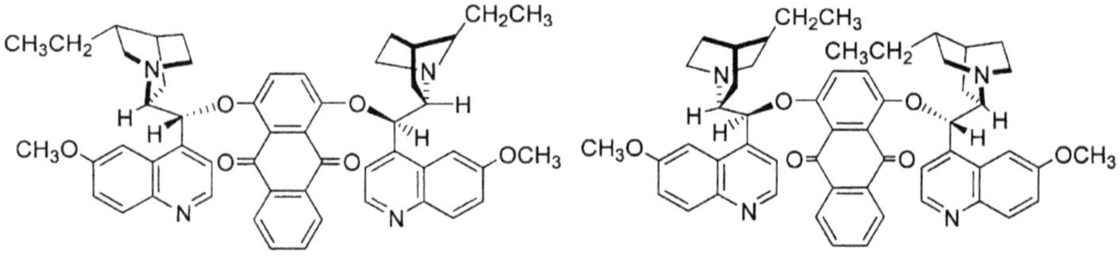

Fig. 5 Anthraquinone derivatives of cinchona alkaloids

Styrenes and cinnamate esters are among the most successful types of alkenes to undergo AA. The AA protocols now enable selective synthesis of either of the regioisomeric β-aminoalcohols.

The AA of styrenes provides a direct access to both enantiomers of *N*-protected 2-aryl-2-aminoethanols (α-arylglycinols). Styrenes have performed poorly in AA studies using sulfonamides as the nitrogen source [18, 61]. However, a study by Reddy and Sharpless has shown styrenes to be excellent substrates when a carbamate, particularly benzyl carbamate is used as the nitrogen source [19]. This modification converts styrenes to α-arylglycinols in excellent enantioselectivities and high degree of regiochemical control. The best results were obtained by using sodium *N*-chlorobenzylcarbamate (BnOC(O)NNaCl, ~3 equivalents), phthalazine ligands **21** or **22** (5 mol%), $K_2Os_2(OH)_4$ (4 mol%) in *n*-propanol/H_2O (3:2) at 25 °C. Under these conditions, styrenes (**42**) favor the formation of α-arylglycinols **43** over the regioisomeric product **44** (Table 4, entries 1, 3, 12, and 17). The regiochemistry may be significantly altered to favor **44** by changing the solvent to CH_3CN. Using AQN ligands **40** or **41** in combination with CH_3CN/H_2O amplifies this reversal of regioselectivity (Table 4, entries 5, 11, and 15). In spite of the increased regioselectivity in favor of **44**, the reaction is not

Table 4 Examples of AA of styrenes

Ar-CH=CH$_2$ (**42**) → [XNClNa, $K_2Os_2(OH)_4$, Solvent, Ligand] → Ar-CH(NHX)-CH$_2$OH (**43**) + Ar-CH(OH)-CH$_2$NHX (**44**)

entry	X	Ar	Ligand	Solvent	Yield[a] (%)	Product ratio 43 : 44	% ee 43	% ee 44	Ref.
1	Cbz	C_6H_5-	**21**	CH_3CN	40	55 : 45	93	--	[19]
2	Ac	C_6H_5-	**40**	CH_3CN/H_2O	--	1 : 13	--	88	[91]
3	Cbz	3-CH_3O-4-BnO-C_6H_3-	**21**	*n*-PrOH/H_2O	70	77 : 23	98	--	[19]
4	Cbz	3-CH_3O-4-BnO-C_6H_3-	**40**	*n*-PrOH/H_2O	--	33 : 66	--	--	[19]
5	Cbz	3-CH_3O-4-BnO-C_6H_3-	**40**	CH_3CN/H_2O	--	20 : 80	--	58	[19]
6	Boc	3-CH_3O-4-BnO-C_6H_3-	**21**	*n*-PrOH/H_2O	65	75 : 25	99	--	[19]
7	Cbz	3-Pyridyl-	**21**	*n*-PrOH/H_2O	35	50 : 50	96	--	[19]
8	Cbz	4-TsO-C_6H_4-	**21**	*n*-PrOH/H_2O	42	50 : 50	83	--	[19]
9	Cbz	4-TsO-C_6H_4-	**21**	CH_3CN/H_2O	--	14 : 86	--	--	[19]
10	Cbz	4-TsO-C_6H_4-	**40**	*n*-PrOH/H_2O	--	17 : 83	--	--	[19]
11	Cbz	4-TsO-C_6H_4-	**40**	CH_3CN/H_2O	--	<1 : 50	--	0	[19]
12	Cbz	4- BnO-C_6H_4-	**21**	*n*-PrOH/H_2O	76	88 : 12	97	--	[19]
13	Boc	4- BnO-C_6H_4-	**21**	*n*-PrOH/H_2O	68	83 : 17	99	--	[19]
14	Cbz	4- BnO-C_6H_4-	**21**	CH_3CN/H_2O	--	75 : 25	--	--	[19]
15	Cbz	4- BnO-C_6H_4-	**40**	CH_3CN/H_2O	--	25 : 75	--	--	[19]
16	Ac	4- BnO-C_6H_4-	**40**	CH_3CN/H_2O	76	1 : 9	--	86	[91]
17	Cbz	3,5-$(CH_3O)_2$-C_6H_3-	**21**	*n*-PrOH/H_2O	68	75 : 25	90	--	[19]
18	Boc	3,5-$(CH_3O)_2$-C_6H_3-	**21**	*n*-PrOH/H_2O	60	75 : 25	97	--	[19]

a) yield of isolated major product

synthetically useful to prepare these products in enantiopure form. It was determined that, under these conditions, the regioisomer **44** is always obtained with low enantioselectivity, suggesting that they are probably formed in the non-selective secondary catalytic cycle [19]. These regioisomers could be prepared with high regio- and enantioselectivity by the amide variant of AA. When sodium *N*-haloamides are used in place of sodium *N*-halocarbamates [91] in the AA of styrenes, the reaction favors formation of the regioisomer **44** with excellent enantioselectivity (Table 4, entries 2 and 16). A major advantage of using the *N*-

haloamide variant of AA is that only stoichiometric amounts of the amide reagent are needed, making isolation and purification easier; the use of excess reagent does not improve the reaction.

The use of *tert*-butyl carbamate in the AA reaction when applying the same conditions as benzyl carbamate were initially unsatisfactory leading to low yields and poor chemoselectivity (more diol was formed). Using CH_3CN/H_2O as solvent mixture suppressed the diol formation but resulted in low regioselectivity. The use of somewhat more ligand (6 mol% instead of 5 mol%), low temperature (0 °C) and *n*-propanol/H_2O (2:1) as solvent, proved to be beneficial giving good regioselectivity and excellent enantioselectivity (Table 4, entries 6 and 18). In general, the *tert*-butyl carbamate AA give slightly poorer regioselectivity and lower yield as compared to benzyl carbamate. However, the reaction is fast and the enantioselectivity is excellent, it provides another viable and convenient alternative for AA of alkenes.

α-Arylglycinols can readily be oxidized by TEMPO/NaOCl [99–101] or RuO_4/H_5IO_6 [29, 102, 103] to α-arylglycines, found in many biologically active molecules [19 and references listed therein]. Oxidation of the mixture of the regioisomers **43** and **44** generates a mixture of the protected amino acid **45** and the protected amino ketone **46** (Scheme 8). The amino acid is easily separated from the amino ketone by simple base extraction and acidification; hence, chromatography is avoided.

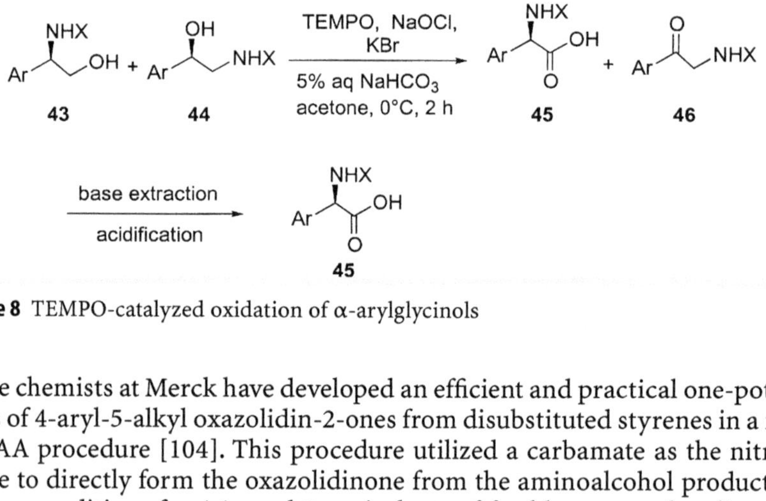

Scheme 8 TEMPO-catalyzed oxidation of α-arylglycinols

The chemists at Merck have developed an efficient and practical one-pot synthesis of 4-aryl-5-alkyl oxazolidin-2-ones from disubstituted styrenes in a modified AA procedure [104]. This procedure utilized a carbamate as the nitrogen source to directly form the oxazolidinone from the aminoalcohol product. The existing conditions for AA used 3 equivalents of freshly prepared sodium *tert*-butyl hypochlorite as the co-oxidant which was unsuitable for large-scale preparations. The chemists found two very efficient alternative co-oxidants, 1,3-dichloro-5,5-dimethyl hydantoin and sodium dichloroisocyanurate, to replace *tert*-butyl hypochlorite. These two co-oxidants are commercially available, economic, stable and convenient to use solids. Both chlorine atoms (in either compound) are available for maximum efficiency. The optimized conditions for the AA reaction [NaOH (3.05 equivalents), urethane (3.08 equivalents), 1,3-dichloro-5,5-dimethyl hydantoin (1.53 equivalents), **21** or **22** (0.025 equivalents) and

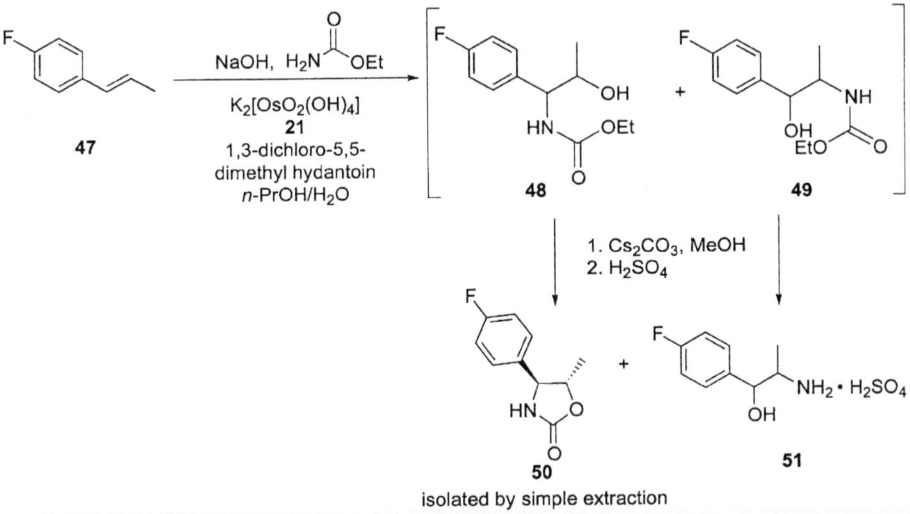

Scheme 9 One-pot preparation of 2-oxazolidinone

$K_2OsO_2(OH)_4$ (0.02 equivalents) in *n*-PrOH/H_2O (1:1)] were successfully applied to several styrenes including *p*-fluorophenylstilbene **47**. The reactions favored the formation of the benzylic amine derivatives **48** over the regioisomeric benzylic alcohol derivatives **49** (Scheme 9). It was also discovered that under basic conditions the benzylic amine **48** cyclizes to the desired oxazolidinone **50** faster than the benzylic alcohol **49**. Thus, following the AA reaction, the crude product mixture is treated with a suitable base (Cs_2CO_3 in MeOH) to affect the cyclization of the major product **48**. Treatment with an acid selectively deprotects and removes the uncyclized isomer as the amine salt **51** leaving the oxazolidinone, which is isolated by simple extraction. This procedure allows an easy separation of the desired product without chromatography, e.g., oxazolidinone **50** is isolated in good yield (71%) and high regio and enantioselectivity (90–93% ee).

The AA of cinnamate esters (**52**, Eq. 8) usually proceeds with better regiochemical control of the addition than that shown in styrenes. Dramatic influences on the regioisomeric ratio can be effected by choice of ligand. Sharpless has shown that benzylic amine product **53**, is the major regioisomer employing the PHAL ligands **21** and **22** [98]. Regioisomer **54** predominates when using the AQN ligands **40** and **41**. In general, moderate to good yields and ees are obtained for each of the regio isomeric AA products **53** and **54** using either ligand type.

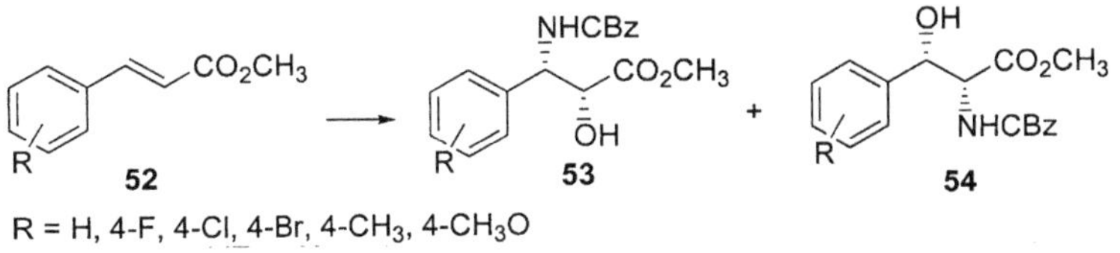

(8)

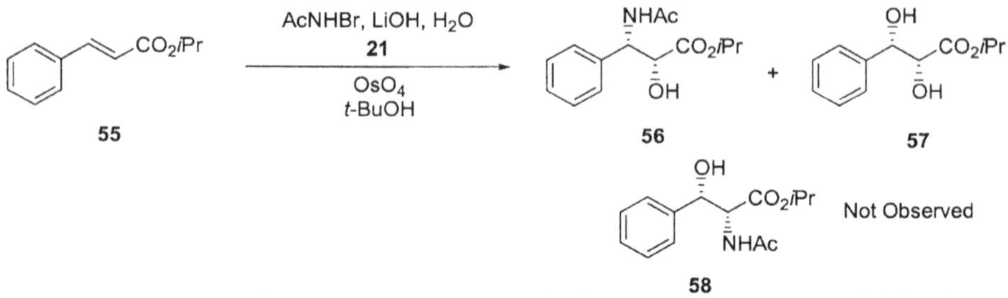

Scheme 10 AA of isopropyl cinnamate

An example that demonstrates this high regiochemical control is the AA of isopropyl cinnamate (**55**) in which only one regioisomer (**56**) was observed together with the diol **57** but none of the regioisomer **58** (Scheme 10) [20]. The reaction was used by *Pharmacia* chemists in the direct synthesis of the phenylisoserine derivative **56** needed to introduce the side chain of Paclitaxel (Taxol). Attempts to carry out the reaction on a large-scale led to the discovery of a very interesting concentration dependence [20]. When the reaction was carried out at higher concentrations than the typical literature AA concentration of 0.014 g/ml, it resulted in the formation of larger amounts of the undesired diol **57**. For example, the ratio of **56**/**57** fell from 95/5 at a concentration of 0.014 g/ml down to 45/55 at a concentration of 0.1 g/ml. The chemists discovered that formation of the diol **57** was suppressed by addition of 1 equivalent acetamide to the reaction mixture. This addition restored the ratio of **56**/**57** to 95/5 and the desired amidoalcohol was isolated in 60% yield and 99% ee after recrystallization. It is unclear why acetamide has the observed beneficial effect of reducing the level of diol formation since it has been shown experimentally that acetamide does not react with OsO_4 [20]. While the addition of acetamide to the reaction mixture provides conditions to carry out the reaction on a large-scale with better volume efficiency, the mechanism by which this occurs remains unknown. It is interesting to note that addition of propionamide (instead of acetamide) resulted in formation of a mixture of the propionyl and acetyl derivatives. Addition of methanesulfonamide didn't improve the reaction while addition of trifluoroacetamide or urea completely shuts down the reaction: no diol or amidoalcohol is formed.

6 The Use of Solid Supports

Many examples of the AD and AA have demonstrated excellent reactivity and selectivity under low catalyst loading [1–10]. In spite of these successes, large-scale applications continued to be hindered by the expense, toxicity, volatility, and disposal of osmium and the expense of the alkaloid ligands. In an attempt to address these problems, a variety of solid-supported catalysts have been examined. Ideally, a solid-supported catalyst would have the same activity as its solution phase counterpart and allow for its complete recovery and reusability without contamination. These developing methodologies have not yet been reported in the literature for large-scale synthesis but clearly focus on issues associated with the scale up of the AD and AA.

Sharpless first used a solid-supported ligand **59** in 1990 to perform the AD of stilbene with good success (Table 5, entry 1) [105]. Since then a vast number of

Table 5 The use of solid-supported ligands in AD and AA

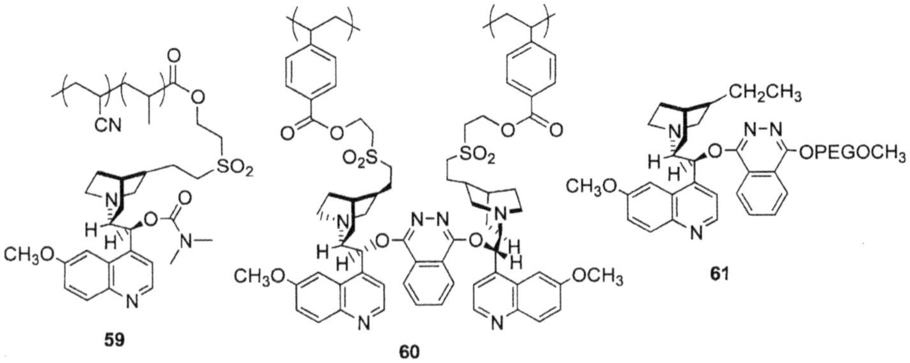

59 **60** **61**

Entry	Alkene	Ligand	Conditions	Yield	% ee
1	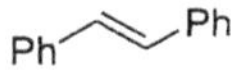 Ph Ph	59	A	96	87
2	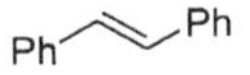 Ph Ph	60	B	90	>99
3	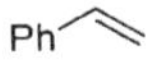 Ph	60	B	86	91
4	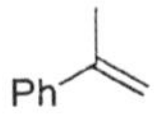 Ph	60	B	88	94
5	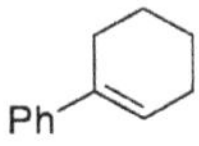 Ph	60	B	85	97
6	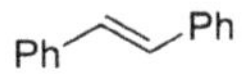 Ph Ph	61	C	93	98
7	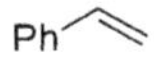 Ph	61	C	87	82
8	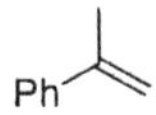 Ph	61	C	89	79
9	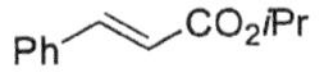 Ph CO_2iPr	60	D	98	87

Conditions: A. 0.25 equiv **59**, 1.25 mol % OsO_4, $K_3Fe(CN)_6$, K_2CO_3, rt, 18 h, 0.07 M *t*-BuOH/H_2O (1:1). B. 0.1 to 0.25 equiv **60**, 0.5 to 1 mol % OsO_4, $K_3Fe(CN)_6$, K_2CO_3, 0 °C, 20 h, *t*-BuOH/H_2O (1:1). C. 0.1 equiv **61**, 0.4 mol % $K_2OsO_2(OH)_4$, 3 equiv $K_3Fe(CN)_6$ and K_2CO_3, 0 °C, 18-24 h, 0.09 M *t*-BuOH/H_2O (1:1). Note 1 equiv $CH_3SO_2NH_2$ added for trans olefins. D. 0.1 equiv **60**, 4 mol % $K_2OsO_2(OH)_4$, 3 equiv Chloramine-M, 20 °C, 20 h, 0.07 M *n*-PrOH/H_2O (1:1).

solid-supports have been attached to alkaloid ligands [106, 107]. Among these ligands polystyrene modified ligand **60** appears to be one of the most useful to date. This ligand has been used to perform the AD (Table 5, entries 2–5) [108, 109] and AA (Table 5, entry 9) [110] reaction providing comparable results to those of unbound systems. In another example a soluble poly(ethylene glycol) (PEG) based ligand **61** was attached to the PHAL core. This ligand has been very successful at performing AD of a wide range of classes of alkenes (Table 5, entries 6–8) [111].

6.1 Microencapsulated Osmium Catalyst

To address the issues associated with the volatility, toxicity, and expense of osmium tetroxide, Kobayashi and coworkers have studied its microencapsulation of osmium tetroxide on solid-support [112–114]. Microencapsulation has been described as a method for immobilizing a catalyst onto a polymer by physical envelopment or by electron interactions between the π-electrons of the polymer support and the vacant orbitals of the catalyst [115]. Initially an acrylonitrile-butadiene-styrene copolymer (ABS) was successfully utilized to form a microencapsulated OsO_4 catalyst (ABS-MC OsO_4). This catalyst was used in conjunction with 5 mol% of **22** and NMO as co-oxidant to perform the AD efficiently on several alkenes. Slow addition (24 h) of the alkene was necessary to enhance selectivity and prepare the desired diols in yields and enantioselectivities comparable to those obtained with "free" OsO_4. The osmium catalyst was completely

Table 6 AD of alkenes using PEM-MC OsO_4

$R_1R_2C{=}CR_3R_4$ → (**62** (5 mol %), **22** (5 mol %); acetone/H_2O = (1:1), $K_3Fe(CN)_6$, K_2CO_3, 30 °C) → $R_2R_1C(OH){-}C(OH)R_3R_4$

Entry	Alkene	% Yield	% ee
1	Ph‑CH=CH₂ (styrene)	85	78
2	Ph‑CH=CH‑CH₃	86	94
3	Ph‑C(CH₃)=CH₂	85	76
4	1-Phenylcyclohexene	85	95
5	C_4H_9‑CH=CH‑C_4H_9	41	91
6	Ph‑CH=CH‑Ph	66	> 99
7	Ph‑CH=CH‑CO_2iPr	51	> 99

recovered after workup and was used up to five times without loss of activity. Recently [112] a new polymer, phenoxyethoxymethyl-polystyrene (PEM) [5% phenoxyethoxymethyl polystyrene+95% polystyrene], was prepared and used to microencapsulate osmium tetroxide. The resulting microencapsulated catalyst (PEM-MC OsO_4) (**62**) was used successfully to perform the AD of a variety of alkenes. This catalyst did not require slow addition of the alkene to obtain the desired diols in high selectivities (Table 6). The encapsulated osmium catalyst was recovered quantitatively and was reused up to three times without loss of activity or selectivity. Typical reaction conditions employ 5 mol% of **62** with an equal amount of ligand, and multiple equivalents of $K_3Fe(CN)_6$ as the co-oxidant and K_2CO_3 as a base in 1:1 acetone/H_2O at 30 °C. The procedure was introduced as a possible means for industrial scale AD reactions.

6.2 A Continuous Flow AD

The dilute reaction conditions required to obtain optimal results is a significant scale up concern which effects the throughput in the catalytic AD reaction. In an attempt to address this problem, Woltinger [116] has utilized a chemzyme membrane reactor (CMR) (Fig. 6) to perform a continuous flow catalytic AD of alkenes. This stainless steel vessel utilizes a semi-permeable membrane as to retain the PEG linked AQN bis-cinchona alkaloid ligand **65** in the reactor. The membrane allows passage of lower molecular weight species thereby allowing continuous dosing of the vessel. The AD reaction was performed on (E)-*tert*-butyl homocinnamate(**63**). The CMR reaction vessel used has a volume of 10 ml and a residence time of 85 min. Two pumps were used to deliver either reagents or starting materials along with the appropriate solvent(s) to the reactor (Fig. 6). Fractions of the reaction mixture were collected as to determine chronological order of conversion and ee.

Analysis of the initial fractions collected gave results comparable to batch experiments of 80% conversion (ee of 80%). However, after six residence times the conversion had dropped to 18% without change in the % ee. The decrease in reactivity appears to result from osmium leaching from the ligand allowing its passage through the membrane. Upon continuous dosing of osmium, the conversion and ee rose to the initial values. It was concluded that potassium osmate has a very weak attachment to the ligand, thus effective retention of osmate was not achieved in the current system.

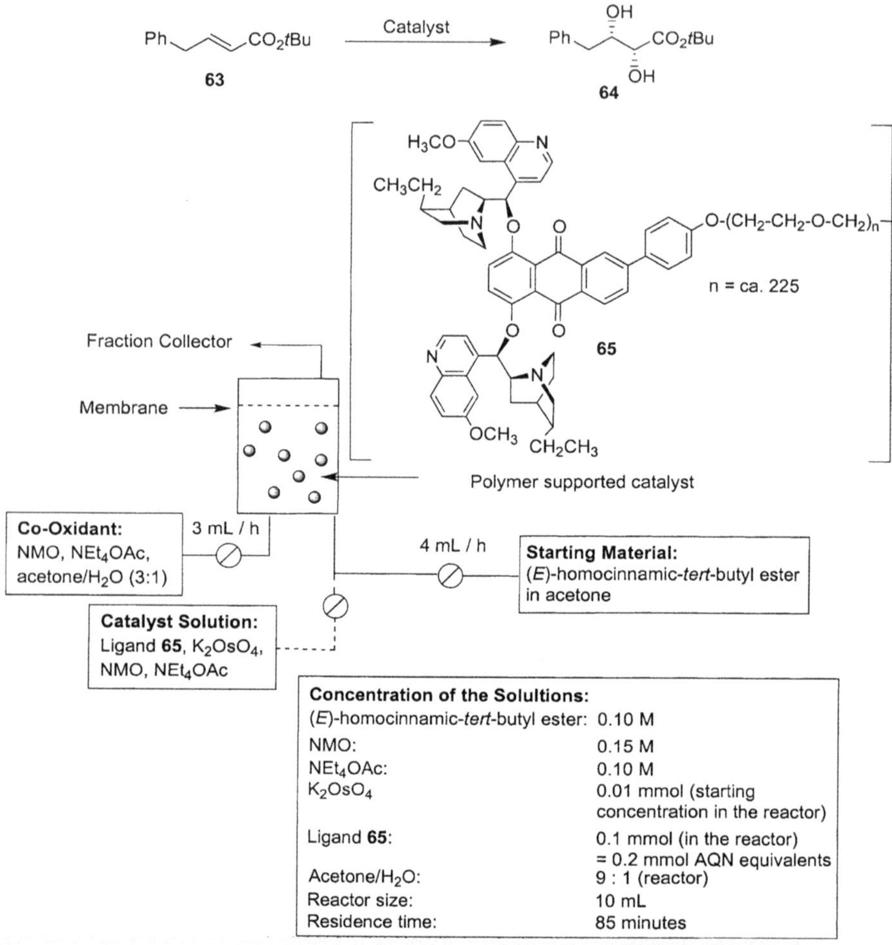

Fig. 6 Continuous AD

7 AD of Chiral Non-Racemic Alkenes

Chiral non-racemic alkenes may be dihydroxylated in high selectivity without the need for chiral ligands. An example is the dihydroxylation of **66** with OsO_4/trimethylamine *N*-oxide ($(CH_3)_3NO$)/pyridine in acetone/H_2O [117]. Heating the mixture at reflux temperature for 18 h afforded a mixture of the diols **67** and **68** in 39:1 ratio (95% de) and isolated yield of 78% (Scheme 11). Applying Sharpless reaction conditions OsO_4/NMO/**7b** in acetone/H_2O showed some improvements in the diastereoselectivity but no marked enhancement in the reaction rate. The slow rate of the reaction was attributed to steric hindrance about the double bond. This reaction was carried out on a multi kilogram scale.

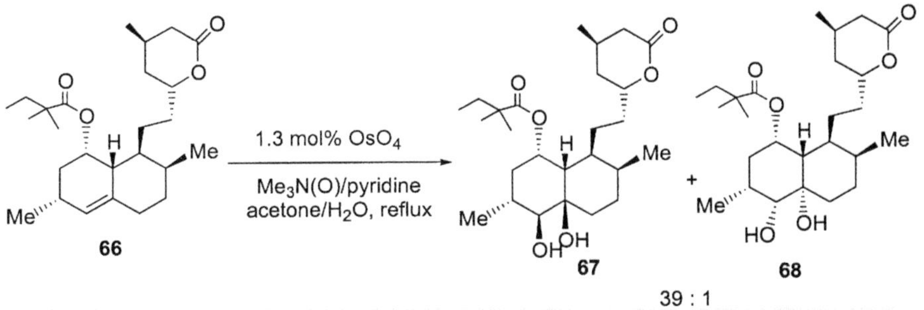

66 → 67 + 68 (1.3 mol% OsO_4; $Me_3N(O)$/pyridine, acetone/H_2O, reflux); 39 : 1

Scheme 11 Dihydroxylation of **66** with non-chiral ligand

Another example is the diastereomeric dihydroxylation of commercial α-pinene (**69**) (93.5% ee) (Eq. 9). As in the previous case, the Sharpless procedure did not offer a real advantage in rate or selectivity [118]. The reaction could only occur at elevated temperature (~56 °C) and there was no enantiomeric enrichment. The authors explained the lack of enantiomeric enrichment by steric factors which hindered the approach of the osmium-ligand complex to alkene **69** such that the catalyst was unable to differentiate the two enantiomers of α-pinene. The use of pyridine as a ligand with 0.002 equivalent of osmium and 60% aqueous NMO in acetone/H_2O solvent mixture at reflux temperature for 44 h afforded the diol **70** in about 90% yield without loss of enantiomeric purity (Eq. 9).

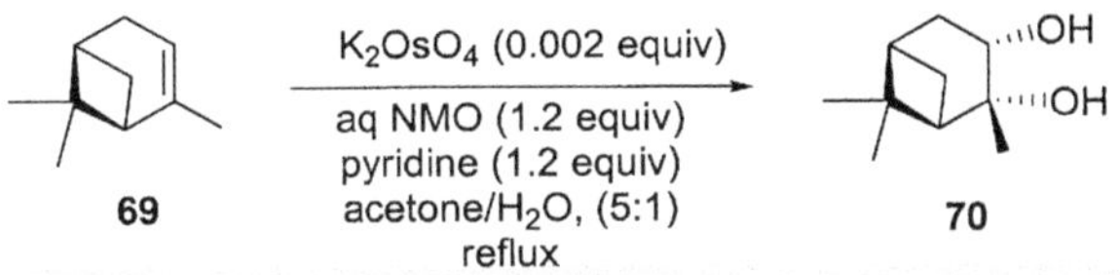

69 → 70 (K_2OsO_4 (0.002 equiv); aq NMO (1.2 equiv), pyridine (1.2 equiv), acetone/H_2O, (5:1), reflux) (9)

8 Conclusion

The discovery of catalytic AD by Sharpless and co-workers in the late 1970s and the AA in the 1990s, as well as the improvements in catalyst/ligand combinations and reaction conditions, have added new very efficient tools to organic synthesis of complex molecules. The value of these two transformations is ever increasing and their applications have been extended to include a large variety of alkenes. The real test for the practicality of a chemical reaction is in its ability to work equally well on large-scale. These reactions have been applied by several process development groups to large-scale syntheses, a proof of their utility.

Process development work to make these reactions of even greater utility is ongoing which will enhance the value of these novel transformations and aid in the development of new and scalable organic reactions. In spite of all the outstanding research and applications of these reactions, this is still a work in progress and their potential is expanding in both fronts, academic and industrial.

References

1. Beller M, Sharpless KB (1996) In: Cornils B, Herrmann WA (eds) Applied homogeneous catalysis with organometallic compounds, vol 2. Wiley-VCH, Weinheim, p 1009
2. Lohray BB (1992) Tetrahedron Asymmetry 3:1317
3. Johnson RA, Sharpless KB (2000) In: Ojima I (ed) Catalytic asymmetric synthesis, 2nd edn. Wiley-VCH, Weinheim, p 357
4. Bolm C, Hildebrand JP, Muñiz K (2000) In: Ojima I (ed) Catalytic asymmetric synthesis, 2nd edn. Wiley-VCH, Weinheim, p 399
5. Markó IE, Svendsen JS (1999) In: Jacobsen EN, Pfaltz A, Yamamoto H (eds) Comprehensive asymmetric catalysis II, vol 2. Springer, Berlin Heidelberg New York, p 713
6. Kolb HC, Sharpless KB (1998) In: Beller M, Bolm C (eds) Transition metals in organic synthesis, vol 2. Wiley-VCH, Weinheim, p 243
7. Kolb HC, Sharpless KB (1998) In: Beller M, Bolm C (eds) Transition metals in organic synthesis, vol 2. Wiley-VCH, Weinheim, p 219
8. Sánchez-Delgado RA, Rosales M, Esteruelas MA, Oro LA (1995) J Mol Catal A Chem 96:231
9. Kolb HC, VanNieuwenhze MS, Sharpless KB (1994) Chem Rev 94:2483
10. O'Brien P (1999) Angew Chem Int Ed Engl 38:326
11. Ahrgren L, Sutin L (1997) Org Process Res Dev 1:425
12. Tan L, Chen C-Y, Larsen RD, Verhoeven TR, Reider PJ (1998) Tetrahedron Lett 39:3961
13. Beck G (2002) Synlett 837
14. Koskinen AMP, Karvinen EK, Siirilä JP (1994) J Chem Soc, Chem Commun 21
15. Wang Z-M, Kolb HC, Sharpless KB (1994) J Org Chem 59:5104
16. Denis J-N, Correa A, Greene AE (1990) J Org Chem 55:1957
17. Lu X, Xu Z, Yang G (2000) Org Process Res Dev 4:575
18. Li G, Sharpless KB (1996) Acta Chem Scand 50:649
19. Reddy KL, Sharpless KB (1998) J Am Chem Soc 120:1207
20. Wuts PGM, Anderson AM, Goble MP, Mancini SE, VanderRoest RJ (2000) Org Lett 2:2667
21. Griesbach RC, Hamon DPG, Kennedy RJ (1997) Tetrahedron Asymmetry 8:507
22. Harada N, Ozaki K, Yamaguchi T, Arakawa H, Ando A, Oda K, Nakanishi N, Ohashi M, Hashiyama T, Tsujihara K (1997) Heterocycles 46:241
23. Horwell DC, Osborne S (1998) WO Patent 9 849 158
24. Jayamma Y, Nandanan E, Lohray BB (1994) J Indian Inst Sci 74:309
25. Josien H, Ko S-B, Bom D, Curran DP (1998) Chem Eur J 4:67
26. Lee K-H, Kashiwada Y, Huang L, Lee TT, Cosentino M, Snider J, Manax M, Xie L (1997) US Patent 5 637 589
27. Sinha SC, Sinha-Bagchi A, Keinan E (1993) J Org Chem 58:7789
28. Albarella L, Piccialli V, Smaldone D, Sica D (1996) J Chem Res, Synop 400
29. Carlsen PHJ, Katsuki T, Martin VS, Sharpless KB (1981) J Org Chem 46:3936
30. Courtney JL (1986) In: Mijs WJ, de Jonge CRHI (eds) Organic synthesis of oxidized metal compounds. Plenum Press, New York, p 445
31. Fatiadi AJ (1987) Synthesis 85
32. Haines AH (1985) Methods for the oxidation of organic compounds: alkanes, alkenes, alkynes, and arenes. Academic Press, London
33. Haines AH (1992) In: Trost BM, Fleming I (eds) Comprehensive organic synthesis, vol 7. Pergamon, Oxford, UK, p 437
34. Lempers HEB, Ripollès i Garcia A, Sheldon RA (1998) J Org Chem 63:1408
35. Milas NA, Sussman S (1936) J Am Chem Soc 58:1302
36. Milas NA (1937) J Am Chem Soc 59:2342
37. Schröder M (1980) Chem Rev 80:187
38. Sharpless KB, Akashi K (1976) J Am Chem Soc 98:1986

39. Sheldon RA, Kochi JK (1981) Metal-catalyzed oxidations of organic compounds. Academic Press, New York
40. Singh HS (1986) In: Mijs WJ, De Jonge CRHI (eds) Organic synthesis of oxidized metal compounds. Plenum Press, New York, p 633
41. Taylor JE, Janini TE, Elmer OC (1998) Org Process Res Dev 2:147
42. Varma RS, Naicker KP (1998) Tetrahedron Lett 39:7463
43. Pearlstein RM, Davison A (1988) Polyhedron 7:1981
44. Balavoine GGA, Manoury E (1995) Appl Organomet Chem 9:199
45. Makowka O (1908) Ber 41:943
46. Criegee R, Marchand B, Wannowius H (1942) Ann 550:99
47. Criegee R (1938) Angew Chem 51:519
48. Criegee R (1936) Ann 522:75
49. Berrisford DJ, Bolm C, Sharpless KB (1995) Angew Chem Int Ed Engl 34:1059
50. Ray R, Matteson DS (1980) Tetrahedron Lett 21:449
51. Schneider WP, McIntosh AV Jr (1956) US Patent 2 769 824
52. VanRheenen V, Kelly RC, Cha DY (1976) Tetrahedron Lett 23:1973
53. Sharpless KB, Patrick DW, Truesdale LK, Biller SA (1975) J Am Chem Soc 97:2305
54. Hentges SG, Sharpless KB (1980) J Am Chem Soc 102:4263
55. Hentges SG, Sharpless KB (1980) J Org Chem 45:2257
56. Griffith WP, Skapski AC, Woode KA, Wright MJ (1978) Inorg Chim Acta 31:L413
57. Jacobsen EN, Markó I, Mungall WS, Schröder G, Sharpless KB (1988) J Am Chem Soc 110:1968
58. Wai JSM, Markó I, Svendsen JS, Finn MG, Jacobsen EN, Sharpless KB (1989) J Am Chem Soc 111:1123
59. Shibata T, Gilheany DG, Blackburn BK, Sharpless KB (1990) Tetrahedron Lett 31:3817
60. Becker H, Sharpless KB (1996) Angew Chem Int Ed Engl 35:448
61. Li G, Chang H-T, Sharpless KB (1996) Angew Chem Int Ed Engl 35:451
62. Patrick DW, Truesdale LK, Biller SA, Sharpless KB (1978) J Org Chem 43:2628
63. Reiser O (1996) Angew Chem Int Ed Engl 35:1308
64. Lohray BB, Bhushan V, Reddy GJ, Reddy AS (2002) Indian J Chem 41B:161
65. Göbel T, Sharpless KB (1993) Angew Chem Int Ed Engl 32:1329
66. Rudolph J, Sennhenn PC, Vlaar CP, Sharpless KB (1996) Angew Chem Int Ed Engl 35:2810
67. Kwong H-L, Sorato C, Ogino Y, Chen H, Sharpless KB (1990) Tetrahedron Lett 31:2999
68. Sharpless KB, Amberg W, Beller M, Chen H, Hartung J, Kawanami Y, Lübben D, Manoury E, Ogino Y, Shibata T, Ukita T (1991) J Org Chem 56:4585
69. Minato M, Yamamoto K, Tsuji J (1990) J Org Chem 55:766
70. Sharpless KB, Amberg W, Bennani YL, Crispino GA, Hartung J, Jeong K-S, Kwong H-L, Morikawa K, Wang Z-M, Xu D, Zhang X-Y (1992) J Org Chem 57:2768
71. Mehltretter GM, Döbler C, Sundermeier U, Beller M (2000) Tetrahedron Lett 41:8083
72. Andersson MA, Epple R, Fokin VV, Sharpless KB (2002) Angew Chem Int Ed 41:472
73. Dupau P, Epple R, Thomas AA, Fokin VV, Sharpless KB (2002) Adv Synth Catal 344:421
74. Pringle W, Sharpless KB (1999) Tetrahedron Lett 40:5151
75. Fokin VV, Sharpless KB (2001) Angew Chem Int Ed Engl 40:3455
76. Rubin AE, Sharpless KB (1997) Angew Chem Int Ed Engl 36:2637
77. McKee BH, Gilheany DG, Sharpless KB (1992) Org Synth 70:47
78. Wang Z-M, Sharpless KB (1994) J Org Chem 59:8302
79. Oi R, Sharpless KB (1992) Tetrahedron Lett 33:2095
80. Song CE, Oh CR, Roh EJ, Choi JH (2001) Tetrahedron Asymmetry 12:1533
81. Gao Y, Zepp CM (1993) WO Patent 9 317 150
82. Döbler C, Mehltretter GM, Sundermeier U, Beller M (2000) J Am Chem Soc 122:10,289
83. Döbler C, Mehltretter G, Beller M (1999) Angew Chem Int Ed Engl 38:3026
84. Döbler C, Mehltretter GM, Sundermeier U, Beller M (2001) J Organomet Chem 621:70
85. Herranz E, Biller SA, Sharpless KB (1978) J Am Chem Soc 100:3596

86. Backväll J-E, Oshima K, Palermo RE, Sharpless KB (1979) J Org Chem 44:1953
87. Gold EH, Babad E (1972) J Org Chem 37:2208
88. Fukuyama T, Jow C-K, Cheung M (1995) Tetrahedron Lett 36:6373
89. Weinreb SM, Demko DM, Lessen TA, Demers JP (1986) Tetrahedron Lett 27:2099
90. Li G, Angert HH, Sharpless KB (1996) Angew Chem Int Ed Engl 35:2813
91. Bruncko M, Schlingloff G, Sharpless KB (1997) Angew Chem Int Ed Engl 36:1483
92. Campbell MM, Johnson G (1978) Chem Rev 78:65
93. Herranz E, Sharpless KB (1978) J Org Chem 43:2544
94. Wallis ES, Lane JF (1946) Organic reactions III, p 267
95. Han H, Cho C-W, Janda KD (1999) Chem Eur J 5:1565
96. Chuang C-Y, Vassar VC, Ma Z, Geney R, Ojima I (2002) Chirality 14:151
97. Morgan AJ, Masse CE, Panek JS (1999) Org Lett 1:1949
98. Tao B, Schlingloff G, Sharpless KB (1998) Tetrahedron Lett 39:2507
99. Anelli PL, Biffi C, Montanari F, Quici S (1987) J Org Chem 52:2559
100. Inokuchi T, Matsumoto S, Nishiyama T, Torii S (1990) J Org Chem 55:462
101. Miyazawa T, Endo T, Shiihashi S, Okawara M (1985) J Org Chem 50:1332
102. Caron M, Carlier PR, Sharpless KB (1988) J Org Chem 53:5185
103. Chong JM, Sharpless KB (1985) J Org Chem 50:1560
104. Barta NS, Sidler DR, Somerville KB, Weissman SA, Larsen RD, Reider PJ (2000) Org Lett 2:2821
105. Kim BM, Sharpless KB (1990) Tetrahedron Lett 31:3003
106. Bolm C, Gerlach A (1998) Eur J Org Chem 21
107. Salvadori P, Pini D, Petri A (1999) Synlett 1181
108. Petri A, Pini D, Salvadori P (1995) Tetrahedron Lett 36:1549
109. Salvadori P, Pini D, Petri A (1997) J Am Chem Soc 119:6929
110. Mandoli A, Pini D, Agostini A, Salvadori P (2000) Tetrahedron: Asymmetry 11:4039
111. Kuang Y-Q, Zhang S-Y, Wei L-L (2001) Tetrahedron Lett 42:5925
112. Kobayashi S, Ishida T, Akiyama R (2001) Org Lett 3:2649
113. Kobayashi S, Endo M, Nagayama S (1999) J Am Chem Soc 121:11,229
114. Nagayama S, Endo M, Kobayashi S (1998) J Org Chem 63:6094
115. Kobayashi S, Nagayama S (1998) J Am Chem Soc 120:2985
116. Wöltinger J, Henniges H, Krimmer H-P, Bommarius AS, Drauz K (2001) Tetrahedron Asymmetry 12:2095
117. DeCamp AE, Mills SG, Kawaguchi AT, Desmond R, Reamer RA, DiMichele L, Volante RP (1991) J Org Chem 56:3564
118. Krishnamurthy V, Landi J Jr, Roth GP (1997) Synth Commun 27:853

Topics Organomet Chem (2004) 6: 181–203
DOI 10.1007/978-3-540-36966-0

Palladium-Catalyzed Heck Arylations in the Synthesis of Active Pharmaceutical Ingredients

Mahavir Prashad

Process Research and Development, Chemical and Analytical Development, Novartis Institute for Biomedical Research, One Health Plaza, East Hanover, New Jersey 07936, USA
E-mail: Mahavir.prashad@pharma.novartis.com

Abstract The utility of inter- and intramolecular Heck arylations in the synthesis of intermediates for the preparation of active pharmaceutical ingredients and in the synthesis of natural products of pharmaceutical importance is described.

Keywords Heck arylation · Active pharmaceutical ingredients (APIs) · Palladium

1 Introduction

The Heck reaction is an old and versatile method for palladium-catalyzed arylation and vinylation of alkenes (Scheme 1) [1–3]. The traditional mechanism [4] of the Heck reaction is well-known and involves an oxidative addition of aryl halide to a PdL_2 complex to form an organopalladium halide ($ArPdXL_2$) intermediate. The *syn*-addition of the organopalladium halide ($ArPdXL_2$) to the olefin followed by the *syn*-elimination of hydropalladium halide ($HPdXL_2$) from the

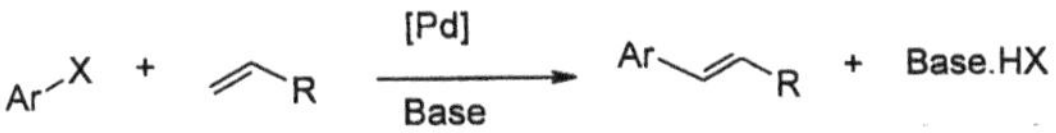

Ar–X + CH_2=CH–R —[Pd], Base→ Ar–CH=CH–R + Base.HX

Scheme 1

resulting adduct affords the olefin. The active PdL_2 complex is regenerated by the addition of a base to remove HX from the inactive $HPdXL_2$.

Since its discovery about three decades ago, the Heck reaction has enjoyed an extraordinary growth, which can be attributed to its excellent functional group tolerability. However, it was only in the mid-eighties when it was recognized that the synthetic potential of this reaction was far from being fully exploited. Several review articles have been written on various aspects of the Heck reaction [5–15] including intramolecular Heck reaction [6], use of tetraalkylammonium salts in the Heck reaction [9], asymmetric Heck reaction [10], mechanistic aspects of the Heck reaction [8, 11, 12], and the Heck reaction on a solid support [13]. Some of recent advances on this reaction include development of phosphine-imidazolium salts [16], phosphites as ligands [17], phosphine-free conditions [18–20], milder reaction conditions [21], conditions using aryl chlorides [22, 23], use of palladacycles [24–26], use of additives [27], reaction in aqueous medium with and without phase-transfer catalysts [28–30], reaction in ionic liquids [31–33], in supercritical carbon dioxide [34, 35], use of supported palladium catalysts [36–38], palladium-modified zeolites [39], microwave-mediated conditions [40–42], and ultrasound promoted reaction in ionic liquids [43]. The use of the Heck reaction in the production of fine chemicals (Scheme 2), e.g., Prosulfuron (a herbicide by former Ciba-Geigy, now Syngenta), 2-ethylhexyl-*p*-methoxy cinnamate (a UV-B sunscreen agent), and monomers for coatings of electronic components (known as Cyclotene), has been reviewed recently [44, 45]. The purpose of this review is to highlight the utility of the Heck arylation reaction in

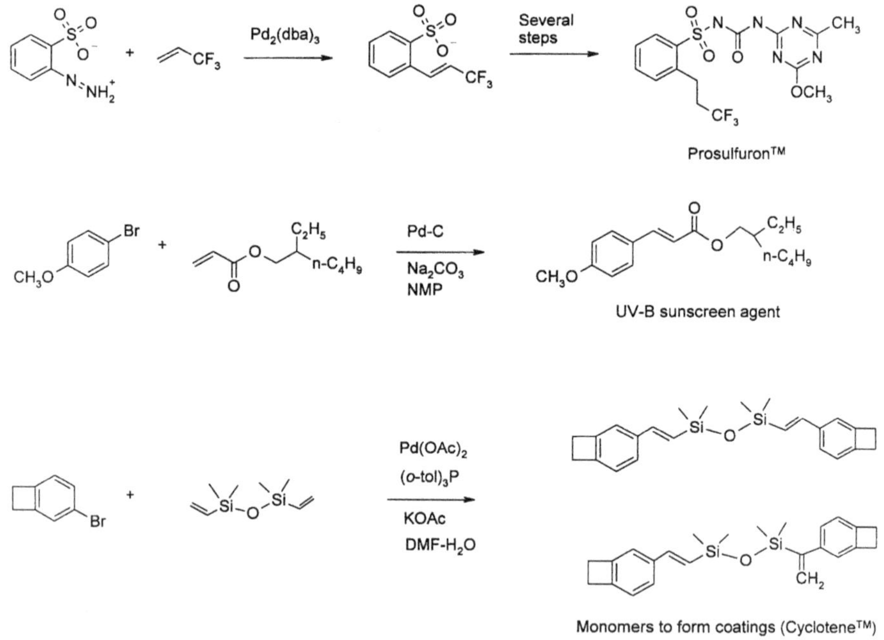

$Pd_2(dba)_3$

Several steps

Prosulfuron™

Pd-C

Na_2CO_3

NMP

UV-B sunscreen agent

$Pd(OAc)_2$

$(o\text{-tol})_3P$

KOAc

$DMF\text{-}H_2O$

Monomers to form coatings (Cyclotene™)

Scheme 2

the pharmaceutical industry in the synthesis of intermediates for the preparation of active pharmaceutical ingredients, both for development phases and marketing, and in the synthesis of some natural products of pharmaceutical importance.

2
Intermolecular Heck Arylations

4-(2-Methoxyethyl)phenol (**3**), which is an important intermediate in the preparation of metoprolol (**4**), a β_1-blocker for the treatment of hypertension, was synthesized using the Heck arylation [46]. The Heck arylation of methyl vinyl ether with 4-bromonitrotoluene (**1**) in the presence of palladium on charcoal and triethylamine at 120 °C in toluene followed by hydrogenation in the same pot afforded 4-methoxyethylaniline (**2**), not isolated, but converted to **3** directly (Scheme 3). A much faster reaction was achieved in acetonitrile, DMF, and DMSO, but with loss in regioselectivity. Elimination of the solvent resulted in a considerably slower reaction. Palladium acetate performed as well as palladium on carbon, whereas the presence of triarylphosphine favored arylation on the oxygen-bearing carbon.

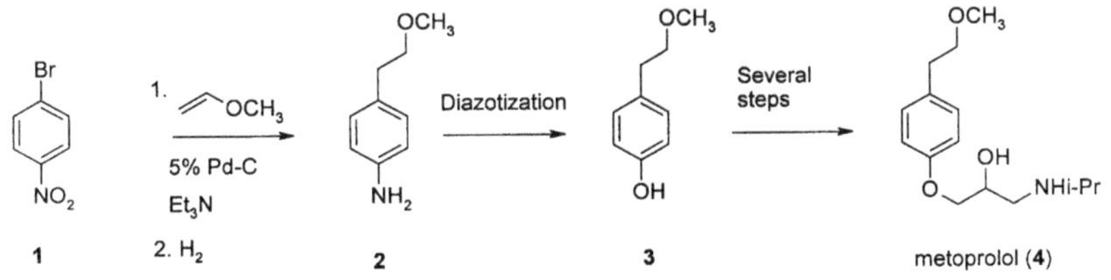

Scheme 3

The main framework of Merck's Singulair (**8**), an LTD_4 antagonist for the treatment of asthma, was formed by the Heck arylation (Scheme 4) [47–49]. The Heck arylation of allylic alcohol **5** with methyl 2-iodobenzoate (**6**) was performed in refluxing acetonitrile in the presence of 1.5 equiv of triethylamine and only 1 mol% of palladium acetate (compared to 2.5–5 mol% normally used). No ligand, phase-transfer catalysts, or other salts were required. Use of 1 mol% of catalyst required only 1 h for this reaction with the isolated allylic alcohol, while with 0.5 mol% catalyst the reaction time was 12 h. The keto ester **7** crystallized from the reaction mixture upon cooling to 22 °C and was isolated in 83% yield. A minor by-product resulted from the arylation of the C-2 carbon of the allylic alcohol, which was removed easily in the mother liquor. The same keto ester **7** was also a key intermediate in the synthesis of another LTD_4 antagonist, L-699,392 (**9**).

A new process for the production of naproxen (**12**), a nonsteroidal anti-inflammatory drug, was developed by Albemarle Corporation utilizing the Heck arylation of ethylene with 2-bromo-6-methoxynaphthalene (**10**) to afford 6-methoxy-2-vinylnaphthalene (**11**), followed by hydroxycarbonylation

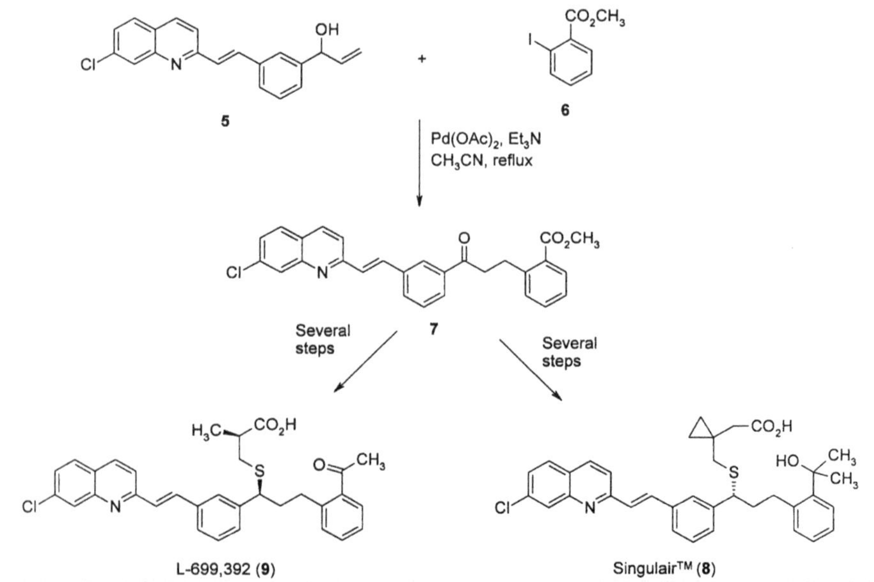

Scheme 4

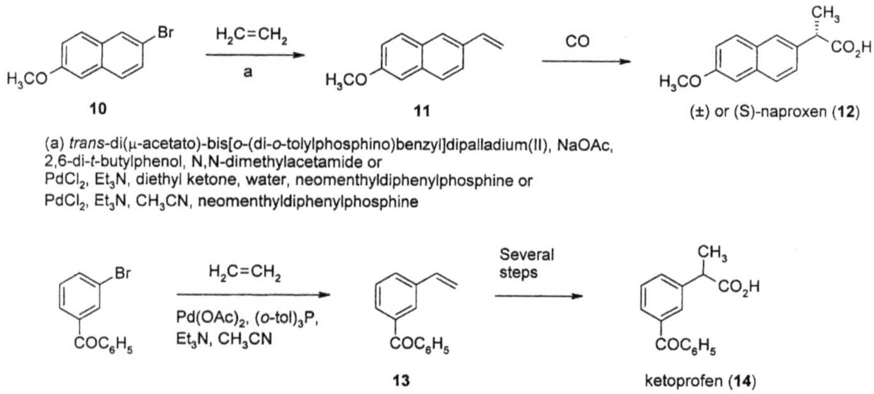

(a) *trans*-di(μ-acetato)-bis[*o*-(di-*o*-tolylphosphino)benzyl]dipalladium(II), NaOAc, 2,6-di-*t*-butylphenol, N,N-dimethylacetamide or
$PdCl_2$, Et_3N, diethyl ketone, water, neomenthyldiphenylphosphine or
$PdCl_2$, Et_3N, CH_3CN, neomenthyldiphenylphosphine

Scheme 5

(Scheme 5) [50, 51]. The Heck reaction conditions involved the use of palladium chloride, triethylamine, diethyl ketone, water, and neomenthyldiphenylphosphine as the ligand. The reaction typically took 4–6 h to give >95% conversion (yield 85–95%) at 95 °C with an ethylene pressure of 425–450 psig. A very high ratio of substrate to catalyst (1870:1) and substrate to ligand (330:1) was used. Recently, Aventis has developed the Heck conditions for the arylation of ethylene with **10** utilizing (*trans*-di(μ-acetato-bis[*o*-(di-*o*-tolylphosphino)benzyl]dipalladium(II), a palladacycle, in *N*,*N*-dimethylacetamide in the presence of sodium acetate and 2,6-di-*tert*-butylphenol [52, 53]. The reaction took 10–16 h to give >95% conversion (yield 89%) at 140 °C with an ethylene pressure of 20 bar. Again a very high ratio of substrate to palladacycle (1876:1) was used. The Heck methodology has also been used to prepare several "profens", including ketoprofen (**14**) [50, 54]. The conditions for ketoprofen utilized palladium acetate, triethylamine, tri-*o*-tolylphosphine, and 20 atm of ethylene in acetonitrile at 125 °C to afford an 80% yield of 3-vinyl benzophenone (**13**).

Resveratrol (**18**), a small molecule naturally occurring in grape skins, mulberries, peanuts, and other plants, has been shown to have potential as a therapeutic for a range of diseases. Its disease preventive qualities in humans have been attributed as the reason for the "French Paradox" (although the Mediterranean diet contains high levels of fat and alcohol, the expected increase in rates of cancer and heart disease is not observed). The key step in the synthesis of resveratrol involved the Heck arylation of 4-acetoxystyrene (**16**) with 3,5-dimethoxybenzoyl chloride (**15**) in the presence of palladium acetate and triethylamine in *p*-xylene at 120–130 °C for 18 h (Scheme 6) [55]. The desired (*E*)-4-acetoxy-3′,5′-dimethoxystilbene (**17**) was obtained in 75% yield. These conditions for the decarbonylative aromatic-olefin Heck reaction are a variation of the classical aryl halide-olefin reaction with the loss of carbon monoxide.

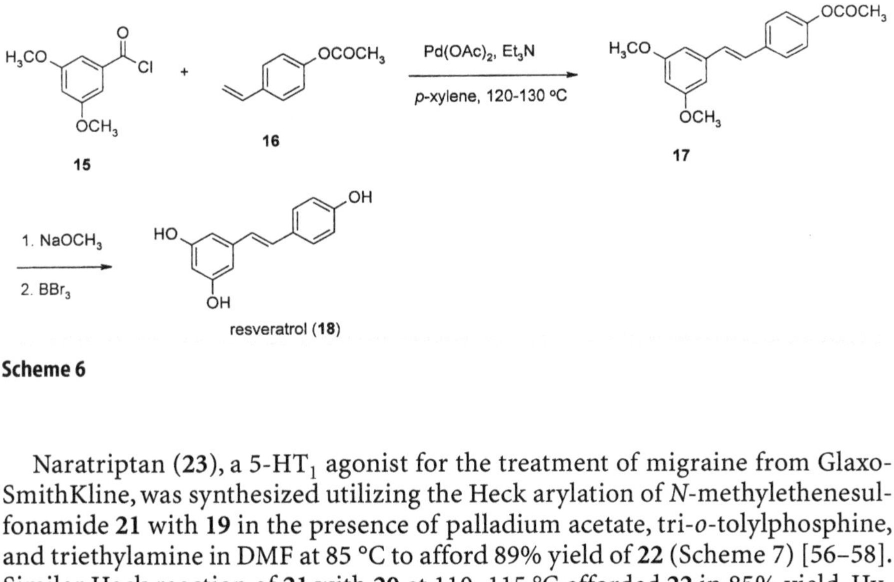

Pd(OAc)$_2$, Et$_3$N; *p*-xylene, 120-130 °C; 1. NaOCH$_3$; 2. BBr$_3$; resveratrol (**18**)

Scheme 6

Naratriptan (**23**), a 5-HT$_1$ agonist for the treatment of migraine from GlaxoSmithKline, was synthesized utilizing the Heck arylation of *N*-methylethenesulfonamide **21** with **19** in the presence of palladium acetate, tri-*o*-tolylphosphine, and triethylamine in DMF at 85 °C to afford 89% yield of **22** (Scheme 7) [56–58]. Similar Heck reaction of **21** with **20** at 110–115 °C afforded **22** in 85% yield. Hydrogenation of this intermediate then yielded naratriptan (**23**).

Another antimigraine agent and 5-HT$_{1D}$ agonist, eletriptan (**27**) from Pfizer, was also synthesized using the Heck arylation of phenylvinyl sulfone (**25**) with **24** in the presence of palladium acetate, tri-*o*-tolylphosphine, and triethylamine in DMF at 85 °C to afford **26**, which was hydrogenated to eletriptan (Scheme 8) [59, 60]. Because phenylvinyl sulfone (**25**) is an irritant to the eyes, respiratory tract and skin, handling of this compound is of concern. This problem could be

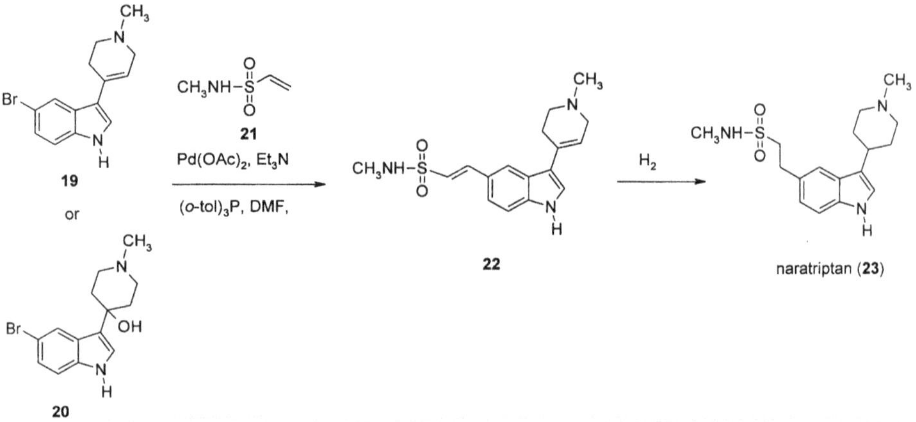

Scheme 7

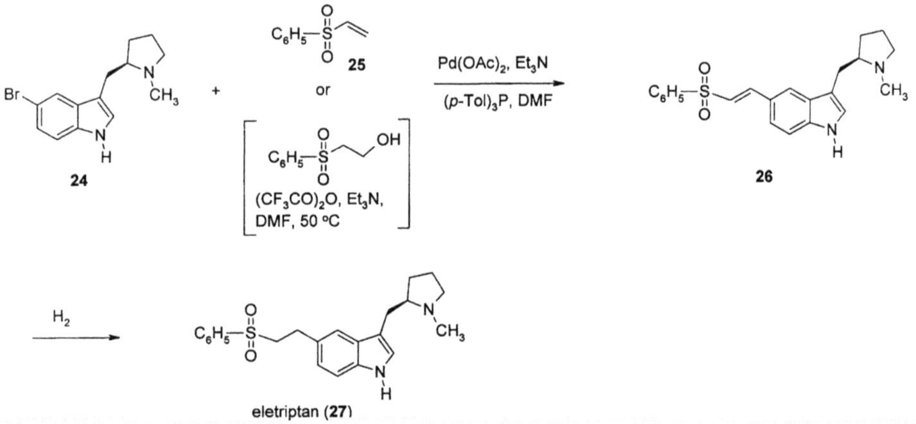

Scheme 8

overcome by generating phenylvinyl sulfone in situ by treatment of non-toxic 2-phenylsulfonylethanol with trifluoroacetic anhydride and triethylamine in DMF at 50 °C followed by the Heck reaction in the same pot.

Acrivastine (**30**) is a non-sedative antihistamine agent from GlaxoSmithKline. The Heck arylation of ethyl acrylate with 2-bromo-6-(*p*-toluoyl)pyridine (**28**) in the presence of palladium acetate, triphenylphosphine, and triethylamine in acetonitrile at 150 °C for 6 h afforded **29** in 85% yield, which was then converted to acrivastine (Scheme 9) [61–63].

The quinoline group in ABT-773, a novel ketolide having potent activity against multidrug-resistant respiratory tract pathogens and excellent in vivo efficacy in experimental animal infection models, was introduced by the Heck arylation of the 6-*O*-allyl ketolide **31** with 3-bromoquinoline in the presence of palladium acetate, triethylamine, tri-*o*-tolylphosphine in acetonitrile (Scheme 10)

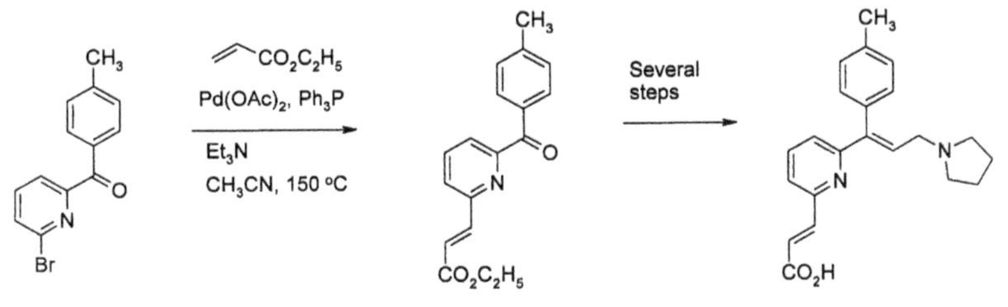

$CO_2C_2H_5$
$Pd(OAc)_2$, Ph_3P
Et_3N
CH_3CN, 150 °C

Several steps

28 29 acrivastine (30)

Scheme 9

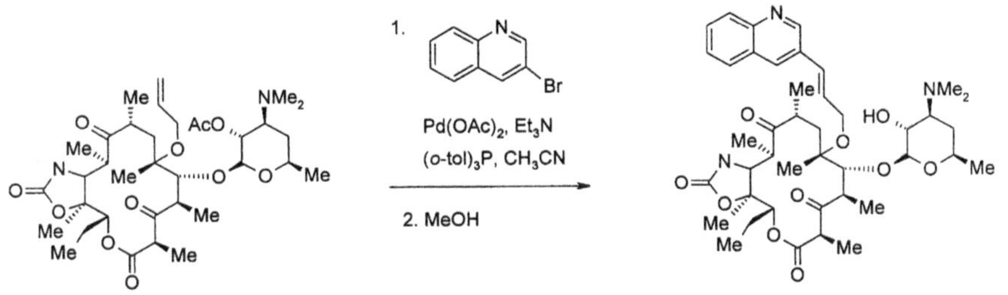

1. Pd(OAc)$_2$, Et_3N
(*o*-tol)$_3$P, CH_3CN
2. MeOH

31 ABT-773

Scheme 10

[64, 65]. These results clearly demonstrated that a variety of functional groups are well tolerated under Heck reaction conditions.

A number of approaches are described for the synthesis of LY231514 (Pemetrexed disodium or Alimta) from Eli Lilly, which represents a new generation of folate-requiring enzyme inhibitors with improved antitumor potency and spectrum [66, 67]. An earlier approach utilized the Heck arylation of alkene **32** with diethyl 4-iodobenzoylglutamate (**33**) in the presence of palladium acetate, tri-*o*-tolylphosphine, and triethylamine to give the desired ethano-bridged pyrrolopyrimidine **34** in 68% yield. None of the anticipated vinyl-bridged pyrrolinopyrimidine was observed (Scheme 11). The unexpected migration of the double bond obviated the anticipated need for reduction of the unsaturated bridge and subsequent oxidation of the pyrroline ring. A more practical synthesis of LY231514 utilized the Sonogashira reaction of methyl-4-bromobenzoate (**35**) with 3-butyn-1-ol (**36**) in the presence of palladium chloride, triphenylphosphine, cuprous iodide, and diethylamine in ethyl acetate at 40 °C for 4 h to afford methyl 4-(4-hydroxy-1-butynyl)benzoate (**37**) in 83% yield on a multikilogram scale [68].

Towards the development of a rapid and enantioselective synthesis of an NK-1 receptor antagonist **41**, an ambitious regio- and stereo-selective Heck arylation of the spirodiene **38** was explored, which in theory could result in eight isomeric products, provided that only single addition occurred. In the presence of palladium acetate, tetra-*n*-butylammonium chloride, potassium formate, lithium

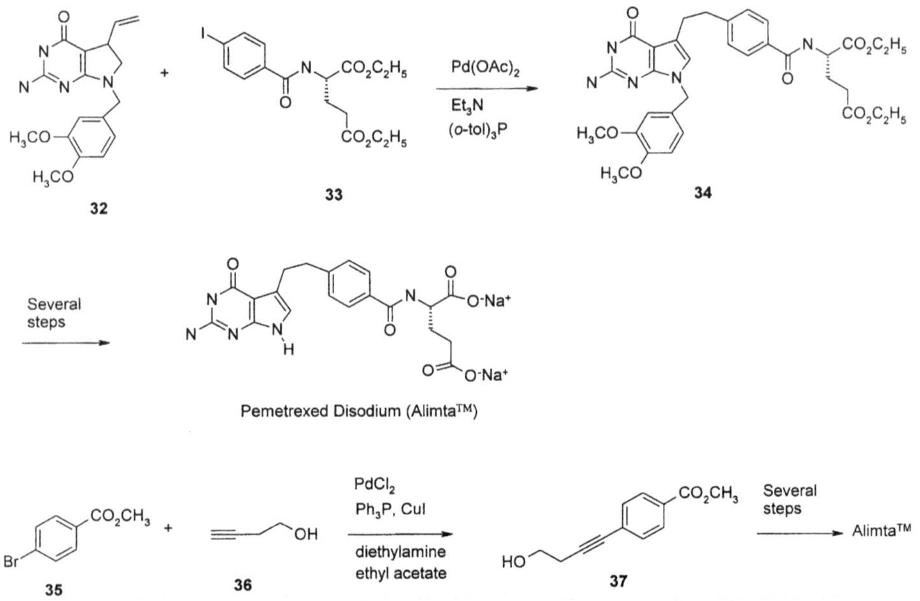

Pemetrexed Disodium (Alimta™)

Scheme 11

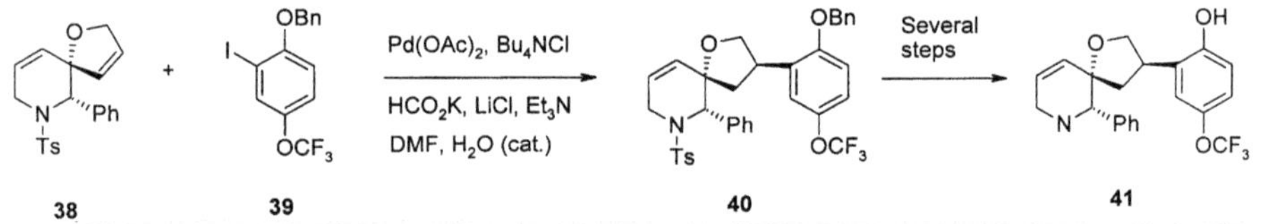

Scheme 12

chloride, and triethylamine in mixture of DMF and water (95/5) the Merck chemists achieved the coupling of aromatic iodide **39** with spirodiene **38** to afford **40** in 60% yield and 90% ds (Scheme 12) [69]. While the role of water (5%) is not clear, significant quantities of regioisomeric products were observed in its absence, suggesting that addition of water in this Heck reaction plays a critical role for obtaining high selectivities.

Nabumetone [4-(6′-methoxy-2′-naphthyl)butan-2-one] (**44**) is a non-steroidal anti-inflammatory drug, which is manufactured by the hydrogenation of ketone **42**. Aventis has developed a process for the manufacture of this intermediate utilizing the Heck arylation of methyl vinyl ketone with 2-bromo-6-methoxynaphthalene (**10**) in the presence of potassium carbonate, palladium dichloride, triphenylphosphine in DMF containing 0.5 wt% of water (Scheme 13). Formation of diarylated by-product **43** was observed [70]. The optimum conditions required 0.255 mol% of palladium chloride, which at both 120 and 132 °C gave 98% of **42** and only 2% of the diarylated by-product **43** at complete conversion of **10**. The results with larger quantities of the catalyst were not significantly bet-

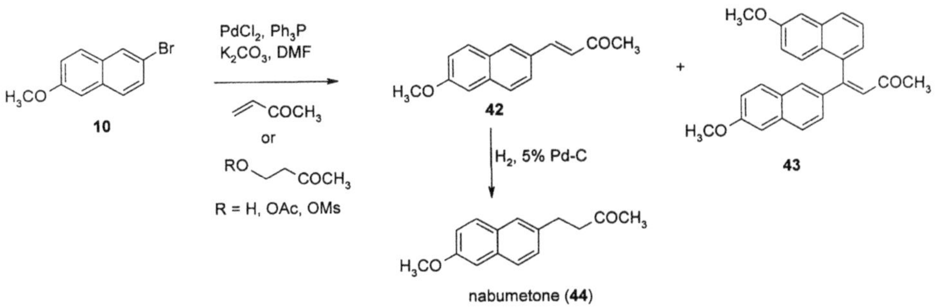

$PdCl_2$, Ph_3P
K_2CO_3, DMF
$\diagup\!\!\diagdown COCH_3$ or $RO\diagdown\!\!\diagup COCH_3$
R = H, OAc, OMs
10 → 42 + 43
H_2, 5% Pd-C
nabumetone (**44**)

Scheme 13

ter or worse. Smaller catalyst quantities (e.g., 0.129 mol%) produced higher amounts of diarylated product (**43**; 53% at 85% conversion). This Heck reaction in DMF in the presence of potassium carbonate and palladacycle, (*trans*-di(μ-acetato-bis[*o*-(di-*o*-tolylphosphino)benzyl]dipalladium(II), produced large amounts of diarylated product **43**, as did the reaction with the comparable amount of palladium chloride [71]. When the reaction was carried out in *N*,*N*-dimethylacetamide instead of DMF and at 140 °C instead of 132 °C, the diarylated product **43** was nearly the exclusive product. Either 4-acetoxy or 4-hydroxy-2-butanone, both of which generate methyl vinyl ketone in situ and are cheaper, less toxic, and more stable than methyl vinyl ketone, can also be used as a replacement for methyl vinyl ketone. However, the use of either of these substitutes requires control of the reaction rates with an appropriate amount of palladium catalyst. With less than optimum catalyst amounts, the methyl vinyl ketone generated in situ decomposes faster than it can react with the halide. With greater than optimal catalyst concentration, the desired monoaryl product **42** is converted to the diarylated by-product **43**. In either case, the deficiency of available methyl vinyl ketone results in larger amounts of diarylated product **43**. Optimum amounts of palladium chloride were 50% (0.128 mol%) and 6% (0.0154 mol%) of that for the reactions with methyl vinyl ketone itself. Thus, such a replacement yielded the desired product with no loss in yield and offered an additional advantage in that less palladium catalyst can be used.

Merck chemists utilized the Heck arylation of *tert*-butyl acrylate with the 2-bromopyridine derivative **45** in the presence of $PdCl_2$(dppf) (CH_2Cl_2 complex) and sodium acetate trihydrate in *N*,*N*-dimethylacetamide at 80 °C to afford the α,β-unsaturated ester **46** in 85% yield. This intermediate is an important Michael acceptor in the practical asymmetric synthesis of a selective Endothelin A receptor antagonist **47** (Scheme 14) [72].

Synthesis of kilogram quantities of 2,3-dihydro-5-hydroxy-6-[3-(2-hydroxymethylphenyl)-2-propenyl]benzofuran (**49**), a topical anti-inflammatory, required an efficient synthesis of (*E*)-3-(2-methoxycarbonylphenyl)propenoic acid (**48**). One of the routes used to prepare this acid utilized the Heck arylation of acrylic acid with methyl-2-bromobenzoate in the presence of 0.2 mol% of palladium acetate, tri-*n*-butylamine, and 0.4 mol% of tri-*o*-tolylphosphine in refluxing toluene for 6 h to obtain a 73% yield (Scheme 15′) [73].

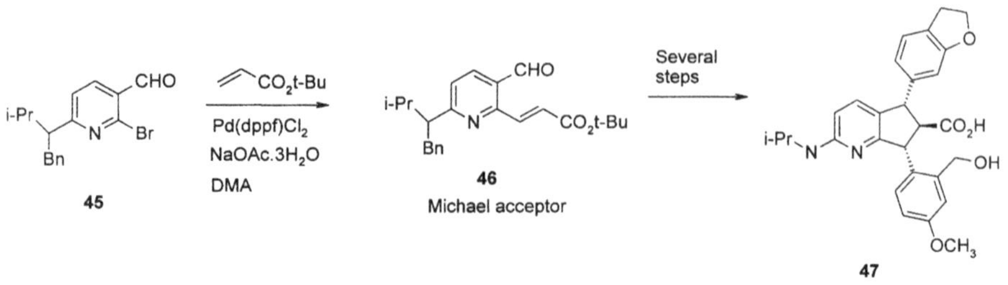

Scheme 14

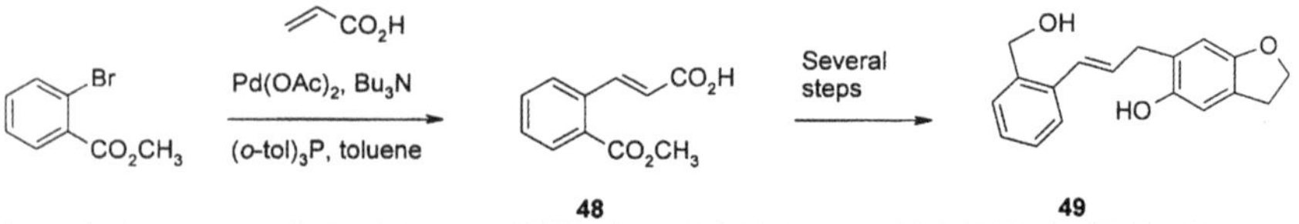

Scheme 15

The Heck reaction was utilized efficiently in the synthesis of natural product plicatin B (**53**), which is the anti-microbial principle of *Psoralea juncea* [74]. The Heck arylation of methyl acrylate with 2-prenylated-4-bromophenol under a variety of conditions failed to give any product. This would have been a straightforward synthesis of plicatin B. Because bromophenols are moderate substrates for the Heck reaction, the additional alkyl group seems to be sufficiently electron donating to suppress the reaction totally. However, the reaction of 2-prenylated-4-bromo-acetoxybenzene (**50**) with methyl acrylate in the presence of 5 mol% of palladium acetate, 10 mol% of tri-*o*-tolylphosphine, and triethylamine in toluene at 100 °C gave 60% yield of the plicatin B acetate (**52**, Scheme 16). Use of triphenylphosphine or a larger amount of tri-*o*-tolylphosphine led to reduced yields. The Heck arylation of methyl acrylate with 2-prenylated-4-iodo-acetoxybenzene (**51**) was tried next in the presence of 5 mol% of palladium acetate and triethylamine in toluene to afford plicatin B acetate **52** in 62% yield. Surprisingly, in the presence of 10 mol% of tri-*o*-tolylphosphine, plicatin B acetate **52** was obtained in 96% yield. These results were in contrast to the reports that addition of a ligand inhibits the Heck reaction with iodides.

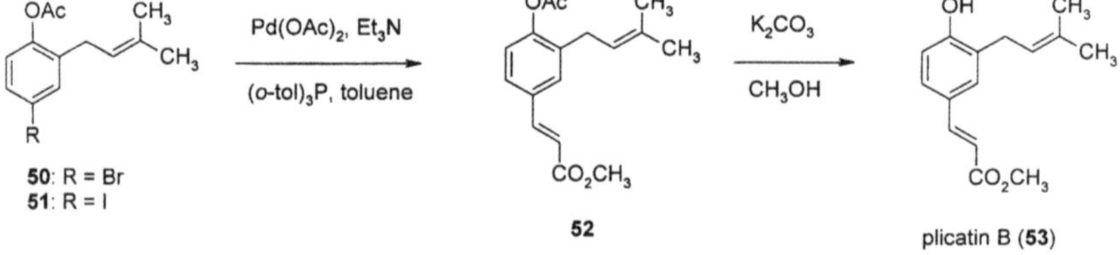

Scheme 16

A new, efficient and much shorter synthesis of DX-9065a, a potent orally active inhibitor of the blood coagulation enzyme factor XA (fXa), was developed using the Heck arylation as the key step [75]. Arylation of **54** with sulfonate **55** in the presence of 2.5 mol% of the stable palladacycle (*trans*-di(μ-acetato-bis[*o*-(di-*o*-tolylphosphino)benzyl]dipalladium(II) and triethylamine in DMF at 120 °C for 32 h afforded the desired alkene **56** in 43% yield as 2:1 mixture of *E*/*Z* isomers (Scheme 17). This isomer ratio was inconsequential to the synthesis of the target molecule.

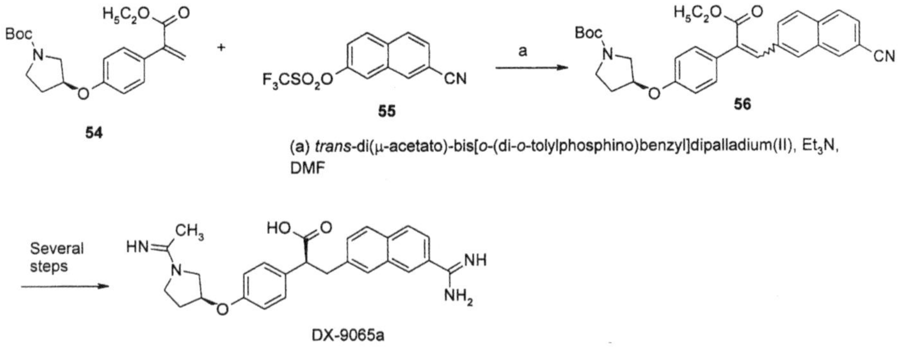

(a) *trans*-di(μ-acetato)-bis[*o*-(di-*o*-tolylphosphino)benzyl]dipalladium(II), Et_3N, DMF

Scheme 17

An analog of ceramide (**59**), which is an important signaling molecule involved in a number of physiological events, including the regulation of cell growth and differentiation, inflammation and apoptosis, is synthesized utilizing the Heck arylation of ethyl acrylate with 1-bromo-3-dodecylbenzene (**57**) in the presence of triethylamine and catalytic tetrakis(triphenylphosphine)Pd(0) at 120 °C in DMF to give ethyl *m*-dodecylcinnamate (**58**) in 92% yield (Scheme 18) [76].

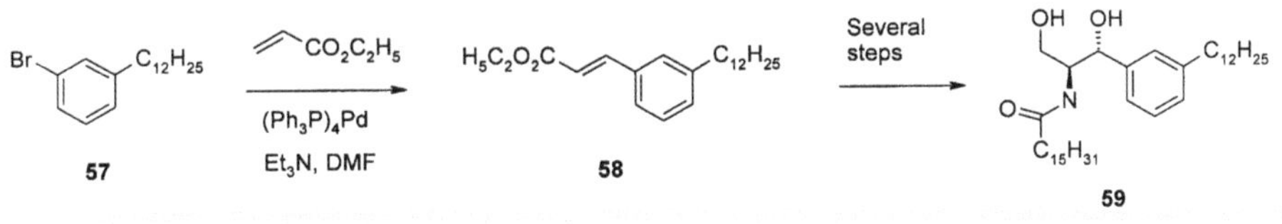

Scheme 18

Novartis chemists utilized the Heck arylation of 2,5-dimethoxystyrene with methyl 5-bromosalicylate in the presence of 3 mol% of palladium acetate in refluxing acetonitrile to give a >90% yield of an *E*/*Z* mixture of stilbene **60**. The olefin **60** which was hydrogenated to afford LAP977 (Scheme 19), a potential therapeutic agent for the treatment of hyperproliferative and anti-inflammatory disorders and cancer [77]. Crude 2,5-dimethoxystyrene was found to be unstable and polymerized on several occasions. This prevented the practical use of this route.

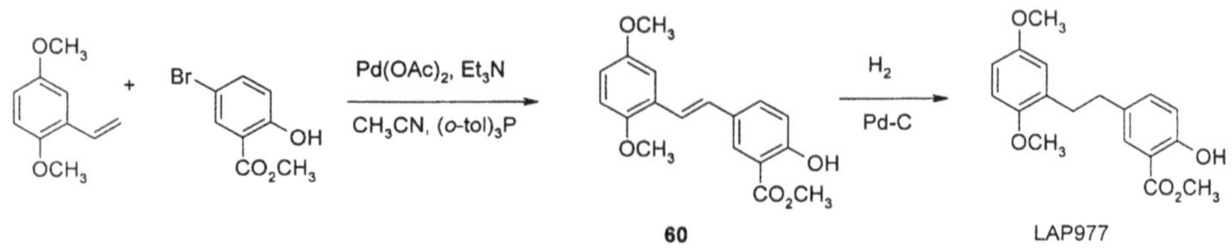

OCH3 + Br OH CO2CH3 — Pd(OAc)2, Et3N / CH3CN, (o-tol)3P → 60 — H2 / Pd-C → LAP977

Scheme 19

The intermolecular Heck arylation was also utilized to prepare a number of drug candidates; such as 5-methoxy-2-[*N*-(2-benzamidoethyl)-*N*-*n*-propylamino]tetralin with weak dopamine D2 and 5-HT_{1A} receptor binding properties [78], a series of bisaryl cyclobutenes as COX-2 inhibitors by Merck-Frost [79], a series of *N,N′*-diphenyl ureas as ACAT inhibitors by Pfizer [80], antitumor anthracyclines [81, 82], indanone analogs of pterosins as a potent smooth muscle relaxants [83], imidazo[4,5-*b*)pyridin-2(3*H*)-ones and thiazolo[4,5-*b*]pyridin-2(3*H*)-ones as novel cAMP PDE III inhibitors [84], HMG-CoA reductase inhibitors by Pfizer [85], (+)-1,2,3,4-tetrahydro-5-(2-phosphonoethyl)-3-isoquinolinecarboxylic acid as a competitive NMDA antagonist with anticonvulsant activity [86], several antifungal triazoles [87] by Pfizer, tryptamines by Boehringer-Ingelheim [88], and the isocoumarin moiety of the rubromycins, e.g., γ-rubromycin, which exhibits activity against the reverse transcriptase of human immunodeficiency virus-1 and human telomerase [89].

3 Intramolecular Heck Cycloarylations

Ondansetron or Zofran (**63**) from GlaxoSmithKline, a 5-HT_3 antagonist for the treatment of emesis, pain disorders, and withdrawal syndrome, utilized an intramolecular Heck cycloarylation reaction of 3-(2-bromo-*N*-methylanilino)-cyclohex-2-en-1-one (**61**) in the presence of 2 mol% of palladium acetate, 4 mol% of triphenylphosphine, and sodium bicarbonate in DMF to afford **62** in 32% yield (Scheme 20) [90, 91]. Usually, the reaction ceased after 25–30 h with ~50% of the starting material unchanged.

Adosar and CC-1065 from PharmaciaUpjohn are potent antitumor agents [92, 93]. One of the syntheses of the cyclopropylpyrroloindole (CPI) fragment **67** of these agents utilized an intramolecular Heck cycloarylation of **64** (Scheme 21) [94, 95]. The Heck cycloarylation of **64** in the presence of tetrakis(triphenyl-

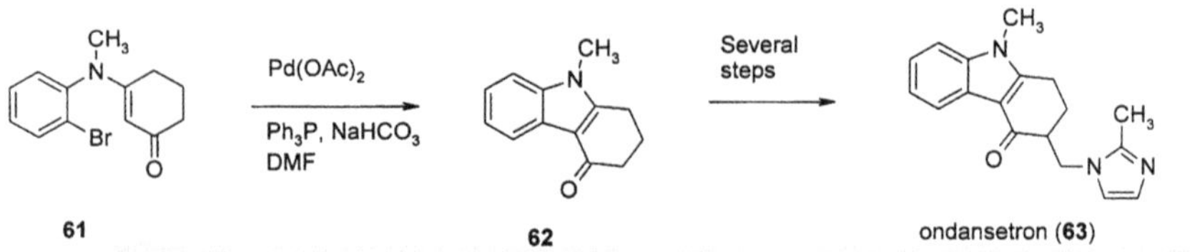

61 — Pd(OAc)2 / Ph3P, NaHCO3, DMF → 62 — Several steps → ondansetron (63)

Scheme 20

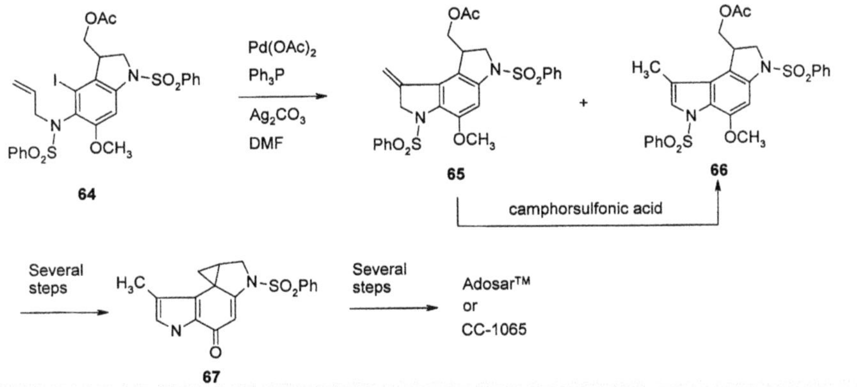

Scheme 21

phosphine)palladium and triethylamine in acetonitrile for 18 h yielded a 17:1 mixture of undesired **65** and desired *seco*-CPI **66** in 90% yield. However, this cyclization in the presence of palladium acetate, triphenylphosphine, and silver carbonate in DMF at room temperature for 3 h afforded exclusively the undesired product **65** in 90% yield. It was noteworthy that lower concentrations of palladium acetate (0.03 mol/l) gave a 1.3:1 ratio of **65** and desired *seco*-CPI **66** and at higher concentrations (0.3 mol/l) only undesired product **65** was formed. However, the selectivity of the Heck cycloarylation was of little importance since the undesired product **65** was quantitatively isomerized to *seco*-CPI (**66**) by a treatment with camphorsulfonic acid in dichloromethane at 20 °C.

The tricyclic structural core of another potent antitumor antibiotic, duocarmycin (**70**), was constructed using an intramolecular Heck cycloarylation [96]. The *trans*-isomer **68** underwent the cyclization with 5 mol% of palladium acetate, 13 mol% of tri-*o*-tolylphosphine, and triethylamine in acetonitrile at 110 °C for 26 h to afford the desired tricyclic compound **69** in 82% yield (Scheme 22) along with a double bond regioisomer in 11% yield. This regioisomeric ratio was inconsequential. Interestingly, no cyclization was observed with the corresponding *cis*-isomer under similar conditions and only the starting material was recovered.

Synthesis of SDZ SER082, a selective and potent 5-$HT_{2C/2B}$ receptor antagonist from Novartis utilized an intramolecular Heck cycloarylation as one of the

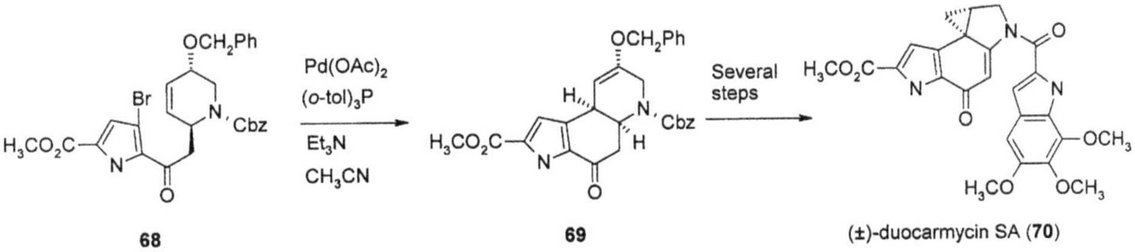

Scheme 22

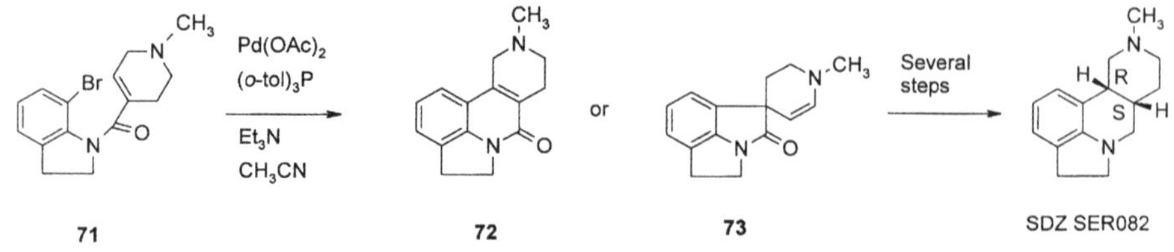

Scheme 23

initial approaches [97]. The Heck cyclization of 7-bromo-2,3-dihydro-1-[(1,2,5,6-tetrahydro-1-methyl-4-pyridyl)carbonyl]-*1H*-indole (**71**) in the presence of a catalytic amount of palladium acetate, tri-*o*-tolylphosphine, and triethylamine in acetonitrile at 90 °C gave only <10% of the desired product **72** via a 6-*endo*-trig cyclization (Scheme 23). No spirocyclic compound **73**, arising from the 5-*exo*-trig cyclization, was observed. Use of stoichiometric amounts of palladium acetate afforded the desired compound **72** in 75% yield.

Syntheses of CP-122,288 and CP-122,638, antimigraine agents from Pfizer, utilized an intramolecular Heck cycloarylation as the key step [98, 99]. Treatment of the monobromo derivative **74** with palladium acetate and triethylamine in refluxing DMF afforded *N*-Cbz-protected CP-122,288 (**75**) in 81% yield, which was converted to the desired drug substance CP-122,288 (Scheme 24). The Heck cycloarylation of the dibromo derivative **76** in the presence of 10 mol% of palladium acetate, triethylamine, tetra-*n*-butylammonium chloride in DMF-DME at 80 °C in 1 h afforded a 76% yield of the 7-bromo-indole **77**, which was converted to CP-122,638 by hydrogenation and to CP122,288 by lithium aluminum hydride reduction followed by hydrogenation. The presence of the second bromine – the bromine "passenger" – was not significantly deleterious to the Heck cyclization.

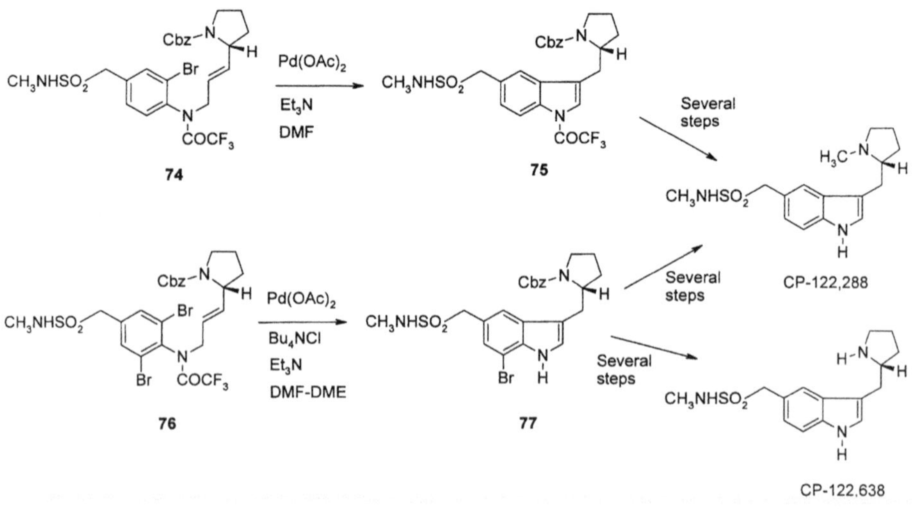

Scheme 24

A palladium-catalyzed annulation between iodoanilines and ketones, via the formation of enamine followed by intramolecular Heck cycloarylation, was also utilized in the synthesis of another antimigraine agent MK-0462 from Merck [100, 101]. The coupling of iodoaniline **78** with C-silylated 3-butyn-1-ol revealed that the more stable C-protection gave better results (Scheme 25). The bulky *tert*-butyldimethylsilyl protected butyn-1-ol coupled considerably slower, hence, triethylsilyl was the preferred protecting group. Protection of the hydroxyl group also played a key role in this reaction. The *bis*-triethylsilyl butyn-1-ol (**79**) was chosen as the substrate because it offered a suitable coupling rate and stability. Thus, the reaction of iodoaniline **78** with **79** in the presence of 2 mol% of palladium acetate and sodium carbonate in DMF at 100 °C afforded an 80% yield of a mixture of desilylated **80** and 2-silylated **81** indoles. This mixture was of no consequence as the silyl group was removed in the next step with methanol and HCl, which was followed by several steps to yield MK-0462.

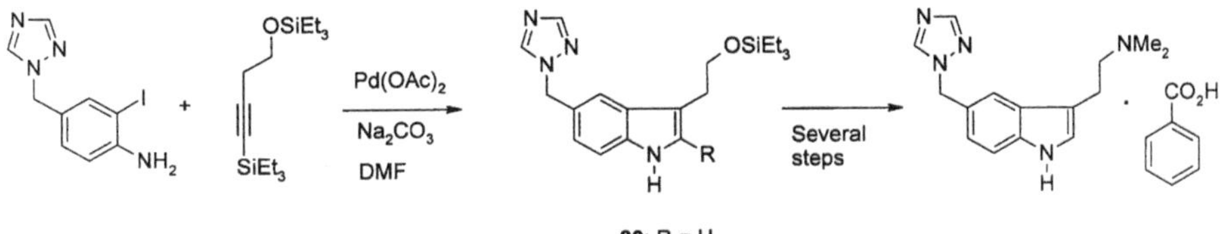

Scheme 25

FR 900482, a potent antitumor agent, utilized an intramolecular Heck cycloarylation of the iodide **82** as the key step in the presence of triethylamine and catalytic tetrakis(triphenylphosphine)Pd(0) in acetonitrile at 80 °C for 10 h to afford the tetracycle **83** in 90% yield (Scheme 26) [102].

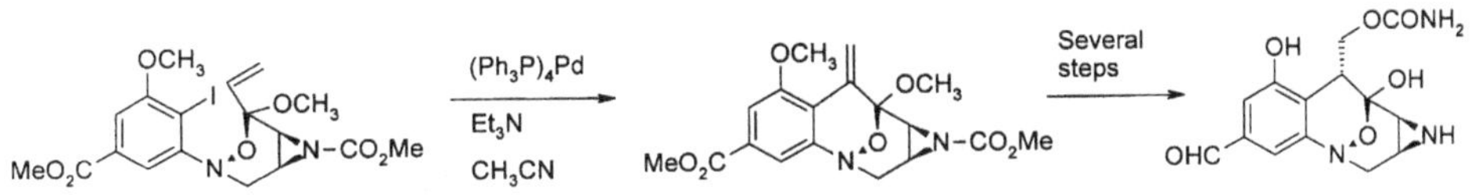

Scheme 26

The intramolecular Heck cycloarylation was utilized to construct the B-ring of the steroid structure of esterone (**86**) [103, 104]. The Heck cyclization of the bromide **84** in the presence of 13 mol% of palladium acetate, 27 mol% of triphenylphosphine, and silver phosphate in DMF at 115 °C for 60 h gave a 63% yield of the tetracyclic steroid skeleton (**85**, Scheme 27). However, the same cyclization in the presence of 2.5 mol% of palladacycle (*trans*-di(μ-acetato-bis[*o*-(di-*o*-tolylphosphino)benzyl]dipalladium(II) and tetra-*n*-butylammonium acetate in a mixture of DMF, acetonitrile, and water at 115 °C for 4.5 h afforded a 99% yield

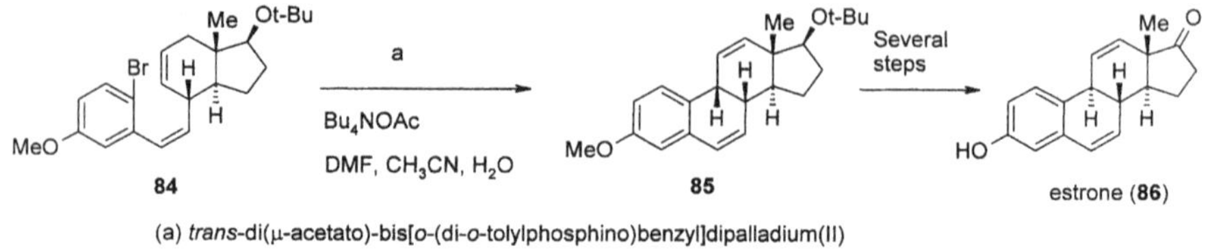

(a) *trans*-di(μ-acetato)-bis[*o*-(di-*o*-tolylphosphino)benzyl]dipalladium(II)

Scheme 27

of the desired product as a single diastereomer with a *cis*-junction of the rings B and C. In this case, the addition of water led to an increase in the reaction rate.

Cephalotaxine (**89**), a parent compound of the antileukamic-active harringtonines, was synthesized by an intramolecular Heck cycloarylation of alkene-bromide **87** in the presence of 4 mol% of palladacycle (*trans*-di(μ-acetato-bis[*o*-(di-*o*-tolylphosphino)benzyl]dipalladium(II) and tetra-*n*-butylammonium acetate in a mixture of DMF, acetonitrile, and water at 110–120 °C for 7 h to afford the pentacyclic core **88** in 81% yield (Scheme 28) [105, 106]. This cyclization was highly stereoselective. No reaction was observed in the presence of tetrakis(triphenylphosphine)palladium.

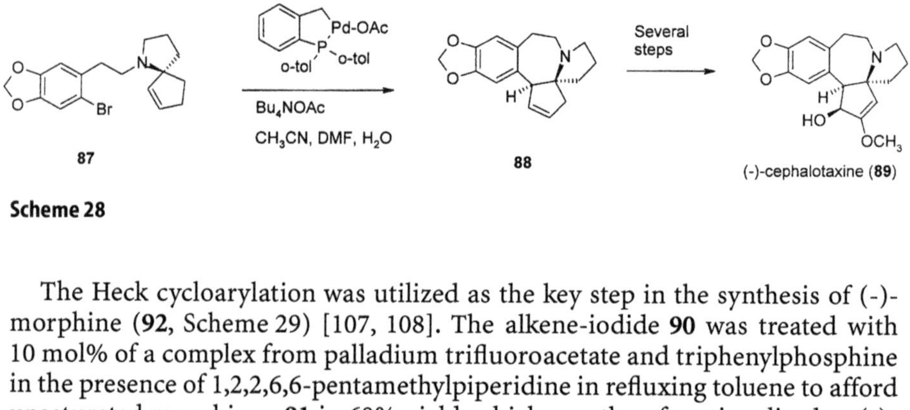

Scheme 28

The Heck cycloarylation was utilized as the key step in the synthesis of (-)-morphine (**92**, Scheme 29) [107, 108]. The alkene-iodide **90** was treated with 10 mol% of a complex from palladium trifluoroacetate and triphenylphosphine in the presence of 1,2,2,6,6-pentamethylpiperidine in refluxing toluene to afford unsaturated morphinan **91** in 60% yield, which was then functionalized to (-)-morphine.

An intramolecular Heck cycloarylation reaction was also used in the synthesis of 2-aminoethyl substituted tricycles with NMDA receptor affinity [109], aminochromans as potential dopamine analogs [110], aryl-fused azapolycyclic

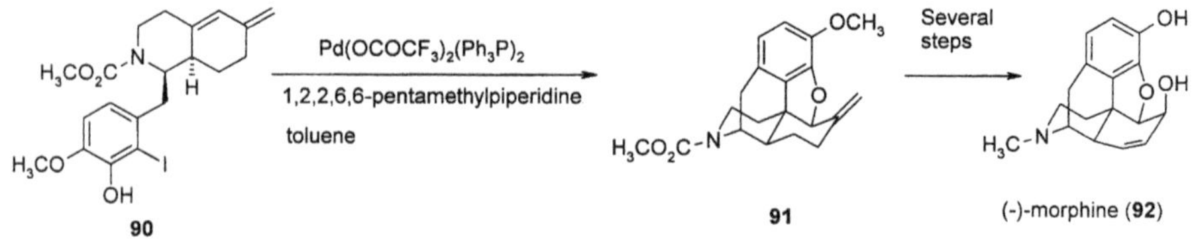

Scheme 29

compounds as nicotine binding inhibitors [111], the northern part of TMC-95A, a potent and selective proteasome inhibitor [112], and analogs of CC-1065 and duocarmycin [113, 114].

4 Use of Multiple Heck Arylations

Novartis chemists synthesized (1α,2β,5β)-2,5-bis(3,4,5-trimethoxyphenyl)-1-methoxycyclopentane (**95**) as a potent platelet activating factor (PAF) antagonist in only three steps, of which the key step was the double Heck arylation of cyclopentene with **93** (Scheme 30) [115, 116]. Double Heck arylation of cyclopentene with **93** under classical conditions involving palladium acetate, tri-*o*-tolylphosphine, and triethylamine in acetonitrile at 100 °C furnished a mixture of regioisomers, 1,3-diarylcyclopentene **94** and the corresponding 1,4-diarylcyclopentene (ratio 2:1). However, none of the 1,4-diarylcyclopentene was observed when the Heck reaction was carried out under phase-transfer conditions involving palladium acetate, tetra-*n*-butylammonium chloride, and potassium acetate in DMF at 80 °C affording 1,3-diarylcyclopentene **94** in >80% yield. Originally, the synthesis of the equipotent analog (1α,2α,5β)-2,5-bis(3,4,5-trimethoxyphenyl)-1-methoxycyclopentane (SDZ 64–688, **96**) was much longer [117].

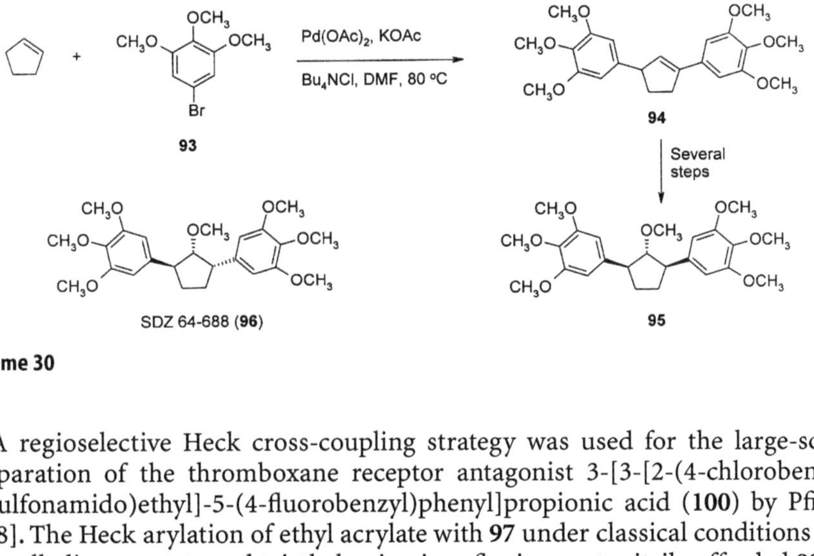

$Pd(OAc)_2$, KOAc

Bu_4NCl, DMF, 80 °C

93

94

Several steps

SDZ 64-688 (**96**)

95

Scheme 30

A regioselective Heck cross-coupling strategy was used for the large-scale preparation of the thromboxane receptor antagonist 3-[3-[2-(4-chlorobenzenesulfonamido)ethyl]-5-(4-fluorobenzyl)phenyl]propionic acid (**100**) by Pfizer [118]. The Heck arylation of ethyl acrylate with **97** under classical conditions using palladium acetate and triethylamine in refluxing acetonitrile afforded **98** in 79% yield (Scheme 31). Only 0.4% of bis(cinnamate) could be detected by GC/MS. Thus, an excellent selectivity was achieved in this phosphine-free Heck reaction of aryl iodides compared with aryl bromides. The second Heck arylation involved the reaction of **98** with *N*-vinylphthalimide, which was accomplished on a small scale in the presence of diisopropylamine, tri-*o*-tolylphosphine, and palladium acetate in toluene at 100 °C in a sealed tube or in refluxing xylene at atmospheric pressure to afford a 67% yield of **99**. A scale-up of the refluxing xylene

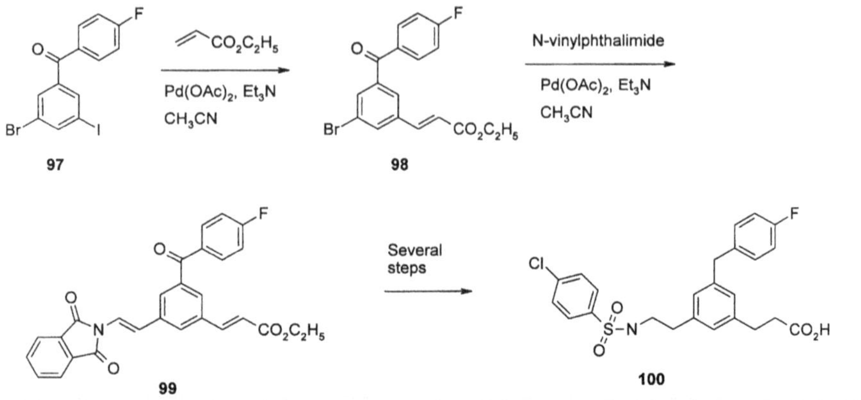

Scheme 31

conditions completely failed. A careful investigation suggested that the batch of tri-*o*-tolylphosphine ligand used in the scale-up was the cause of the problem, because using a fresh source of tri-*o*-tolylphosphine allowed smooth Heck reaction again in 65% yield. Surprisingly, this Heck arylation also proceeded smoothly in the absence of the phosphine ligand source with only a modest decrease in the isolated yield (58%) of **99**. The phosphine-free conditions were used to carry out this Heck arylation on a multi-kilogram scale due to their reliability.

GI147211C from GlaxoSmithKline is a potent and novel water-soluble topoisomerase I inhibitor for the treatment of a variety of tumor types [119]. Two intramolecular Heck cycloarylations were utilized in a practical synthesis of this camptothecin analog. The first Heck cycloarylation was used to prepare **103** (Scheme 32). Thus an intramolecular Heck reaction of crotyl ether **101** in the

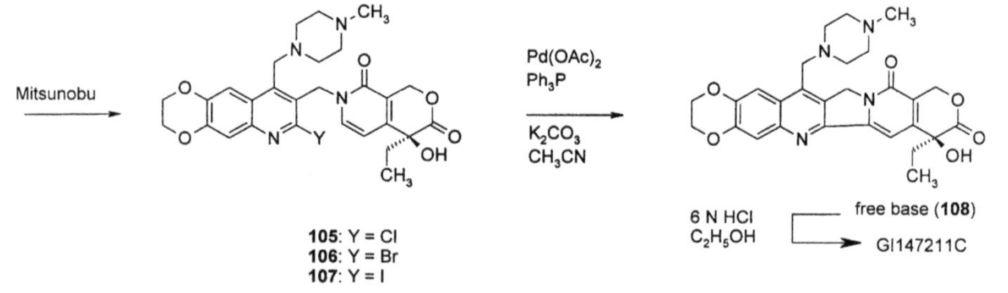

Scheme 32

presence of palladium acetate, potassium carbonate, and tetra-*n*-butylammonium bromide in DMF at 90 °C for 24 h afforded an 8:1 regioisomeric mixture of the desired six-membered enol ether **102** and corresponding undesired allylic ether. The desired enol ether **102** could be obtained in a 79% yield after a fractional crystallization of the mixture. The allylic ether could also be isomerized to the enol ether **102** upon treatment with Wilkinson's catalyst in refluxing *n*-propanol. The second key intramolecular Heck cycloarylation of the haloquinolines **105–107** was achieved in the presence of palladium acetate, triphenylphosphine, and potassium carbonate in refluxing acetonitrile. Under these conditions all three quinolines underwent cyclization smoothly to afford the free base **108** of GI147211C in 34, 71, and 55% yields, respectively. The cyclization of the chloroquinoline **105** was unexpected as chlorides usually involve forcing conditions. Traces of palladium from the product were removed by recrystallizations from dichloromethane, methanol, and acetone containing triphenylphosphine to give the metal-free drug substance as the free base **108.**

A combination of an inter and intramolecular Heck arylations was used to prepare (-)-7-chloro-3-(4-aminomethyl-2-(carboxymethoxy)-phenyl)aminocabonylmethyl-1,3,4,5-tetrahydrobenz-[*c,d*]indole-2-carboxylic acid hydrochloride (**113**), a potent NMDA-glycine antagonist from Sumitomo for the treatment of stroke and neurodegenerative disorders such as Alzheimer's and Hungtington's diseases [120]. The first Heck reaction was used to achieve the formylethylation at C-4 of **109** with allyl alcohol in the presence of palladium acetate, sodium bicarbonate, and benzyltriethylammonium chloride in DMF at 50 °C for 4 h to afford **110** in 90% yield (Scheme 33). A Wittig-Horner-Emmons olefination of this intermediate followed by iodination at C-3 afforded the precursor **111** for the second Heck reaction. Intramolecular Heck cycloarylation of **111** under standard conditions using palladium acetate, triphenylphosphine, and triethylamine led to a sluggish reaction and a significant amount of de-iodinated product along with small amounts of the desired tricyclic product **112.** Addition of silver phosphate (2.0 equiv) and triethylamine (0.5 equiv) to the palladium acetate-triphenylphosphine system in DMF co-operatively improved the yield of **112** to 80%. Use of silver carbonate and silver sulfate instead of silver phosphate

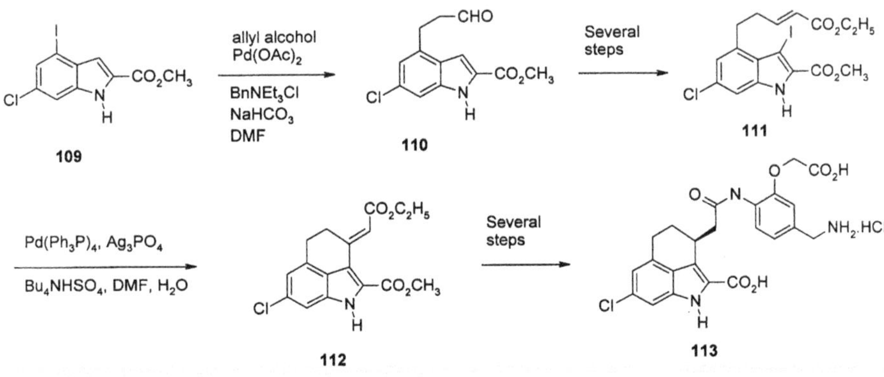

allyl alcohol
$Pd(OAc)_2$
$BnNEt_3Cl$
$NaHCO_3$
DMF
Several steps
$Pd(Ph_3P)_4$, Ag_3PO_4
Bu_4NHSO_4, DMF, H_2O
Several steps

109 **110** **111** **112** **113**

Scheme 33

afforded poorer yields. Replacement of triethylamine with tetrabutylammonium hydrogensulfate was also effective. Interestingly, addition of water significantly increased the reaction rate. Tetrakis(triphenylphosphine)palladium(0) was a suitable catalyst in this reaction. The reaction proceeded smoothly with excellent yield regardless of the presence of tetrabutylammonium hydrogensulfate. Anhydrous conditions resulted in the retardation of the reaction. This Heck cycloarylation was performed on a large scale under optimized conditions using 0.8 equiv silver phosphate, 2 mol% of tetrakis(triphenylphosphine)palladium(0), and 20 vol.% of water in DMF at 90 °C for 4 h to afford the desired tricycle **112** in 81% isolated yield.

5
Conclusion

Since its discovery about three decades ago, the Heck arylation reaction has emerged as a powerful "atom-economic" reaction, which tolerates a variety of functional groups. Utility of this reaction in the pharmaceutical industry in the synthesis of intermediates for the preparation of active pharmaceutical ingredients (APIs) has been demonstrated with a number of examples. Recent developments on the use of aromatic chlorides instead of iodides, development of stable and active catalysts towards achieving higher reaction rates, development of milder, ligand-free conditions, and the possibility of recovering and recycling the palladium catalyst make this reaction even more practical. In contrast to the fine-chemicals industry, the use of palladium-catalyzed reactions, such as the Heck arylation reaction, in the pharmaceutical industry presents a challenge because the palladium has to be removed from intermediates or the API to achieve a 2 ppm specification in the bulk API. This challenge, which is not special to the Heck reaction but common to all the palladium-catalyzed reactions, has been met successfully by developing various methods for palladium removal. Thus, the Heck reaction will continue to be a safe, ecologically friendly, efficient, and practical methodology for process chemistry in the pharmaceutical industry.

Acknowledgements I would like to thank Dr. Oljan Repič and Dr. Thomas J. Blacklock for their helpful suggestions.

References

1. Heck RF (1968) J Am Chem Soc 90:5518
2. Heck RF (1972) J Organomet Chem 37:389
3. Heck RF, Nolley JP (1972) J Org Chem 37:2320
4. Heck RF (1982) Org React 27:345
5. Heck RF (1991) Vinyl substitution with organopalladium intermediates In: Trost BM, Fleming I (eds) Comprehensive organic synthesis, vol 4. Pergamon Press, Oxford, UK, p 833
6. Overman LE (1994) Pure Appl Chem 66:1423
7. De Meijere A, Meyer FE (1994) Angew Chem Int Ed Engl 33:2379
8. Cabri W, Candiani I (1995) Acc Chem Res 28:2
9. Jeffery T (1996) Tetrahedron 52:10113

10. Shibasaki M, Boden CDJ, Kojima A (1997) Tetrahedron 53:7371
11. Crisp GT (1998) Chem Soc Rev 27:427
12. Amatore C, Jutand A (2000) Acc Chem Res 33:314
13. Franzen R (2000) Can J Chem 78:957
14. Beletskaya IP, Cheprakov AV (2000) Chem Rev 100:3009
15. Whitcombe NJ, Hii KK, Gibson SE (2001) Tetrahedron 57:7449
16. Yang C, Lee HM, Nolan SP (2001) Org Lett 3:1511
17. Beller M, Zapf A (1998) Synlett 793
18. Reetz M, Westermann E, Lohmer R, Lohmer G (1998) Tetrahedron Lett 39:8449
19. Gurtler C, Buchwald SL (1999) Chem Eur J 5:3107
20. Gruber AS, Pozebon D, Monteiro AL, Dupont J (2001) Tetrahedron Lett 42:7345
21. Littke AF, Fu GC (2001) J Am Chem Soc 123:6989
22. Littke AF, Fu GC (1999) J Org Chem 64:10
23. Ehrentraut A, Zapf A, Beller M (2000) Synlett 1589
24. Zapf A, Beller M (2001) Chem Eur J 7:2908
25. Buchmeiser MR, Schareina T, Kempe R, Wurst K (2001) J Organomet Chem 634:39
26. Herrmann WA, Bohm VPW, Reisinger CP (1999) J Organomet Chem 576:23
27. Hartung CG, Kohler K, Beller M (1999) Org Lett 1:709
28. Jeffrey T (1994) Tetrahedron Lett 35:3051
29. Basnak I, Takatori S, Walker RT (1997) Tetrahedron Lett 38:4869
30. Williams DBG, Lombard H, Holzapfel CW (2001) Synthetic Comm 31:2077
31. Carmichael AJ, Earle MJ, Holbrey JD, McCormac PB, Seddon KR (1999) Org Lett 1:997
32. Calo V, Nacci A, Monopoli A, Lopez L, Cosmo AD (2001) Tetrahedron 57:6071
33. Hagiwara H, Shimizu Y, Hoshi T, Suzuki T, Ando M, Ohkube K, Yokoyama C (2001) Tetrahedron Lett 42:4349
34. Shezad N, Oakes RS, Clifford AA, Rayner CM (2001) Chem Ind (Dekker) 82:459
35. Bhanage BM, Ikushima Y, Shirai M, Arai M (1999) Tetrahedron Lett 40:6427
36. Wall VM, Eisenstadt A, Ager D, Laneman SA (1999) Platinum Metals Rev 43:138
37. Kohler K, Wagner M, Djakovitch L (2001) Catal Today 66:105
38. Alper H, Arya P, Bourque SC, Jefferson GR, Manzer LE (2000) Can J Chem 78:920
39. Djakovitch L, Koehler K (2001) J Am Chem Soc 123:5990
40. Larhed M, Hallberg A (1996) J Org Chem 61:9582
41. Diaz-Ortiz A, Prieto P, Vazquez E (1997) Synlett 269
42. Villemin D, Caillot F (2001) Tetrahedron Lett 42:639
43. Deshmukh RR, Rajagopal R, Srinivasan KV (2001) Chem Commun 1544
44. Stephan MS, De Vries JG (2001) Chem Ind (Dekker) 82:379
45. De Vries JG (2001) Can J Chem 79:1086
46. Hallberg A, Westfelt L, Anderson C-M (1985) Synthetic Comm 15:1131
47. Drugs Fut (1997) 22:1103
48. Shinkai I, King AO, Larsen RD (1994) Pure Appl Chem 66:1551
49. King AO, Corley EG, Anderson RK, Larsen RD, Verhoeven TR, Reider PJ, Xiang YB, Belley M, Leblanc Y, Labelle M, Prasit P, Zamboni RJ (1993) J Org Chem 58:3731
50. Wu TC (1996) US Patent 5 536 870
51. Lin RW, Herndon RC, Allen RH, Chockalingham KC, Roy RK (1998) World Patent WO 98/30529
52. Herrmann WA, Beller M, Tafesh A (1999) US Patent 6 005 151
53. Beller M, Tafesh A, Herrmann WA (1996) German Patent DE 195 03 119
54. Ramminger C, Zim D, Lando VR, Fassina V, Monteiro AL (2000) J Braz Chem Soc 11:105
55. Andrus M, Meredith E (2001) World Patent WO 01/60774
56. Drugs Fut (1996) 21:476
57. Blatcher P, Carter M, Hornby R, Owen MR (1995) World Patent WO 95/09166
58. Oxford AW, Butina D, Owen MR (1993) European Patent EP 0 303 507
59. Drugs Fut (1997) 22:221

60. Orita A, Katakami M, Yasui Y, Kurihara A, Otera J (2001) Green Chem 3:13
61. Coker GG, Findlay JWA (1989) European Patent EP 0 085 959
62. Findlay JWA, Coker GG (1987) US Patent 4 650 807
63. Findlay JWA, Coker GG (1985) US Patent 4 501 893
64. Drugs Fut (2000) 25:445
65. Or YS, Clark RF, Wang S, Chu DTW, Nilius AM, Flamm RK, Mitten M, Ewing P, Alder J, Ma Z (2000) J Med Chem 43:1045
66. Drugs Fut (1998) 23:498
67. Taylor EC, Liu B (1999) Tetrahedron Lett 40:5291
68. Barnett CJ, Wilson TM, Kobierski ME (1999) Org Proc Res Dev 3:184
69. Wallace DJ, Goodman JM, Kennedy DJ, Davies AJ, Cowden CJ, Ashwood MS, Cottrell IF, Dolling U-H, Reider PJ (2001) Org Lett 3:671
70. Fritch JR, Aslam M, Rios DE, Smith JC (1996) World Patent WO 96/40608
71. Fritch JR, Rios DE, Kohlpaintner CW, Aslam M (1996) Chem Ind (Dekker) 68:313
72. Song ZJ, Zhao M, Frey L, Lushi T, Chen CY, Tschaen DM, Tillyer R, Grabowski EJJ, Volante R, Reider PJ (2001) Org Lett 3:3357
73. Alabaster RJ, Cottrell IF, Hands D, Humphrey GR, Kennedy DJ, Wright SHB (1989) Synthesis 598
74. Bates RW, Gabel CJ (1993) Tetrahedron Lett 34:3547
75. Kehr C, Neidlein R, Engh RA, Brandstetter H, Kucznierz R, Leinert H, Marzenell K, Strein K, von der Sall W (1997) Helv Chim Acta 80:892
76. Chun J, He L, Byun HS, Bittman R (2000) J Org Chem 65:7634
77. Kucerovy A, Li T, Prasad K, Repič O, Blacklock TJ (1997) Org Proc Res Dev 1:287
78. Homan EJ, Tulp MTM, Nilsson JE, Wikstrom HV, Grol CJ (1999) Bioorg Med Chem 7:2541
79. Dube D, Fortin R, Frenette R, Friesen R, Guay D, Prescott S (1998) US Patent US 5817700
80. Trivedi BK, Stoeber TL, Stanfield RL, Essenburg AD, Hamelehle KL, Krause BR (1993) Bioorg Med Chem Lett 3:259
81. Penco S (1993) Chim Ind (Milan) 75:369
82. Cabri W, Candiani I, DeBernardinis S, Francalanci F, Penco S, Santi R (1991) J Org Chem 56:5796
83. Sheridan H, Frankish N, Farrell R (1999) Eur J Med Chem 34:953
84. Singh B, Bacon ER, Robinson S, Fritz RK, Lesher GY, Kumar V, Dority JA, Reuman M, Kuo GH, Eissenstat MA, Pagani ED, Bode DC, Bentley RG, Connell MJ, Hamel LT, Silver PJ (1994) J Med Chem 37:248
85. Sliskovic DR, Blankley CJ, Krause BR, Newton RS, Picard JA, Roark WH, Roth BD, Sekerke C, Shaw MK, Stanfield RL (1992) J Med Chem 35:2095
86. Bigge CF, Wu JP, Malone TC, Taylor CP, Vartanian MG (1993) Bioorg Med Chem Lett 3:39
87. Bell AS, Stephenson PT (1997) European Patent EP 0 796 858
88. Dong Y, Busacca CA (1997) J Org Chem 62:6464
89. Waters SP, Kozlowski MC (2001) Tetrahedron Lett 42:3567
90. Coates IH, Bell JA, Humber DC, Ewan GB (1987) US Patent 4 695 578
91. Iida H, Yuasa Y, Kibayashi C (1980) J Org Chem 45:2938
92. Drugs Fut (1994) 19:781
93. Wierenge W (1991) Drugs Fut 16:741
94. Tietze LF, Buhr W, Looft J, Grote T (1998) Chem Eur J 4:1554
95. Tietze LF, Grote T (1994) J Org Chem 59:192
96. Muratake H, Abe I, Natsume M (1994) Tetrahedron Lett 35:2573
97. Muhle H, Nozulak J, Cercus J, Kusters E, Beutler U, Penn G, Zaugg W (1996) Chimia 50:209
98. Macor JE, Ogilvie RJ, Wythes MJ (1996) Tetrahedron Lett 37:4289
99. Macor JE, Blank DH, Post RJ, Ryan K (1992) Tetrahedron Lett 33:8011

100. Chen CY, Lieberman DR, Larsen RD, Reamer RA, Verhoeven TR, Reider PJ, Cottrell IF, Houghton PG (1994) Tetrahedron Lett 35:6981
101. Chen CY, Lieberman DR, Larsen RD, Verhoeven TR, Reider PJ (1997) J Org Chem 62:2676
102. McClure KF, Danishefsky S (1993) J Am Chem Soc 115:5094
103. Tietze LF, Nobel T, Spescha M (1998) J Am Chem Soc 120:8971
104. Tietze LF, Nobel T, Spescha M (1996) Angew Chem Int Ed Engl 35:2259
105. Tietze LF, Schirok H (1999) J Am Chem Soc 121:10264
106. Tietze LF, Schirok H (1997) Angew Chem Int Ed Engl 36:1124
107. Hong CY, Overman LE (1994) Tetrahedron Lett 35:3453
108. Hong CY, Kado N, Overman LE (1993) J Am Chem Soc 115:11028
109. Wunsch B, Diekmann H, Hofner G (1995) Tetrahedron: Asymmetry 6:1527
110. Davies AJ, Taylor RJK (1992) Bioorg Med Chem Lett 2:481
111. Coe JW (1999) World Patent WO 99/55680
112. Inoue M, Furuyama H, Sakazaki H, Hirama M (2001) Org Lett 3:2863
113. Boger DL, Turnbull P (1998) J Org Chem 63:8004
114. Boger DL, Turnbull P (1997) J Org Chem 62:5849
115. Prashad M, Tomesch JC, Wareing JR, Larsen D, Fex HD (1992) Eur J Med Chem 27:413
116. Prashad M, Tomesch JC, Wareing JR, Smith HC, Cheon SH (1989) Tetrahedron Lett 30:2877
117. Anderson RC, Houlihan WJ, Smith HC, Villhauer EB (1989) US Patent 4 845 129
118. Waite DC, Mason CP (1998) Org Proc Res Dev 2:116
119. Fang FG, Bankston DD, Huie EM, Johnson MR, Kang MC, LeHoullier CS, Lewis GC, Lovelace TC, Lowery MW, McDougald DL, Meerholz CA, Partridge JJ, Sharp MJ, Xie S (1997) Tetrahedron 53:10,953
120. Katayama S, Ae N, Nagata R (2001) J Org Chem 66:3474

Topics Organomet Chem (2004) 6: 205–245
DOI 10.1007/978-3-540-36966-0

Palladium-Catalyzed Cross-Coupling Reactions in the Synthesis of Pharmaceuticals

Anthony O. King · Nobuyoshi Yasuda

Department of Process Research, Merck & Co. Inc., P.O. Box 2000, Rahway, NJ 07065, USA
E-mail: tony_king@merck.com, nobuyoshi_yasuda@merck.com

Abstract The use of Pd-catalyzed cross-coupling methods for the construction of C-C, C-N, and C-O bonds have increased exponentially over the years. The variety of mild and chemoselective coupling conditions available make these coupling protocols very desirable for the preparation of compounds with highly functionalized and complex molecular structures. Pharmaceutical companies have used these methods for preparing new drug candidates on a small scale as well as the manufacturing of approved drug substances on a commercial scale. In this chapter we will focus our attention mainly on the application of these coupling methods in relatively large-scale pharmaceutical applications.

Keywords Kumada · Negishi · Sonogashira · Migita · Stille · Suzuki · Miyaura · Mizoroki · Heck · Palladium · Cross-coupling

1 Introduction

Since Kumada and Corriu reported the first cross-coupling reaction in the 1970s [1], transition metals catalyzed cross-coupling methods have blossomed and have totally changed the landscape of organic synthetic chemistry [2]. These methods are complimentary to conventional methods and offer several advantages. In material science, for example, new materials such as electron conductive organic polymers and liquid crystals can be selectively and effectively prepared by these cross-coupling methods [3]. It is also true in the pharmaceutical industry. From a medicinal chemistry point of view, the ability to create carbon-carbon bonds under mild conditions provides many new avenues for designing new candidates. This advantage is further enhanced when it is combined with combinatorial approaches [4]. Since some of the palladium-catalyzed cross-coupling reactions are so rapid and mild, even carbon-11 ($t_{1/2}$ is 20.4 min) or fluorine-18 ($t_{1/2}$ is 110 min) labeled compounds can be successfully prepared for PET studies using these cross-coupling methods [5]. From a process chemistry viewpoint, the cross-coupling reactions allow us to develop more convergent processes and provide more flexibility in the process designs. Since convergent processes are generally preferred, the drug candidates can be divided up into smaller building blocks, which have similar complexities and are easier to prepare separately, and then couple together via these coupling methods.

This chapter is mainly devoted to palladium-catalyzed cross-coupling reactions used in industrial applications since 1990. References for this review have been selected from hits obtained from literature and patent searches using Beilstein and CAS. Since a tremendous amount of work has been published, we focused mainly on relatively large-scale pharmaceutical applications. This chapter is divided into 10 sections with the second through eighth sections segregated according to coupling methods. The ninth section focuses on specific issues in pharmaceutical applications using these cross-coupling methods. General information and key reviews are provided at the beginning of each section.

2 Magnesium-Mediated Cross-Coupling (Kumada Reaction)

The magnesium-mediated cross-coupling reaction was the first example of a nickel- and/or palladium-catalyzed cross-coupling reaction with organic halides and is referred to as the Kumada reaction to commemorate his discovery in the 1970s [1]. A lithium-mediated cross-coupling reaction was reported about the same time [1b]. Although still very effective in many synthetic schemes, one of the major drawbacks with the use of Grignard reagents in cross-coupling reactions is their lack of chemoselectivity. Due to the high nucleophilicity of Grignard reagents, limited functional groups can be tolerated on both of the coupling partners. Therefore, the Kumada reaction is not suitable for the synthesis of highly functionalized compounds. In this regard, an interesting high-yielding Kumada cross-coupling was reported with 4-bromobenzonitrile by taking advantage of the reversible nature of the Grignard additions with nitrile

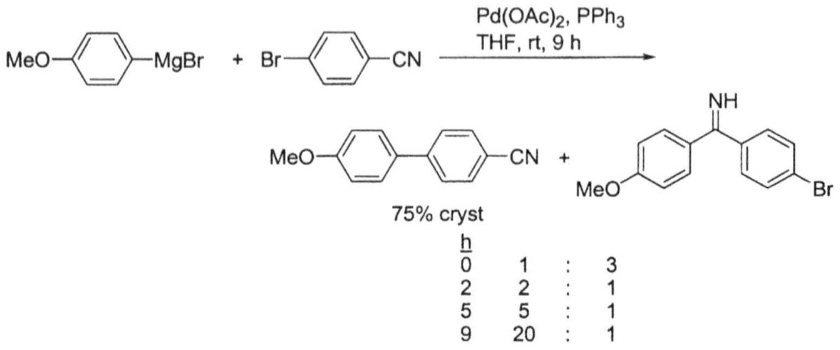

Scheme 1

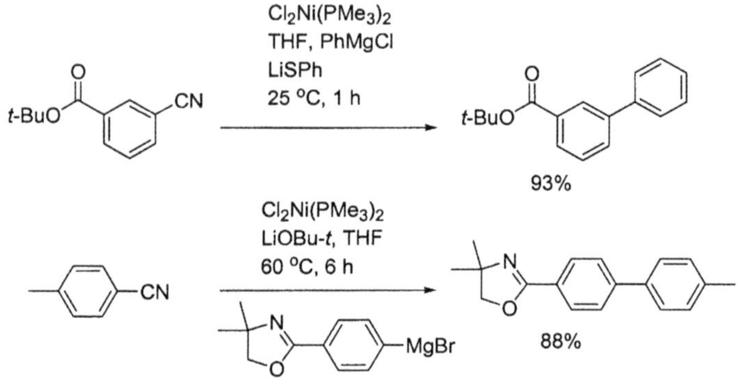

Scheme 2

(Scheme 1) [6]. Excellent yields were reported for the coupling of 4-bromobenzonitrile with a variety of arylmagnesium bromides using <1 mol% of the Pd catalyst. Interestingly, the aryl nitrile can also act as an electrophile under Kumada coupling conditions when $NiCl_2(PMe_3)_2$ is used as a catalyst [7]. This catalyst, in conjunction with the use of Grignard reagents and in the presence of a lithium alkoxide or sulfide, provides high yields of unsymmetrical biaryl products (Scheme 2).

2.1 Diflunisal from Merck

An anti-inflammatory agent, diflunisal, was developed by Merck in the 1970s. There are two patents from Merck for the preparation of this drug (Scheme 3) [8]. In both cases biaryl intermediates were prepared by a Gomberg-Bachmann reaction between the diazonium derivative of 2,4-difluoroaniline and benzene or anisole. Recently, chemists from Zambon Group reported a new process via the coupling of 4-methoxyphenylmagnesium bromide and 2,4-difluorophenyl bromide as a key step. The reaction proceeded well in the presence of 0.05 mol%

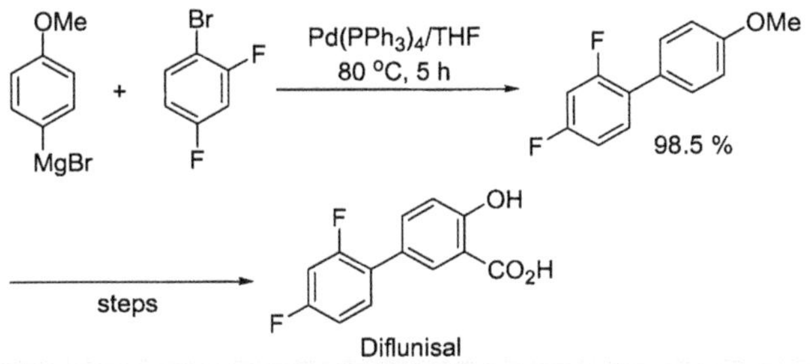

Scheme 3

of $Pd(PPh_3)_4$ and provided the biaryl intermediate in 98.5% isolated yield [9]. Nickel catalysts can also be used effectively for this coupling.

2.2
AG341: An Effective Inhibitor of Thymidylate Synthase from Agouron

A large-scale preparation of 5-methylbenz[*c*,*d*]indol-2(1*H*)-one, which is a key precursor of thymidylate synthase inhibitor AG341, was reported from Agouron (Scheme 4) [10]. The presence of the amide N-H proton did not hinder the coupling reaction.

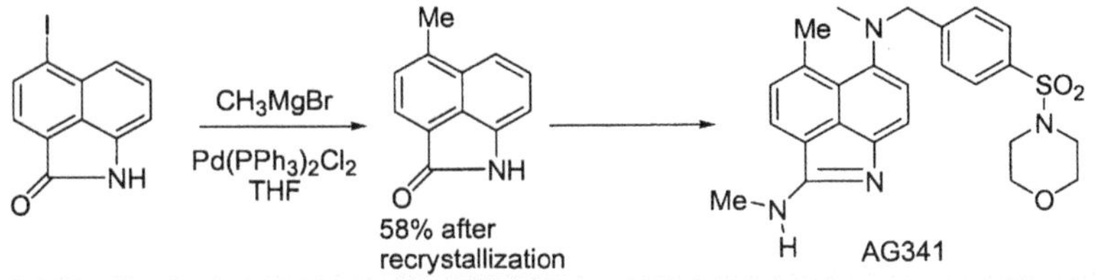

Scheme 4

The low chemoselectivity of Grignard reagents has prompted the development of other organometallics containing Al, Zr, Zn, Cd, Sn, B, and, the latest entry, Si in cross-coupling reactions. These methods are described in the following sections.

3
Zinc-Mediated Cross-Coupling (Negishi Reaction)

Organozinc reagents are one of the earliest carbon nucleophile examples in chemistry history. Compare to organolithiums or Grignard reagents, organozinc reagents are less reactive and can tolerate the presence of many functional groups. However, organozinc reagents still retain some nucleophilicity and exhibit different regioselectivities when used in the presence or absence of transition metals (see Schemes 5 and 6).

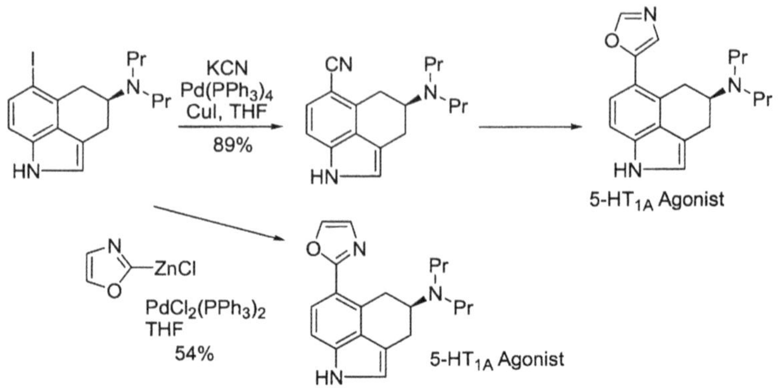

Scheme 5

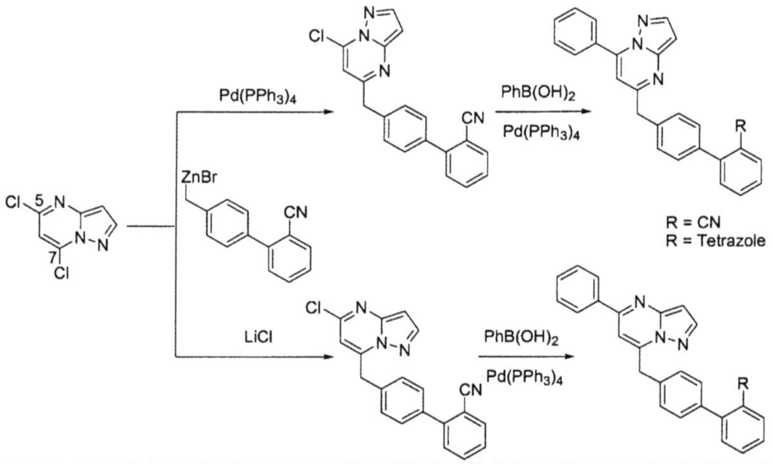

Scheme 6

The zinc-mediated cross-coupling reaction was first reported by Negishi [11]. Extensive studies on zinc reagents, in conjunction with copper and magnesium copper complexes, have been reported by Knochel [12]. For the preparation of organozinc reagents, the most straightforward method is the direct zinc metal insertion into carbon-halide bonds. The halides are typically organic iodides and bromides. In some cases organic chlorides can also be used. To increase the reactivity of zinc, Knochel reported the activation of metallic zinc with dibromoethane and TMSCl. More active metallic zinc was prepared by reduction of zinc salts and is commercially available as Rieke zinc [13]. Transmetalation from other carbon-metal bonds, such as carbon-lithium, carbon-magnesium, carbon-boron, carbon-zirconium, etc., is also a general method. Since organozinc reagents are not stable in the presence of water, the reagents are usually used immediately after their preparation without prior isolation. This could be a disadvantage for organozinc-mediated cross coupling on large scales.

3.1
A Potent 5-HT_{1A} Agonist from Eli Lilly

Preparation of a potent 5-HT_{1A} agonist (2-oxazolyl) was reported from Eli Lilly using Negishi cross coupling as a key reaction (Scheme 5) [14]. Another 5-oxazolyl derivative, which also possessed 5-HT_{1A} antagonist activity, was prepared by the palladium-copper mediated cross coupling with KCN followed by construction of the oxazole. This modified Sekiya-Ishikawa cyanation was dramatically milder than the original Rosenmund-von Brown procedure.

3.2
Angiotensin II Receptor Antagonists from Shionogi

A very interesting regioselectivity was reported on the preparation of angiotensin II receptor antagonists from Shionogi (Scheme 6) [15]. The cross-coupling reaction of the benzylzinc chloride with 5,7-dichloropyrazolo[1,5-*a*]pyrimidine in the presence of $Pd(PPh_3)_4$ provided the 5-benzyl derivative (53%) together with the 7-benzyl derivative (7%). On the other hand, the direct nucleophilic addition reaction in the presence of 2 equivalents of LiCl provided exclusively the 7-benzyl derivative in 54% yield. The products were further converted to drug candidates by a Suzuki-Miyaura reaction.

3.3
DPX-MY926 Cereal Herbicide from DuPont

For the preparation of a cereal herbicide, DPX-MY926, chemists from DuPont reported another interesting example of regioselectivity (Scheme 7) [16]. The palladium-catalyzed cross-coupling reaction between the benzylzinc chloride and 2,4-dichloropyrimidine gave the 4-coupled product as the major product. However, the same coupling with 4-chloro-2-thiomenthylpyrimidine or 2-chloro-4-thiomethylpyrimidine gave the 2-benzyl-4-thiomethylpyrimidine exclusively. According to their GC analysis of the reaction with 4-chloro-2-thiomethylpyrimidine, the intermediate was 2,4-dithiomethylpyrimidine. Oxidation to 2-methylsulfone prevented the methylthiol scrambling and the coupling reaction proceeded as expected on the 4-position.

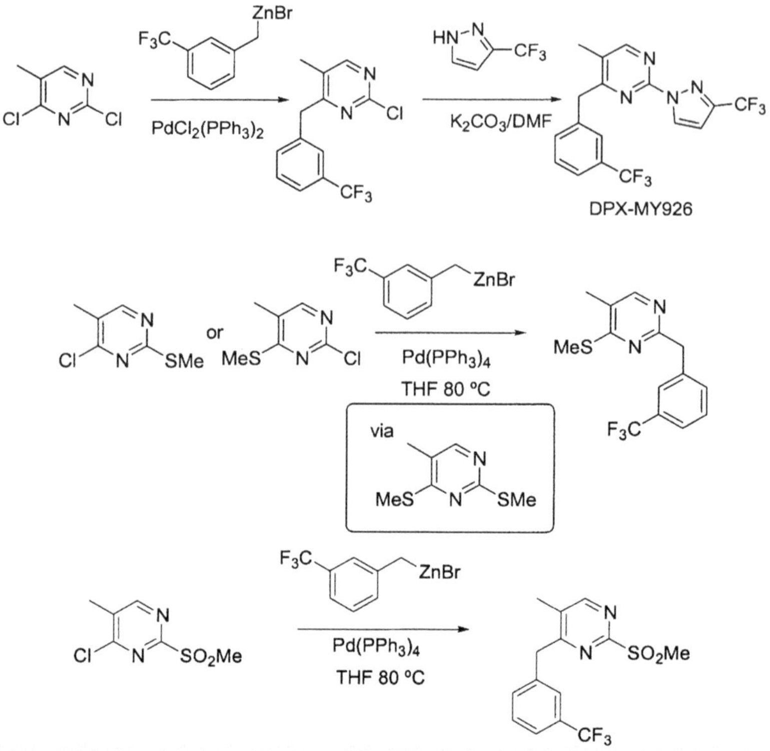

Scheme 7

3.4 Growth Hormone Secretagogue from Merck

The first major cross-coupling reaction undertaken at Merck was the preparation of the tetrazolylmethylbiphenyl fragment for the synthesis of the growth hormone secretagoge candidate (Scheme 8) [17].

The reaction sequence involved the tetrazole-directed *ortho*-lithiation of 2-trityl-5-phenyltetrazole, transmetalation of the resulting aryl lithium to slightly over 0.5 equivalents of $ZnCl_2$, followed by the nickel-catalyzed coupling of the diarylzinc reagent with 4-iodotoluene. This Negishi coupling proceeded excellently, providing the coupled product in 91% isolated yield. Pd catalysts also catalyzed this reaction as effectively. After a radical bromination step, the benzyl bromide was used as an electrophile for the alkylation of the diazepam core nitrogen.

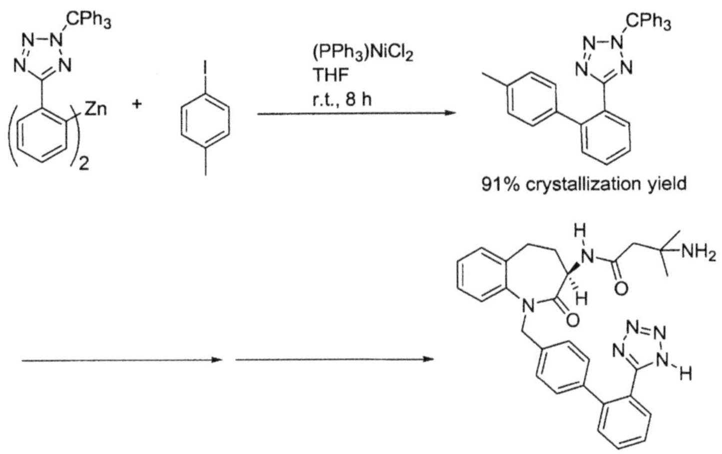

(PPh3)NiCl2
THF
r.t., 8 h

91% crystallization yield

Scheme 8

4 Tin-Mediated Cross-Coupling (Migita-Stille Reaction)

The palladium-catalyzed cross-coupling reaction with organotin reagents was independently reported by Migita-Kosugi [18] and Stille [19]. Many reviews have been published on this coupling [20]. A distinct advantage of this reaction is the mildness of the reaction conditions. The Sn-C bond is one of the most covalent bonds as compared to other cross-coupling partners such as Li-C, Mg-C, Al-C, Zn-C, and B-C. This suggests that Sn-C compounds are the least nucleophilic partners among the group. Organostannanes can be prepared by transmetalation of C-Li or C-MgX with R_3SnX, nucleophilic substitution of R_3SnLi, hydrostannation of olefins with R_3SnH, or palladium-catalyzed coupling of Ar-X with R_3SnSnR_3 [21]. Organostannanes are air and water stable and isolable by standard isolation techniques. Not only are sp and sp^2 carbons transmetalated from organotins but sp^3 carbon can also participate in transmetalations in many cases. This versatile coupling is well suitable in medicinal chemistry where quick assembly of building blocks is one of the more important aspects.

There are some problems associated with the Migita-Stille coupling. The biggest issue is the toxicity of organostannanes which makes it undesirable especially on a process scale. Removal of organostannane by-products (typically R_3SnOH) from the coupled product can sometimes be difficult. When fluorous tin was used for the coupling, the tin residue was effectively extracted into perfluorinated solvents and can be recycled [22]. The third issue is that only one residue out of four is utilized for coupling. The remaining three residues are dummy groups and therefore this reaction is inherently of poor atom economy. In addition, unless the organotins have four identical residues, other carbon residues can competitively transmetalate and create a chemoselectivity issue. Butyl and methyl are generally used as dummy groups since transmetalation of sp^3

carbon is slower than other hybrid carbons. However, some butyl and methyl transfers are observed in many cases. In order to avoid this undesirable chemoselectivity and to increase the transmetalation rate, pentacoordinated (typically amine) organostannanes have been reported to give facile cross coupling [23]. In these cases, the carbon substituent *trans* to the amine moiety is selectively transferred since that particular C-Sn bond has the longest bond length and, therefore, is the weakest bond.

4.1
Anti-MRS Carbapenem from Merck

A new anti-MRS carbapenem candidate, designed based on a releasable side chain concept, was reported (Scheme 9). The key reaction was the hydroxymethylation of the 3-trifluoromethanesulfonyloxy carbapenem with Bu_3SnCH_2OH [24]. The crystalline 2-hydroxymethyl carbapenem was a versatile intermediate and was converted to the target compound via a Tsuji-Trost reaction [25].

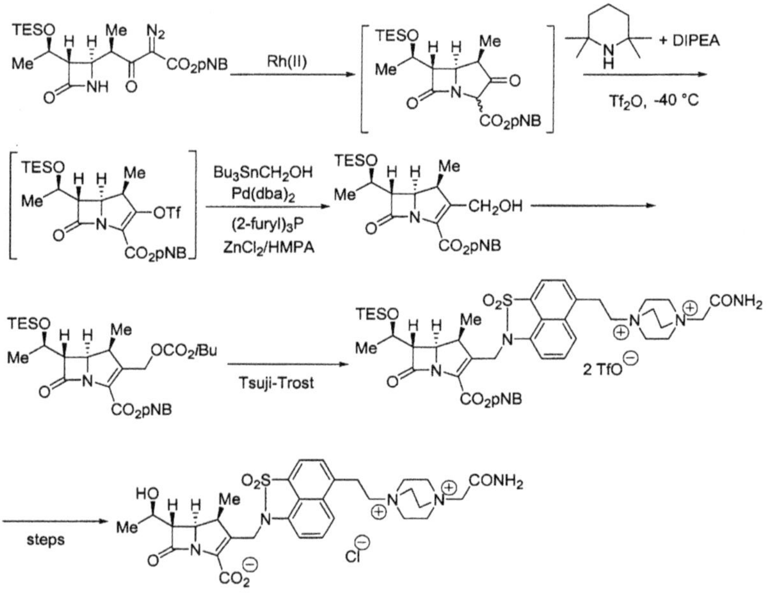

TESO H H Me N₂ CO₂pNB Me NH O — Rh(II) → [TESO H H Me Me N =O CO₂pNB] — + DIPEA, Tf₂O, -40 °C →

[TESO H H Me Me N OTf CO₂pNB] — Bu_3SnCH_2OH, $Pd(dba)_2$, $(2\text{-furyl})_3P$, $ZnCl_2$/HMPA → TESO H H Me Me N CH_2OH CO_2pNB →

TESO H H Me Me N OCO_2iBu CO_2pNB — Tsuji-Trost → TESO H H Me Me N O_2S N CO_2pNB N N CONH₂ 2 TfO⊖

— steps → HO H H Me Me N O_2S N $CO_2^{\ominus}$ N N CONH₂ Cl⊖

Scheme 9

An alternative one step coupling process has also been disclosed (Scheme 10) [26]. This process utilized the weaker carbon-tin bond in the stannatrane derivative. The side chain, naphthasultam, is an electron-withdrawing group such that the bond between the α-carbon of the nitrogen and the tin atom in the stannatrane is longer than that in the corresponding Sn-methyl stannatrane. The coupling proceeded smoothly to provide the product in near quantitative yield, together with quantitative recovery of the crystalline stannatrane chloride after

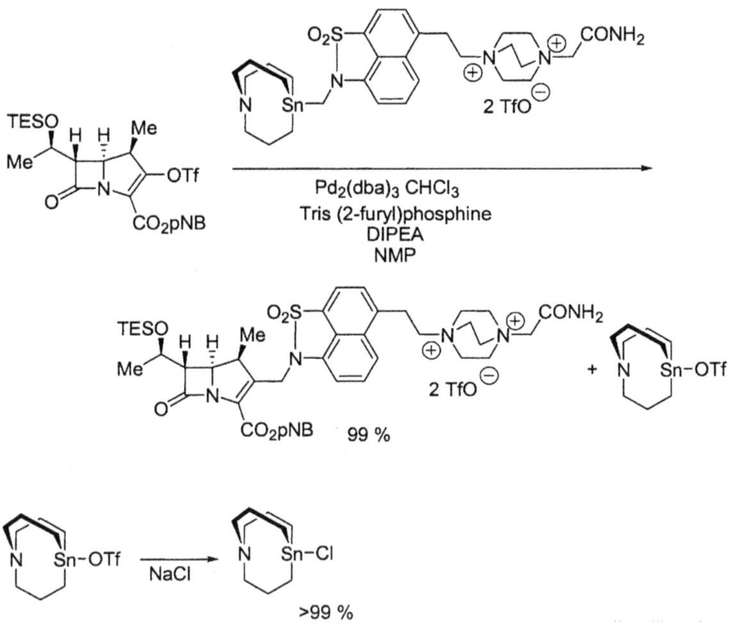

Scheme 10

work-up. This is the first cross coupling example with an sp^3 carbon bearing nitrogen atom at the α-position.

4.2
cis-Cefprozil from Bristol-Meyers Squibb

cis-Cefprozil, an orally active antibiotic, was launched by Bristol-Meyers Squibb (Scheme 11). The first preparation of this drug by Naito was based on the Wittig reaction which had a serious *E*/*Z* selectivity problem [27]. Later, Farina and his co-workers at Bristol-Myers Squibb applied the Migita-Stille coupling reaction,

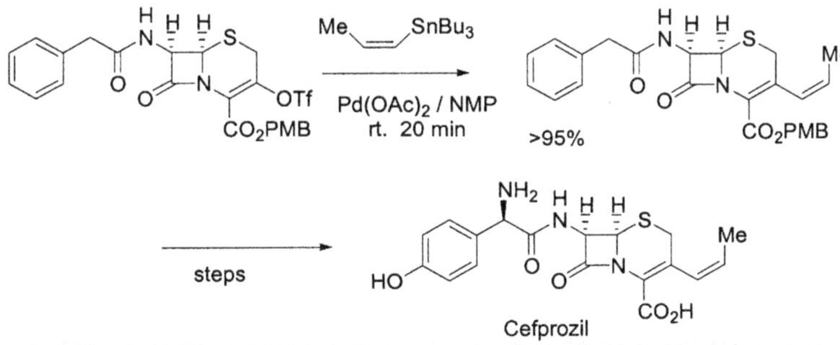

Scheme 11

which took advantage of the retention of *cis* configuration of the vinylstannane reagent. This is the beginning of cross-coupling reactions in the β-lactam field [28]. Ferina's pioneering work led to the discovery of tri-(2-furyl)phosphine as a ligand and his kinetic studies with Liebeskind is the gold standard in cross-coupling reactions [20b]. Subsequently, Roth and his co-workers at Bristol-Myers Squibb reported an effective Migita-Stille coupling using ligandless palladium in NMP [29]. Eventually, *cis*-Cefprozil was prepared by a vinyl cuprate addition to an allenylazetidinone, which was readily prepared from penicillin [30], thus avoiding the use of toxic organostannanes.

4.3
3-(1,3-Butadienyl)cephalosporins from Yamanouchi

A short publication on 3-(1,3-butadienyl)cephalosporins was reported from Yamanouchi in 1994 (Scheme 12) [31], but there were no specific biological activities with these derivatives. In this case, an allyl chloride was the coupling partner. The *E*-configuration of the coupled product was critical for the elimination of the corresponding phosphate, which resulted in the formation of the 1,3-butadienyl functional moiety.

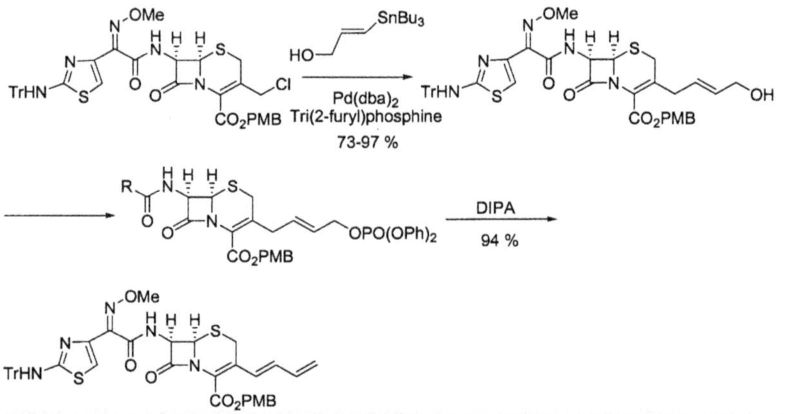

Scheme 12

4.4
2-Aryl and 2-Heteroaryl Penems from SmithKline Beecham

A series of 2-aryl and 2-heteroaryl penems were prepared from the 2-stannyl penem by SmithKline Beecham (Scheme 13). The preparation of the 2-stannyl penem itself was very interesting and worth reviewing here [32].

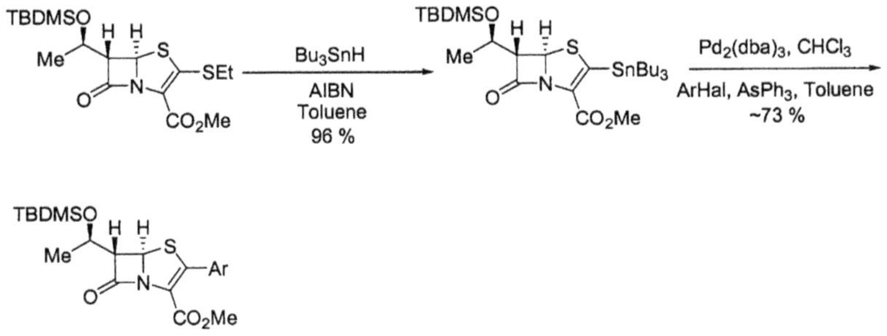

Scheme 13

4.5
C-3 Alkyne-Substituted Cephalosporins from Fujisawa

An acetylene analog of Cefixime was prepared by a tin-mediated alkynyl cross coupling (Scheme 14) [33]. In this case a vinyl mesylate (3-mesyloxy cephalosporin) was the coupling partner. The analog exhibited only low antibacterial activity and poor oral bioavailability. Chemists from Fujisawa also reported the preparation of 3-[(*E*) and (*Z*)-2-substituted vinyl]-cephalosporins and their oral activities [34].

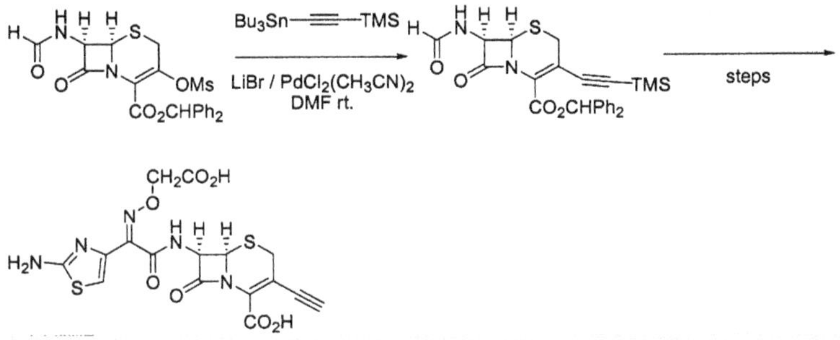

Scheme 14

4.6
A Factor VIIa Inhibitor Discovered by Ono Pharmaceutical Co.

Relying on a Migita-Stille coupling, chemists from Parke-Davis Pharmaceutical Research reported an efficient synthesis of a Factor VIIa Inhibitor which was discovered initially by Ono Pharmaceutical Co. (Scheme 15) [35]. This inhibitor has the potential for treating disease states associated with the extrinsic system such as acute myocardial infarction, stroke, disseminated intravascular coagulation, and thrombolytic disease. The arylstannane was first prepared via a palladium-catalyzed coupling and then utilized in a subsequent cross-coupling.

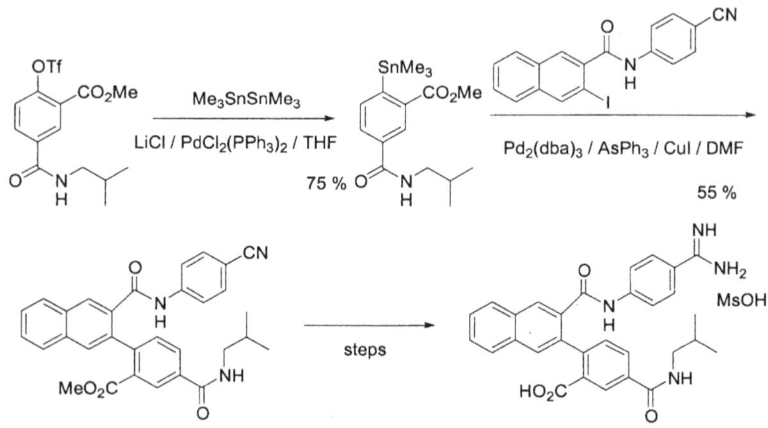

Scheme 15

4.7 ^{11}C Labeled MADAM

PET has become a very effective diagnostic tool. ^{11}C labeled *N,N*-dimethyl-2-(2-amino-4-methylphenylthio)benzylamine (MADAM), used for diagnosing serotonin transporter (5-HTT), was prepared by the Migita-Stille coupling (Scheme 16). With $t_{1/2}$ of ^{11}C being 20.2 min, a short reaction time is a necessity. The reaction was complete in only 7 min at 120 °C [36].

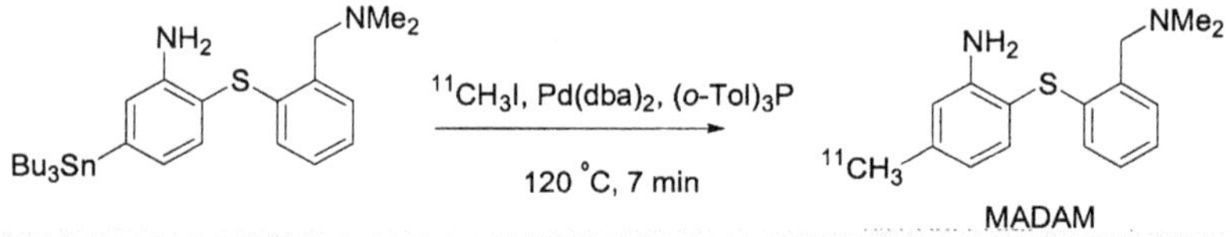

Scheme 16

5 Boron Mediated Cross-Coupling (Suzuki-Miyaura Reaction)

The palladium-catalyzed cross-coupling reaction with borinic and boronic acid derivatives is one of the most well-established cross-coupling methods and has provided number of advantages over other methods. This reaction is referred to as the Suzuki-Miyaura reaction [37].

Boronic acids are generally air and moisture stable and amenable to isolation. In situ generated borate complexes can also be used (see Schemes 26 and 28). The reaction requires excess base or a fluoride to form the borates, which are electronically richer, thus facilitating the transmetalation step in the catalytic cycle. The reaction conditions are usually mild and many functional groups can be tolerated. Purification of products is relatively easy, since the byproduct, boric acid, can be removed by a simple aqueous base extraction. Base selection is critical for the success of the Suzuki-Miyaura reaction [38]. The reaction can be run in homogeneous or biphasic conditions. Furthermore, boronic acids and their boric acid byproduct are believed to be innocuous when compared to organostannanes.

General preparative methods for boronic acids are transmetalation from organometals with trialkylborate, such as $B(O\textit{i}\text{-}Pr)_3$, or hydroboration followed by solvolysis [39]. Recently, Ishiyama and Miyaura reported the preparation of pinacol arylborates from the corresponding arylhalides by a palladium-catalyzed coupling with bis(pinacolato)diboron [40]. The reagent is now commercially available but quite expensive ($93/g from STREM) for production scale. This economical problem was somewhat overcome by the report from Murata and Masuda using the more affordable pinacolborane. [41]. Selection of the appropriate base (KOAc or Et_3N) was critical for these boronic acids syntheses. Tandem couplings are also reported (Scheme 17) [42]. The coupling of an organic electrophile (halide, triflate and others) with diboronate allows the selective cross-coupling of two different organic halides in a stepwise fashion. In many cases the catalytic coupling system for the preparation of the organoboron can be used in the second coupling.

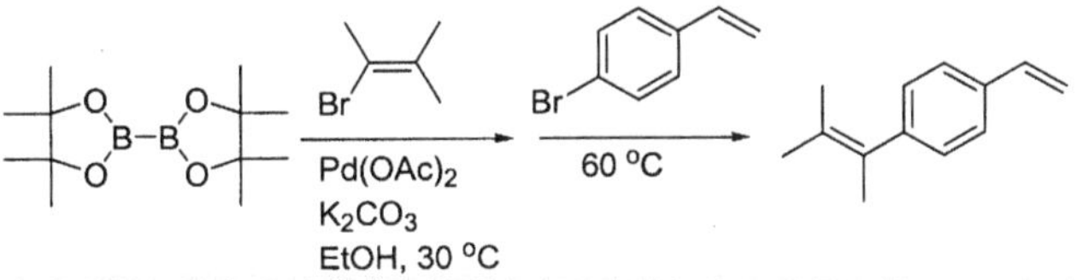

Scheme 17

Iridium complexes can also catalyze the direct borylation of arenes under mild reaction conditions and thus obviate the need for aryl halides or triflates (Scheme 18) [43]. Under the reaction conditions borylation prefers to occur at the *m*- and *p*-positions of both electron-rich and electron-poor monosubstituted arenes in statistical ratios of ca. 2 to 1 and both borons are incorporated into the arenes. Borylation at the ortho positions is negligible.

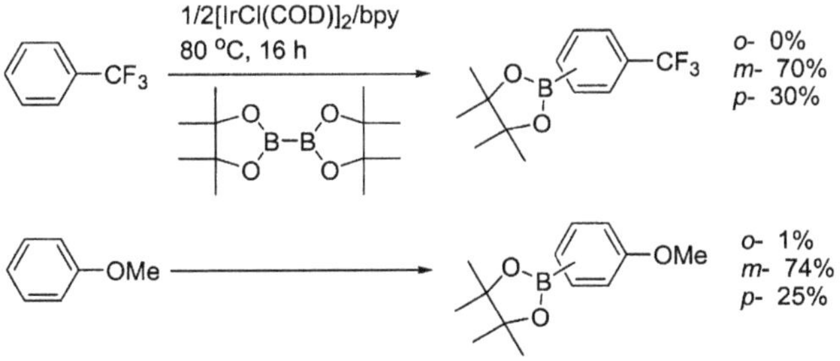

Scheme 18

Most commonly employed coupling electrophiles are aryl, alkenyl, and alkynyl halides and triflates. The boronic acids generally bear sp, sp^2, and sp^3 carbon-boron bonds. The most widely used boronic acids are sp^2 carbon-bearing compounds such as aryl and alkenyl boronic acids. Boronic acids bearing an sp carbon are unstable towards isolation and are typically prepared in situ. Suzuki-Miyaura coupling with sp^3 carbon bearing boronic acids is one of the most difficult classes of coupling. For this coupling, 9-BBN derivatives are used in order to increase electron density at the boron atom [37d]. Generally, organic halides are limited to iodo and bromo compounds. Recently, economically more affordable organic chlorides can also be utilized when sterically bulky phosphine ligands, such as Cy-MAP1 [44] and tris(*tert*-butyl)phosphine, are used as ligands [45]. Fu reported the first successful Suzuki-Miyaura cross coupling between sp^3 carbons using PCy_3 as the ligand [46].

The benefits resulting from the addition of a phase-transfer agent to Suzuki-Miyaura coupling reaction has not been fully investigated. Under two-phase coupling conditions, high rpm mixing may be very critical to obtain an efficient reaction. This is especially important in a large reactor. It has been reported that by using aliquot 336 even slower speed mixing is sufficient for coupling (Scheme 19) [47]. This issue can also be overcome through the use of an organic

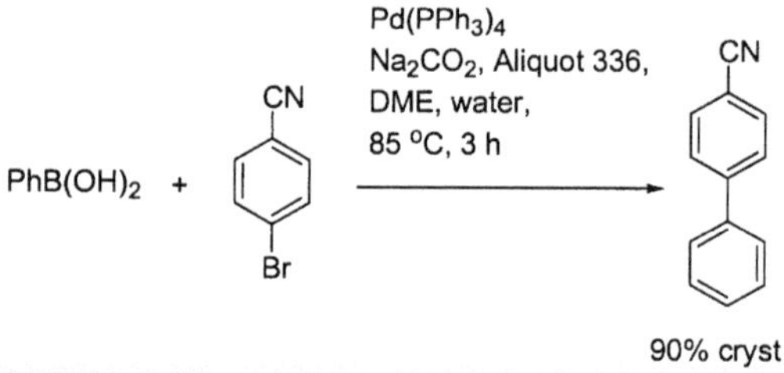

Scheme 19

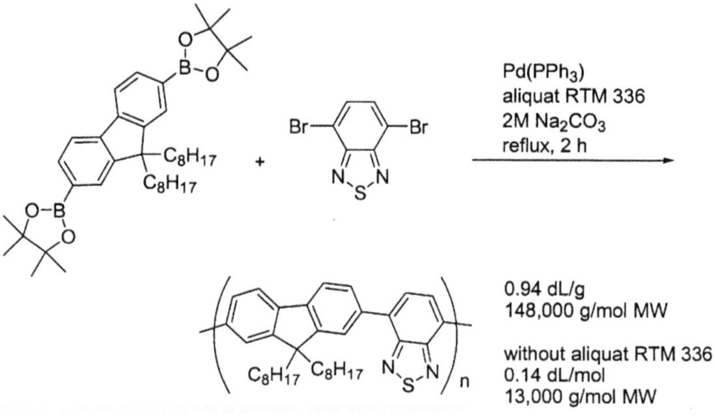

Scheme 20

solvent soluble base, such as tetraalkylammonium carbonate or an amine, and use of water only as a reagent rather than a solvent.

In another report, the presence of aliquot 336 also provides a polymer of substantially higher molecular weight than when it is absent (Scheme 20) [48].

5.1 Losartan and Related Drugs

Losartan is an angiotensin II receptor antagonist developed by DuPont and Merck [49]. Similar to losartan, candesartan cilexetil [50], irbesartan [51], valsartan [52], HR-720 [53] and compounds from Shionogi [15] are members of the sartan family (Fig. 1).

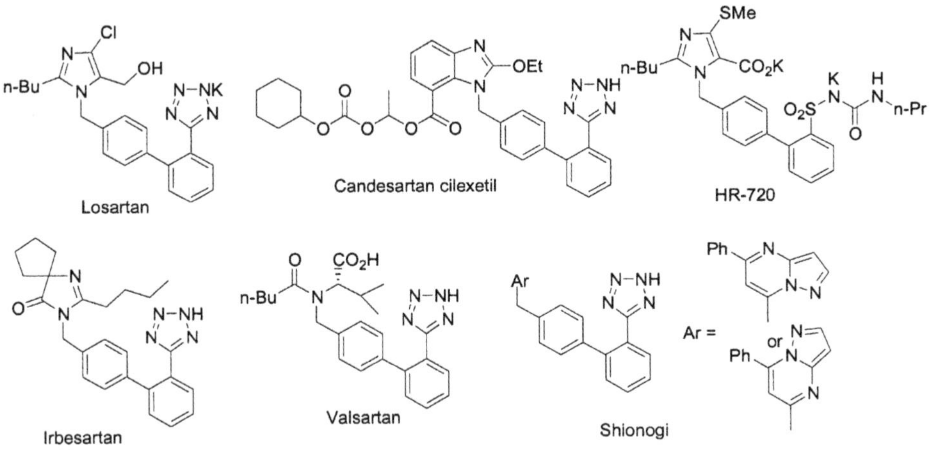

Fig 1 Sartan family

The original approaches to prepare the biaryl bond involved Ullmann coupling or nucleophilic aromatic substitution along with a number of subsequent steps to incorporate the requisite heterocycles. By taking advantage of the newly discovered directed *ortho*-metalation on 5-phenyltetrazole, all of the above steps were avoided by coupling the *ortho*-substituted boronic acid of trityl-protected phenyltetrazole directly with the 4-bromobenzylated imidazole to afford the biphenyl product in one step. This is the first Suzuki-Miyaura application in the pharmaceutical industry (Scheme 21).

HR-720 was prepared a little differently from losartan. The imidazole moiety in HR-720 was constructed subsequent to the Suzuki-Miyaura coupling reaction (Scheme 22).

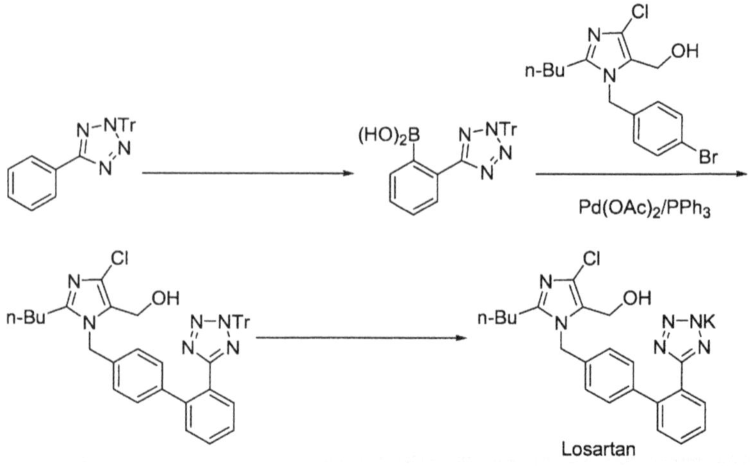

Scheme 21

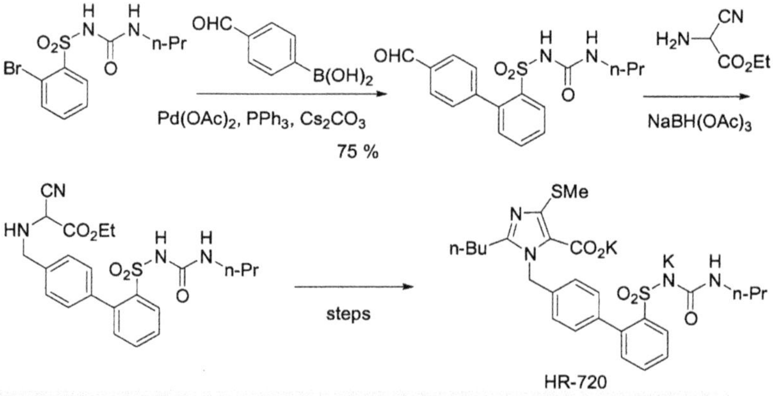

Scheme 22

5.2
Anti-MRS Carbapenem Candidate from Merck

A series of anti-MRS 2-aryl substituted carbapenem candidates were reported from Merck. The latest candidate was prepared via a Suzuki-Miyaura cross coupling as the key step (Scheme 23) [54]. The cross-coupling approach provided a more convergent process where fewer reaction steps were required to complete the synthesis after side chain installation onto the carbapenem skeleton. This is especially critical since the carbapenem intermediates are typically unstable. In this case, the candidate was prepared in high yield via two deprotection steps after the key cross coupling. This cross coupling was one of the most elaborate examples and exhibited the inherent advantage of the Suzuki-Miyaura coupling.

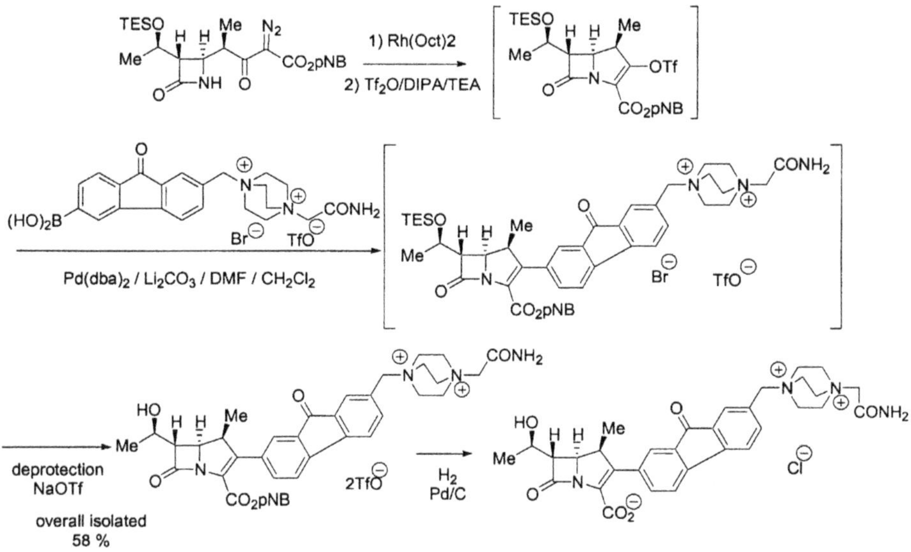

Scheme 23

5.3 Dethiacarba Analogs of Clinically Useful Carbapenems from Shionogi

Shionogi reported dethiacarba analogs of clinically useful carbapenems such as imipenem, panipenem, biapenem, meropenem, and S-4661 (Scheme 24) [55]. These analogs were less potent than the original drugs against both Gram-positive and Gram-negative bacteria. The key step in their syntheses was the hydroboration of olefin derivatives with 9-BBN followed by a Suzuki-Miyaura coupling with a carbapenem triflate.

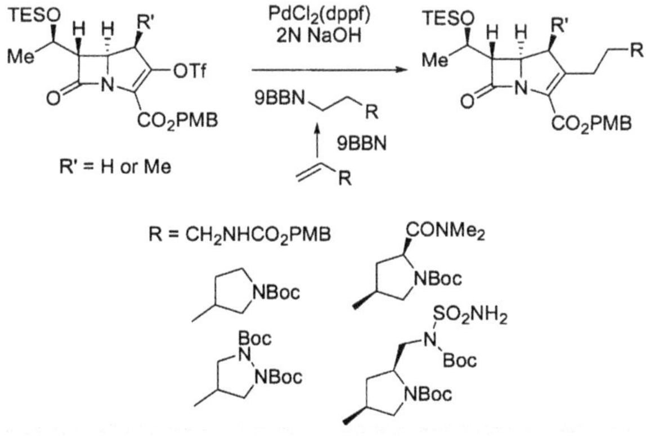

Scheme 24

5.4 SB 242784 from SmithKline Beecham

SmithKline Beecham reported the preparation of SB 242784, a compound in development for the treatment of osteoporosis, by using four sequential palladium-catalyzed reactions (Scheme 25) [56]. The first two reactions were Sonogashira reactions which established a ynenyl boronic ester in high efficiency. The next palladium-catalyzed reaction was the Suzuki-Miyaura coupling. The reaction was observed to give the product without generating double bond isomers and did not require protection of the nitrogen throughout this process. The resulting dienyne was cyclized to an indole in the presence of $Pd(CH_3CN)_2Cl_2$. Further modifications successfully concluded the preparation of SB 22784.

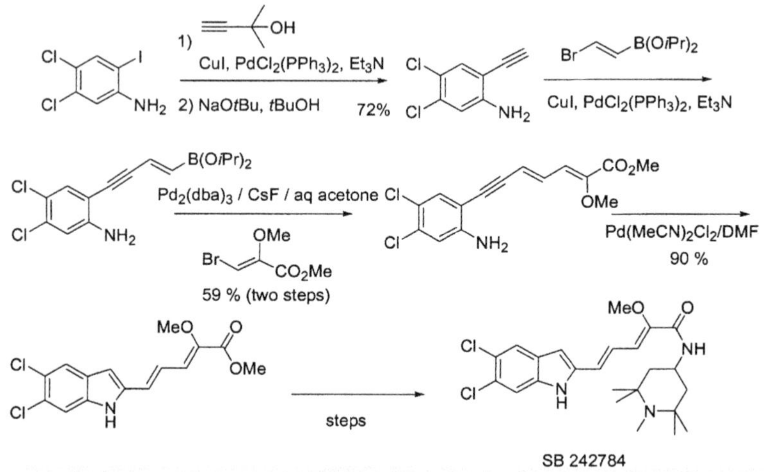

Scheme 25

5.5 NK-104 from Nissan Chemical Industries

Preparation of a highly potent HMG-CoA reductase inhibitor NK-104 was reported from Nissan Chemical Industries (Scheme 26) [57]. The vinyl boron compound was prepared by hydroboration of an alkyne with excess disiamylborane. The excess disiamylborane was quenched with EtONa in EtOH prior to the Suzuki-Miyaura coupling. The reaction conditions were optimized and the best result (99% yield) was obtained when allylpalladium chloride was used in acetonitrile.

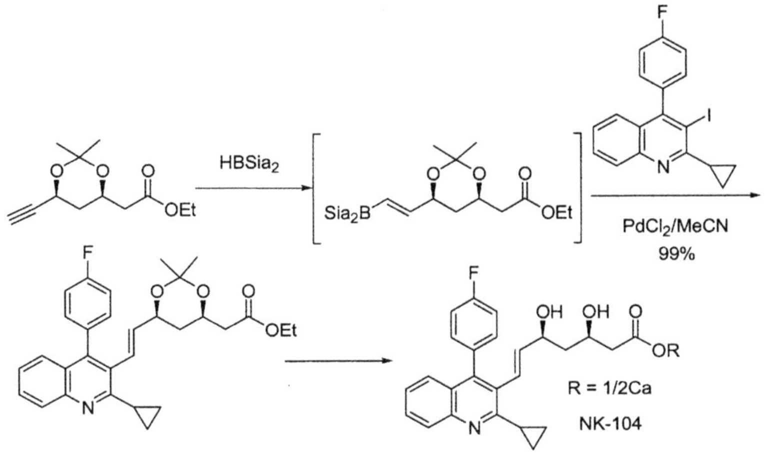
HBSia$_2$
OEt
Sia$_2$B
OEt
F
I
N
PdCl$_2$/MeCN
99%
F
OEt
N
F
OH OH O
OR
R = 1/2Ca
NK-104
N

Scheme 26

5.6 Rofecoxib from Merck

The first generation of COX-II inhibitor, rofecoxib, could be prepared in many ways. One of the processes utilized the Suzuki-Miyaura coupling as the key step (Scheme 27) [58].

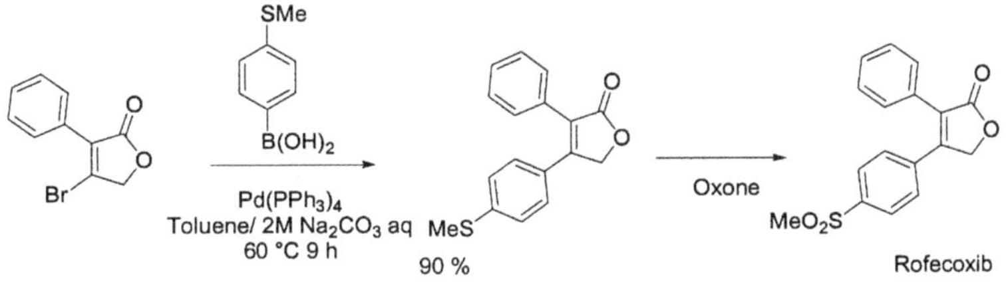
SMe
B(OH)$_2$
Br
Pd(PPh$_3$)$_4$
Toluene/ 2M Na$_2$CO$_3$ aq
60 °C 9 h
MeS
90 %
Oxone
MeO$_2$S
Rofecoxib

Scheme 27

5.7 Etorocoxib from Merck

The original route for the second generation COX-II inhibitor, etorocoxib, utilized the Suzuki-Miyaura reaction twice (Scheme 28) [59]. The first cross coupling with 3-bromo-5-chloro-2-aminopyridine took place at the bromide carbon. The amino group of the product was converted to a bromide under Sandmeyer conditions. The resulting 2-bromopyridine was coupled with 2-methylpyridine-5-borate to provide etorocoxib. An alternative preparation of etorocoxib was also reported [60].

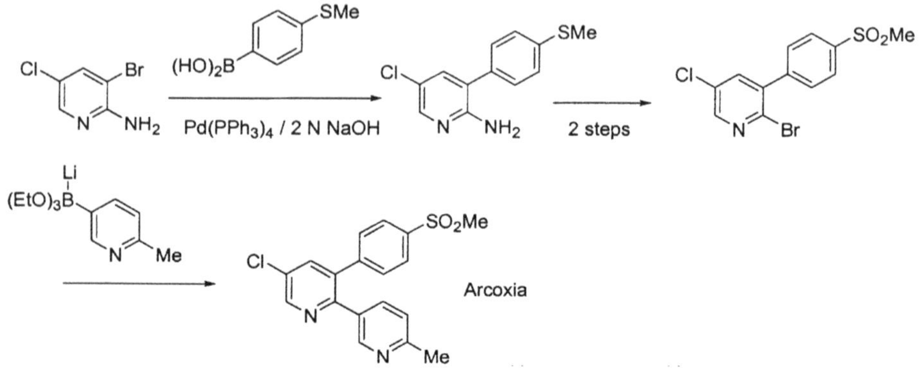

Scheme 28

5.8
Abiraterone Acetate

The steroidal abiraterone acetate, a prodrug for abiraterone, is a potent inhibitor of human cytochrome $P450_{17\alpha}$. Abiraterone acetate has been approved for clinical trials in patients with hormone-dependent prostatic carcinoma. The key reaction is a Suzuki-Miyaura cross coupling between diethyl(3-pyridyl)borane

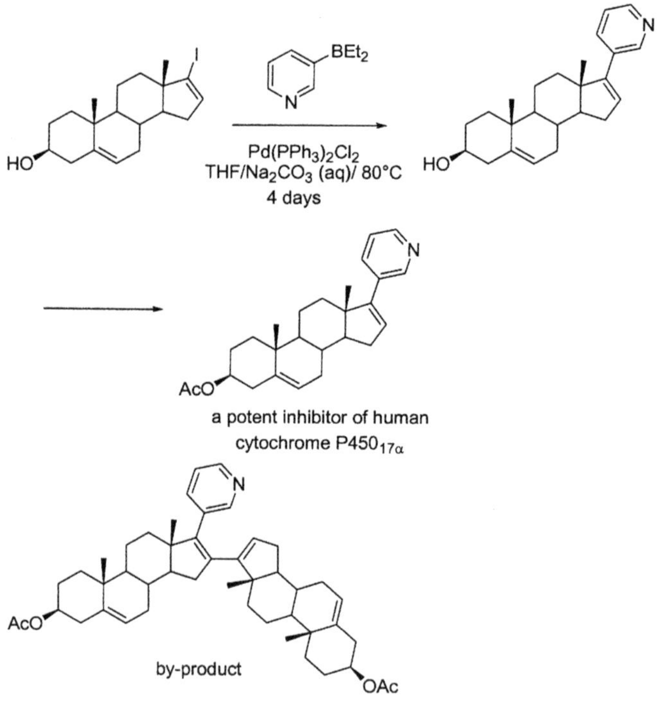

Scheme 29

and the 17-enol iodide (Scheme 29). The reaction required four days at 80 °C. The major by-product was the Heck-Mizorogi product. The by-product was removed from the product after *O*-acylation and reverse phase silica gel chromatography [61].

6 Alkyne Cross-Coupling (Sonogashira Reaction)

Currently, the cross-coupling reaction for forming a sp^2-sp carbon bond is one of the mildest and most successful methods in this field. Previous reaction conditions involving copper acetylides, the Castro-Stephens reaction, are harsh and require a stoichiometric amount of copper [62]. Functional group tolerability is also limited. Furthermore, only aryl iodides can be used in this reaction. The reaction is dramatically improved by the addition of a palladium catalyst. This palladium and copper dual-catalyst system in the presence of an amine base is very efficient for preparing alkynes and is now referred to as the Sonogashira reaction [63]. The success of this coupling relies on the higher acidity of the alkynyl proton. Therefore, unlike other methods, there is no need to activate the nucleophile cross-coupling partner. For instance, Negishi, Suzuki-Miyaura, and Migita-Stille reactions require preparing C-Zn, C-B, and C-Sn bonds, respectively, prior to their coupling reactions. It is interesting to note that copper acetylide is formed in situ during the catalytic cycle. It is recognized that the number of publications for this reaction has increased exponentially since 1993 [64]. This method is also very effective for preparing liquid crystals and electron conductive organic polymers [65]. Practically, this is the easiest method to create carbon-carbon bonds, and the resulting triple bond can be readily transformed to other functional moieties. Cross-coupling using alkynylboronates, alkynylstannanes, and zinc acetylides are complimentary methods [63].

6.1 Tazarotene from Allergan

Tazarotene is a member of the acetylenic retinoids. It has been commercialized by Allergan, Inc. and approved for the treatment of psoriasis and acne. It was prepared via the palladium-catalyzed cross coupling between a zinc acetylide and ethyl 6-chloronictoninate (Scheme 30) [66].

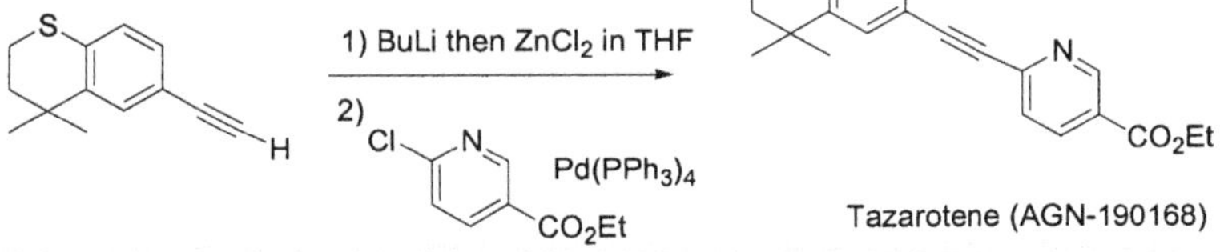

1) BuLi then $ZnCl_2$ in THF
2) Cl, N, CO_2Et, $Pd(PPh_3)_4$

Tazarotene (AGN-190168)

Scheme 30

6.2
An HIV Protease Inhibitor

The key building block of an HIV protease inhibitor candidate, the chlorinated furopyridine, was effectively prepared via the Sonogashira reaction (Scheme 31) [67]. The reaction resembled the Larock indole synthesis [68]. The structure of the candidate is quite similar to that of Crixivan. The only difference is this candidate has the furopyridine instead of a pyridine in Crixivan.

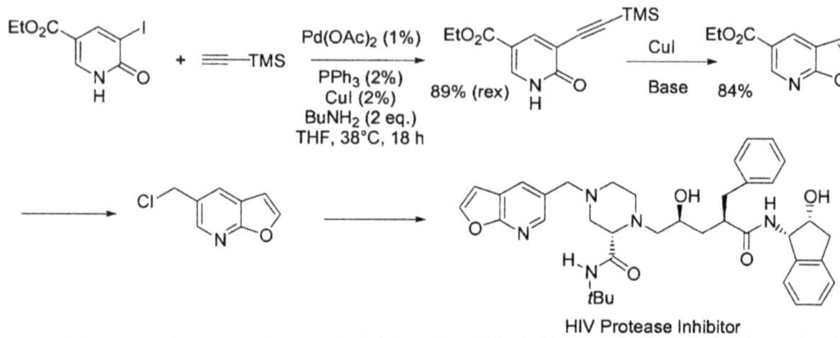
EtO2C
TMS
Pd(OAc)2 (1%)
PPh3 (2%)
CuI (2%)
BuNH2 (2 eq.)
THF, 38°C, 18 h
89% (rex)
CuI
Base
84%
Cl
OH
tBu
HIV Protease Inhibitor

Scheme 31

6.3
SIB-1508Y

SIB-1508Y is a nicotinic acetylcholine receptor antagonist and is developed for the potential treatment of Parkinson's disease, Alzheimer's disease, attention deficit hyperactivity disorder, Tourette's syndrome, and schizophrenia. The structure of SIB-1508Y is 5-ethynylnicotine. The coupling reaction of this preparation is the Sonogashira reaction shown here (Scheme 32). The coupling proceeded in 92% yield by using the masked acetylene. Deprotection provided SIB-1508Y in high efficiency [69].

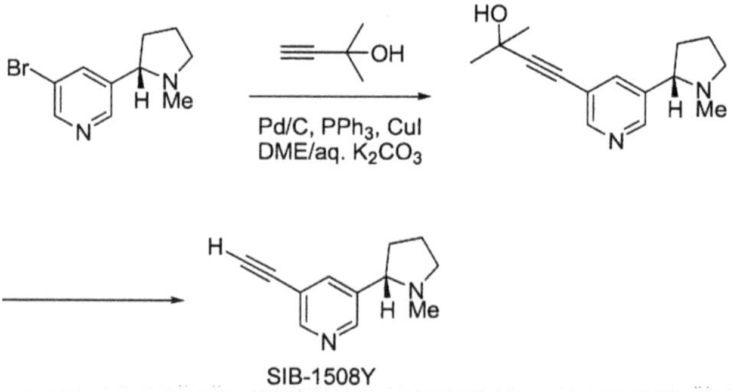
Br
H Me
OH
Pd/C, PPh3, CuI
DME/aq. K2CO3
HO
SIB-1508Y

Scheme 32

6.4
Terbinafine from Sandoz

Terbinafine has been commercialized by Sandoz for oral and topical treatment of mycoses. The original process relied on the *N*-alkylation of a substituted allyl bromide, which provided a 7:3 mixture of *E* and *Z* isomers. Terbinafine, which has the *E*-configuration, was purified as a HCl salt from IPA/Et_2O [70]. Recently, chemists from Sandoz reported a new process using a Sonogashira reaction as the key step (Scheme 33) [71a]. The new process used an *N*-alkylation with 1,3-(*E*)-dichloropropene followed by a Sonogashira reaction to give the pure *E*-isomer selectively. This process was also applied to the synthesis of its carba-analogues at Sandoz [71b].

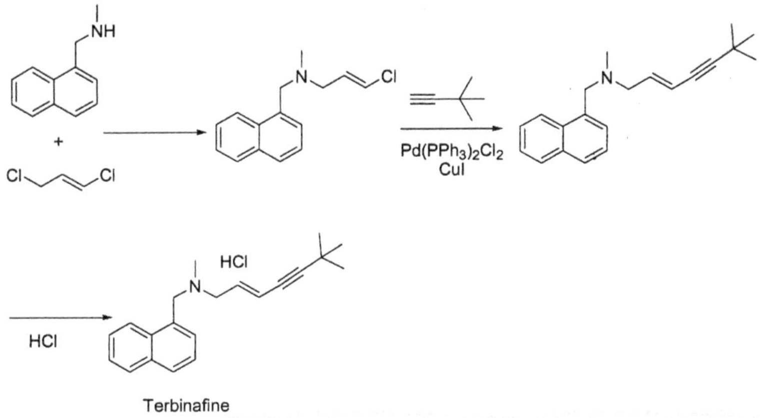

NH
+
Cl Cl
N Cl
$Pd(PPh_3)_2Cl_2$
CuI
N
HCl
N HCl
Terbinafine

Scheme 33

6.5
A-79175 from Abbott

Preparation of multikilogram quantities of the second generation 5-lipoxygenase inhibitor, A-79175, was reported in 1997 from Abbott (Scheme 34) [72]. The key step again was a Sonogashira reaction between two building blocks having equal complexity. This process is therefore very convergent and efficient. They reported that the choice of base was critical for this reaction. With 1.2 equivalents of diisopropylamine, the reaction was completed in 2 h at room temperature yielding 98%. On the other hand, the reaction was sluggish when triethylamine was substituted for diisopropylamine. Even after 8 h at room temperature, the product was formed in only 60% yield along with approximately 20% of unreacted starting materials and 10–15% of isoxazole.

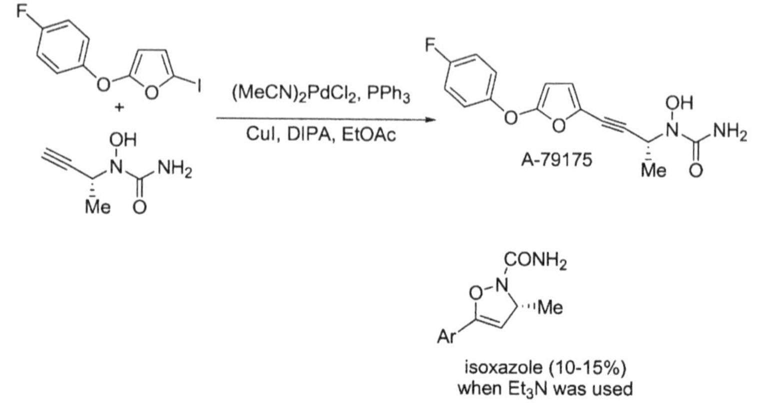

(MeCN)$_2$PdCl$_2$, PPh$_3$
CuI, DIPA, EtOAc
A-79175
isoxazole (10-15%) when Et$_3$N was used

Scheme 34

6.6 Antirhinoviral Agents from Eli Lilly

Large-scale preparations of two antirhinoviral agents were reported from Lilly in 1998 (Scheme 35) [73]. The agents were prepared by a reductive Heck-Mizorogi reaction with 3-aryl-propiolamide, which in turn was prepared by a Sonogashira reaction. Regioselectivity of the reductive Heck-Mizorogi reaction was controlled by the proper choice of catalyst (phosphine free catalysts, such as Pd(dba)$_2$ or Pd(MeCN)$_2$Cl$_2$), and the *E/Z* selectivity of the resulting tri-substituted olefin was easily controlled under the reaction conditions.

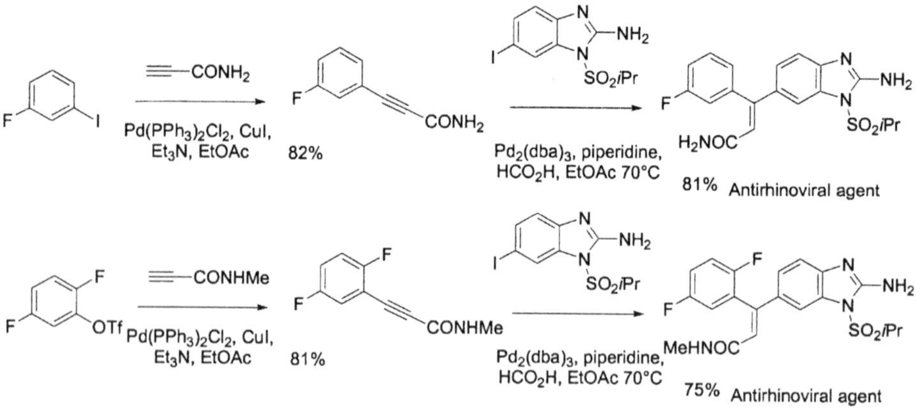

Pd(PPh$_3$)$_2$Cl$_2$, CuI, Et$_3$N, EtOAc 82%
Pd$_2$(dba)$_3$, piperidine, HCO$_2$H, EtOAc 70°C
81% Antirhinoviral agent
Pd(PPh$_3$)$_2$Cl$_2$, CuI, Et$_3$N, EtOAc 81%
Pd$_2$(dba)$_3$, piperidine, HCO$_2$H, EtOAc 70°C
75% Antirhinoviral agent

Scheme 35

6.7
SB 222618 from SmithKline Beecham

SB 222618 has been a target for synthetic chemists at SmithKline Beecham owing to its potential PDE (cAMP-specific cyclic nucleotide phosphodiesterases) IV inhibitor activity against inflammatory diseases such as asthma [74]. The candidate was prepared via two key reactions: Regioselective S_N2' addition of a cuprate to a bromoallene and a Sonogashira reaction (Scheme 36).

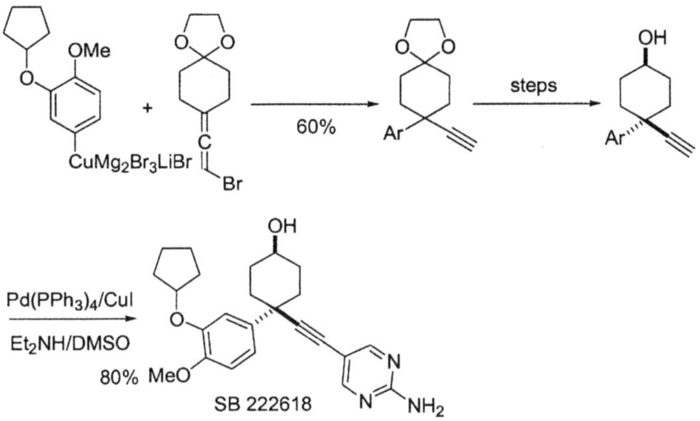

Scheme 36

6.8
Eniluracil

The widely used anticancer drug, Eniluracil, is a potent inactivator of dihydropyrimidine dehydrogenase. The dihydropyrimidine dehydrogenase is the rate-limiting enzyme in the metabolism of 5-fluorouracil (Scheme 37) [75]. GlaxoSmithKline reported the preparation of over 60 kg of Eniluracil after many process issues were overcome.

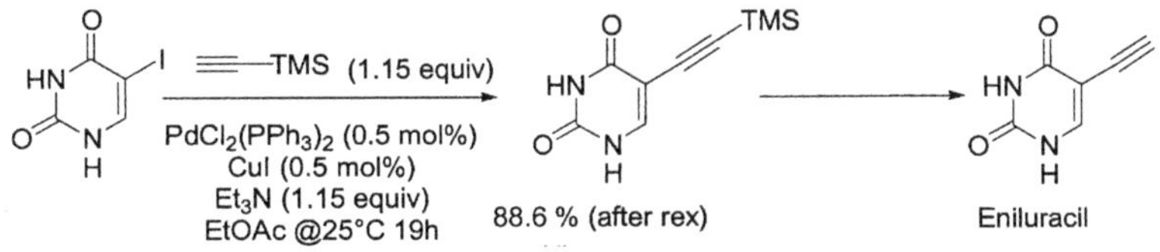

Scheme 37

7
Amination Reaction

From the very beginning of the pharmaceutical industry, aniline derivatives such as indole, quinolone, benzodiazepine, phenothiazine, etc., are key components in many medicines. Traditionally, these compounds have been prepared by multiple chemical steps from aniline derivatives by *N*-modifications, such as *N*-alkylation including reductive amination, *N*-acylation, and *N*-sulfonylation, etc. *N*-Nucleophilic substitutions have been reported in special cases in which aryl halides are highly activated, such as 2-chloropyrimidines. Copper-catalyzed reactions are known but generally are not suitable for scale-up and limited to specific substrates [76]. Kosugi and Migita reported the first preparation of aryl amines from aryl bromide using Bu_3SnNEt_2 in the presence of $PdCl_2[P(o\text{-tol})_3]_2$ in 1983 [77]. The reaction is somewhat limited due to the need of the dialkylamidostannane, which is not stable under aqueous conditions, and only electron-neutral aryl halides can be used. Thus this reaction has not commonly been utilized until Hartwig and Buckwald improved this reaction further. They demonstrated that tin-free aminations of aryl iodides and bromides took place with amines under carefully optimized conditions [78]. The key finding was the choice of base, NaOBu-*t* or Cs_2CO_3, and the use of BINAP as the ligand. Tanaka reported the first successful cross coupling with inexpensive aryl chlorides in 1997 using PCy_3 as a ligand [79]. Later, Buchwald reported an improvement using Cy-MAP1 as a ligand in 1998 [80]. This method has been further improved by the introduction of tri-*tert*-butylphosphine by Yamamoto in 1998 [81].

Similar reaction conditions can also be used for the formation of C-O bonds. However, reductive elimination from C-Pd-O is slower than that from C-Pd-N or C-Pd-C bonds. Therefore, only electron-deficient, highly reactive aryl halides can participate in this *O*-alkylation reaction [78]. At this time copper-catalyzed reaction protocols are better for C-O bond formation [82].

7.1
Norastemizole from Sepracor

The preparation of a potent non sedating histamine H_1-receptor antagonist, Norastemizole, was reported from Sepracor using a palladium-catalyzed *N*-amination reaction (Scheme 38) [83]. An interesting regioselectivity was observed.

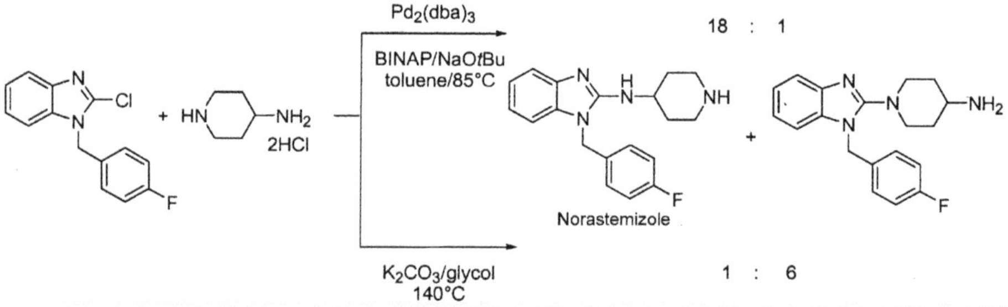

Scheme 38

Palladium-catalyzed amination provided predominantly a secondary amine. On the other hand, base-catalyzed amination provided mainly a tertiary amine. In the palladium-catalyzed amination, steric effect would be a significant factor, but in the base-catalyzed case nucleophilicity of nitrogen played the key role.

7.2 Hydroxyitraconazole

The preparation of an active metabolite of an antifungal and anti-yeast compound, Itraconazole, via a palladium-catalyzed *N*-amination reaction was also reported from Sepracor (Scheme 39) [84]. This convergent method suffered from the competitive Heck-Mizorogi reaction, which gave the desired product in 81% isolated yield together with the Heck-Mizorogi product in 4%.

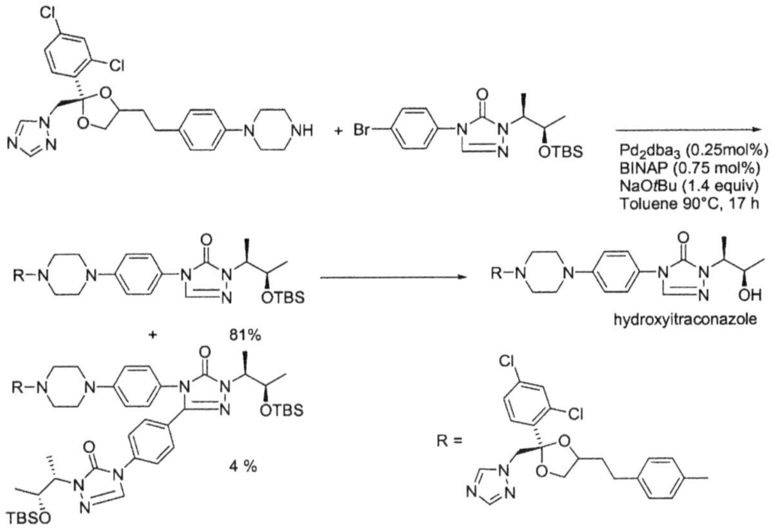

Pd$_2$dba$_3$ (0.25mol%)
BINAP (0.75 mol%)
NaO*t*Bu (1.4 equiv)
Toluene 90°C, 17 h

OTBS 81%

OTBS 4 %

hydroxyitraconazole

Scheme 39

7.3 DuP-721 from DuPont Pharmaceuticals Co.

DuP-721 is an oxazolindione class antibiotics candidate developed by DuPont Co. (Scheme 40) [85]. The original reported method utilized the conventional aniline chemistry. Recently, Abbott reported an alternative method utilizing a palladium-catalyzed *N*-amination (Scheme 41) [86]. The new route is more convergent than the original route.

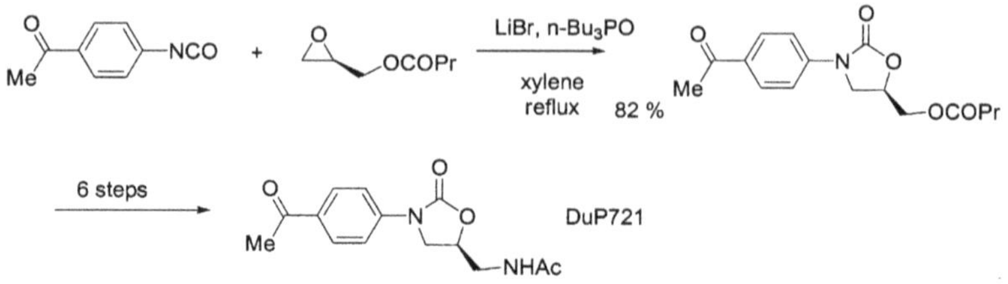

Scheme 40

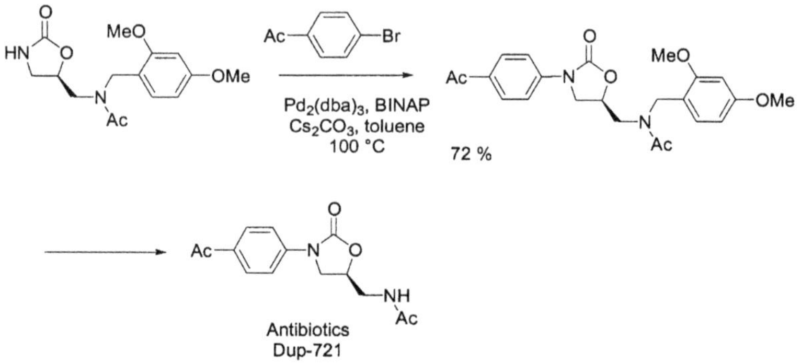

Scheme 41

7.4
A Muscarinic Receptor Antagonist from Banyu and Merck

A process for a Muscarinic receptor antagonist was published from Banyu and Merck recently (Scheme 42) [87]. A palladium-catalyzed amination was used in the preparation of the heterocyclic polyamine portion of the drug candidate. Benzophenone imine was used as an ammonia surrogate since the coupled product could be hydrolyzed readily to the primary amine [88].

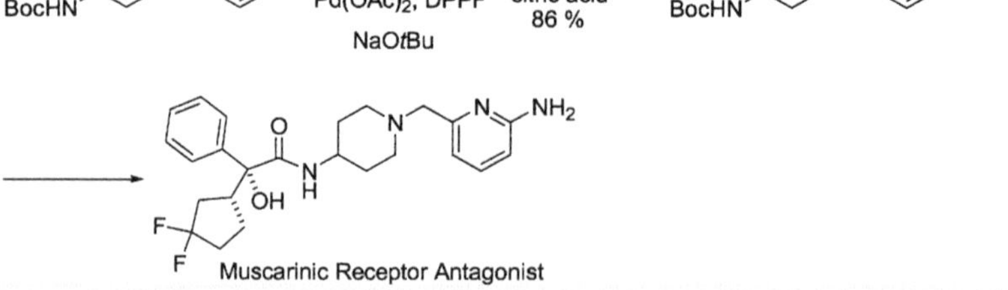

Scheme 42

8
Other Cross Coupling Methods

Although the development of the Pd-catalyzed cross-coupling reaction with organosilicon compounds has been studied extensively by Hiyama and Hatanaka and others [89], no large-scale pharmaceutical application with this method was found in our survey. Likewise, aluminum-, zirconium-, and iron-mediated cross-coupling reactions have been reported and systematically studied but again no pharmaceutical application was found [90]. Economic, productivity concerns and functional groups compatibility may be issues limiting the use of these coupling reactions. Time will tell if this group of coupling methods will supercede the other coupling methods currently in vogue.

Larock's palladium-catalyzed indole synthesis, a Heck-Mizoroki type reaction, could be viewed as a cross-coupling reaction even though this method involves a carbo-palladation step rather than a transmetalation.

8.1
Rizatriptan Benzoate (Maxalt) from Merck

The 5-HT_{1D} receptor agonist, Rizatriptan benzoate, was prepared using the Larock's indole synthesis as the key step (Scheme 43) [91]. Protection of 3-butyn-1-ol with triethylsilyl group was important to prevent coupling at the terminal carbon of the acetylene and minimize de-silylation.

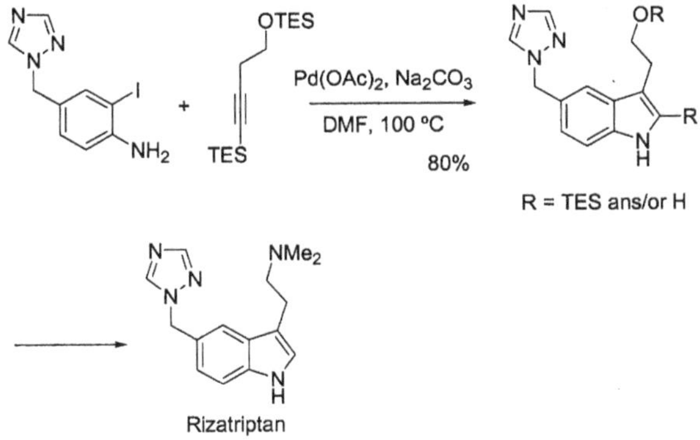

$Pd(OAc)_2$, Na_2CO_3
DMF, 100 °C
80%
R = TES ans/or H
Rizatriptan

Scheme 43

Three other active pharmaceutical ingredients currently on the market, Montelukast Sodium (Merck) [92], Abacavir (Wellcome) [93], and Famciclovir (SmithKline Beecham) [94], may also have utilized palladium chemistry in their manufacturing processes (Fig. 2). The Heck-Mizorogi reaction was used in the Montelukast Sodium process, and the Tsuji-Trost reaction was used in both of the Abacavir and Famciclovir processes.

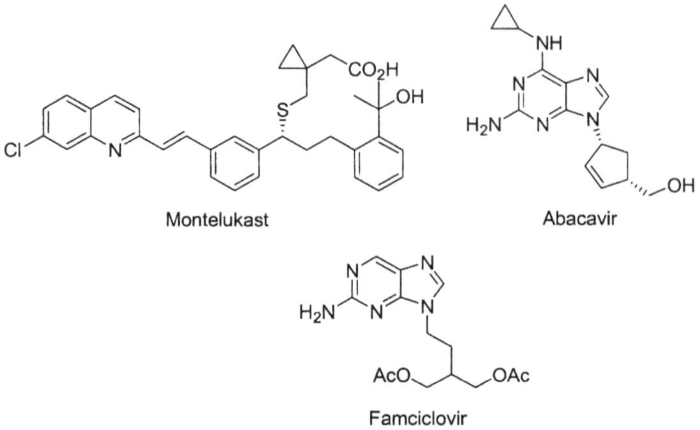

Montelukast

Abacavir

Famciclovir

Fig 2 Drugs on the market which may be utilizing other palladium-mediated reactions as key steps

9 Process Issues

9.1 Heavy Metal Waste Minimization

During the processing of these cross-coupling reactions on a large scale, minimization of heavy metal waste is of utmost importance. Since usage of organostannanes is undesirable due to their toxicity, Zn is generally the second choice behind Suzuki-Miyaura conditions with boron. To minimize the use of Zn, the amount of $ZnCl_2$ used in the transmetalation reaction with organolithium or organomagnesium can be reduced to 0.5 mol equiv vs the organometal reagent. This generates the diarylzinc and both aryl groups can still participate in the coupling. If further minimizing of the zinc waste is necessary, the amount of $ZnCl_2$ can be reduced further to 0.25 mol equiv by forming the tetraarylzinc dianion. All four of the aryl groups transfer effectively during the coupling [95]. Even boron waste can be reduced by using borinic acids and triarylboranes in the Suzuki-Miyaura coupling reaction [96], but unlike boronic and borinic acids which are stable to air and moisture, triorganoboranes are air-sensitive and their isolation and handling could be cumbersome on a large scale.

9.2 Pd Removal, Recycling, and Regeneration

The use of homogeneous transition metal catalysts always poses the problem of residual metal contamination in the isolated product. This can be a serious problem with commercial scale processes for the manufacture of pharmaceuticals, especially if the transition metal-catalyzed method is used at the end of a synthetic route. Minimizing the Pd catalyst loading for the reaction is economically desirable and also leads to the reduction of Pd contamination in the final drug product. In many cases, simply crystallizing the product from the appropriate solvent or solvent mixture can reject essentially all of the residual Pd metal. Passing a solution of the product through a small amount of activated carbon, silica gel, or chelating resin first before crystallization may be necessary to reduce further the amount of residual metal, for which the crystallization alone is not sufficiently effective.

In the case of the coupling step in losartan (Scheme 20), the residual Pd was around 1000 ppm when the trityl-protected losartan coupling product was crystallized directly from diethoxymethane (DEM). Filtration of the THF/DEM solution of the product through a short column of carbon, silica, or chelating resin prior to crystallization still did not provide isolated material having satisfactorily low residual Pd level. The solution to this problem was to increase the solubility of the Pd by complexing it with tributylphosphine. Thus by adding 10 mol equiv of tributylphosphine vs Pd prior to crystallization, the residual Pd in the isolated product was reduced to <25 ppm. After the final de-protection and salt formation steps, the residual Pd was further reduced to 1–2 ppm.

Another reported method was the use of trithiocyanuric acid to precipitate the Pd, which was then filtered off (Scheme 44) [97]. With 25 mol equiv of trithiocyanuric acid vs Pd, the residual Pd in the product was <10 ppm. Increasing the trithiocyanuric acid to 50 equiv further dropped the residual Pd to <1 ppm.

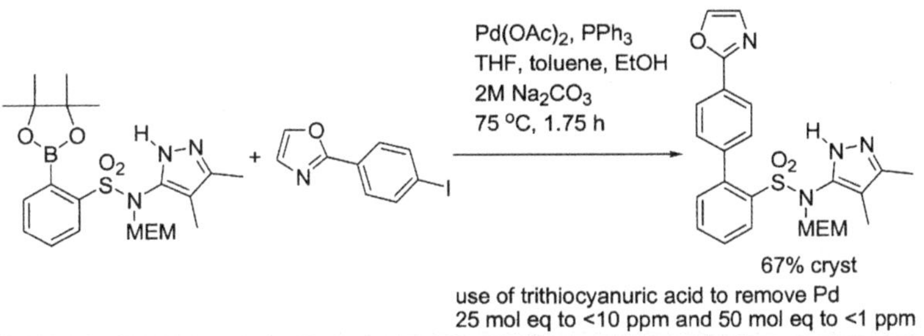

$Pd(OAc)_2$, PPh_3
THF, toluene, EtOH
2M Na_2CO_3
75 °C, 1.75 h

67% cryst
use of trithiocyanuric acid to remove Pd
25 mol eq to <10 ppm and 50 mol eq to <1 ppm

Scheme 44

Even without any added ligand, the use of a slight excess of the electrophiles can also help to stabilize the Pd in the form of $RPd(II)XL_2$ and thus increases the solubility of the Pd in the crystallization solvent. This was indeed accomplished in the Heck-Mizorogi reaction for the preparation of the 1,3-diarylpropanone intermediate for the asthma drug montelucast (Scheme 45) [98]. With a 10% excess of methyl *o*-iodobenzoate, nearly complete rejection of the Pd was achieved by crystallizing the product directly from the reaction mixture by the addition of 15 vol.% of water based on the amount of DMF used.

The ability to recycle and reuse catalyst has a tremendous cost benefit for a commercial process. In the absence of any ligand, the Pd has a tendency to precipitate at the end of some coupling reactions and can be filtered off. To aid in the removal of the Pd, a filter-aid is usually added for the Pd to deposit on. This deposited Pd on the filter-aid can then be re-activated by treatment with I_2 and

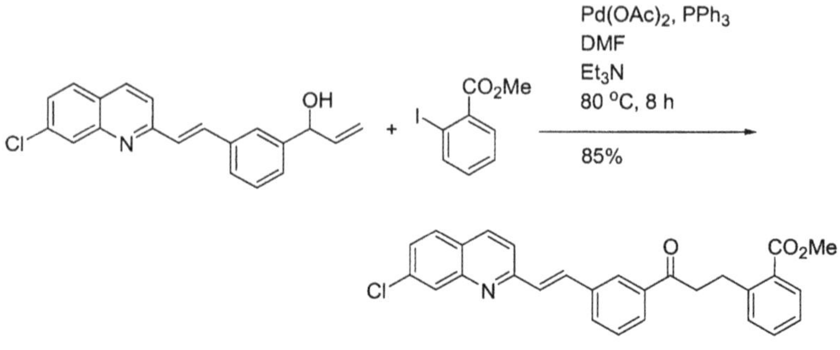
Pd(OAc)2, PPh3
DMF
Et3N
80 °C, 8 h
85%
OH
CO2Me
Cl
N
O

Scheme 45

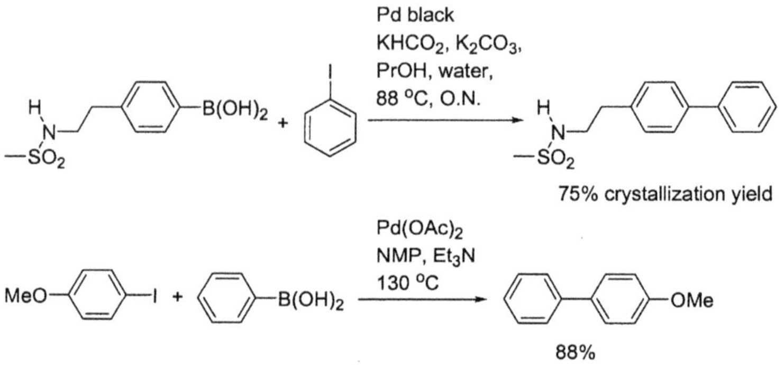
Pd black
KHCO2, K2CO3,
PrOH, water,
88 °C, O.N.
B(OH)2
75% crystallization yield
Pd(OAc)2
NMP, Et3N
130 °C
MeO
OMe
88%

Scheme 46

used again in another coupling reaction with similar effectiveness (Scheme 46) [99]. In this case, the use of excess electrophiles RX is probably undesirable since the resting form of the Pd is RPd(II)X which may be sufficiently stabilized to remain in solution and not precipitate out.

Besides the usual triarylphosphines, such as triphenylphosphine, ionic mono- and di-phosphines have also been used (Scheme 47) [100]. With these water soluble ligands the coupling reactions can even be carried out in water as the

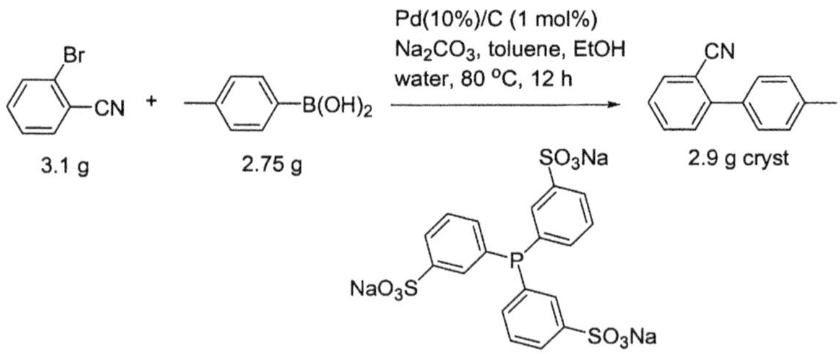
Pd(10%)/C (1 mol%)
Na2CO3, toluene, EtOH
water, 80 °C, 12 h
Br
CN
B(OH)2
3.1 g
2.75 g
2.9 g cryst
SO3Na
NaO3S
P
SO3Na

Scheme 47

sole solvent. This can benefit commercial processing by reducing organic solvents usage and ease of separation of catalyst from the product.

A very interesting 15-membered macrocyclic triolefin Pd(0) complex was reported by Cortes et al. (Fig. 3) [101]. The Pd(0) metal is coordinated to the three double bonds and is very stable to chromatography. After the coupling reaction, this Pd(0) complex can be recovered quantitatively and re-used. The complex has also been anchored onto polystyrene resin further simplifying its recovery.

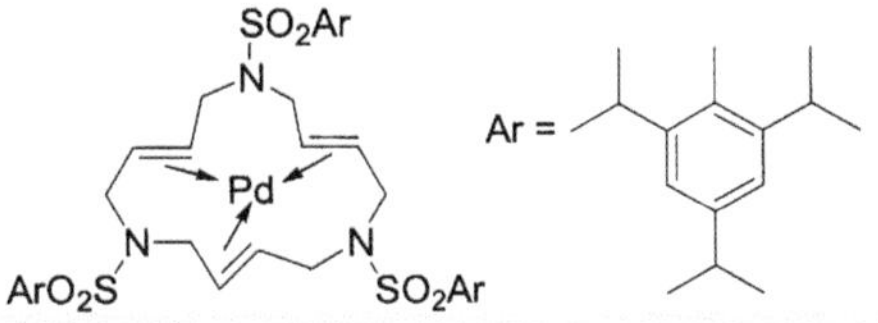

SO2Ar
N
Pd
N N
ArO2S SO2Ar
Ar =

Fig 3

As discussed earlier, the removal of homogeneous catalyst from the product has always been an issue with these reactions. Thus homogeneous catalyst adsorbed or deposited on a support can simplify the catalyst removal process. Activated aryl chlorides were found to couple with boronic acids in the presence of Pd/C [102]. The success of this reaction lies in the use of aqueous DMA as the solvent system, as aqueous EtOH gave homo-coupled products derived from the aryl chloride (Scheme 48) [103]. Neutral and electron-rich aryl chlorides only provided low to moderate yields of the cross-coupling products. Although the addition of a phosphine to this coupling system completely shut down the reaction, others have made use of the Pd/C-PR_3 catalytic system for the coupling of activated aryl chlorides and boronic acids to give excellent yields of biaryl products [104]. In general, homogeneous system is much more active requiring shorter reaction time and lower catalyst charge.

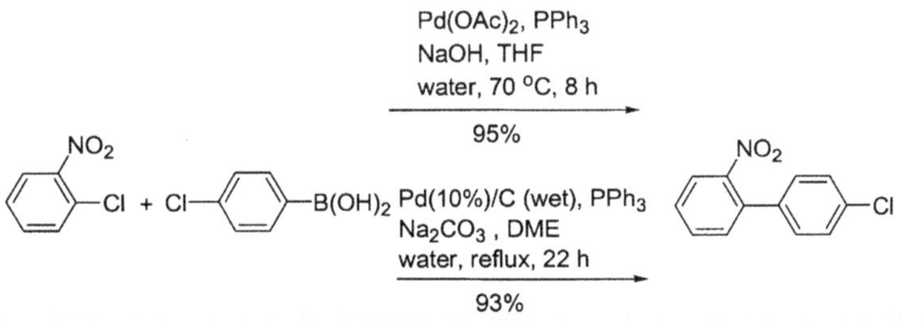

$Pd(OAc)_2$, PPh_3
NaOH, THF
water, 70 °C, 8 h
95%
NO_2
Cl + Cl— —$B(OH)_2$
Pd(10%)/C (wet), PPh_3
Na_2CO_3, DME
water, reflux, 22 h
93%
NO_2
Cl

Scheme 48

The use of Ni/C-PPh_3 catalytic system has also been reported for aryl chloride and alkylzinc coupling (Scheme 49) [105]. Even electron-rich aryl chlorides gave high yields of coupled products. Only a trace amount of Ni was found in the reaction mixture, thus essentially no catalyst bleed occurred during the reaction.

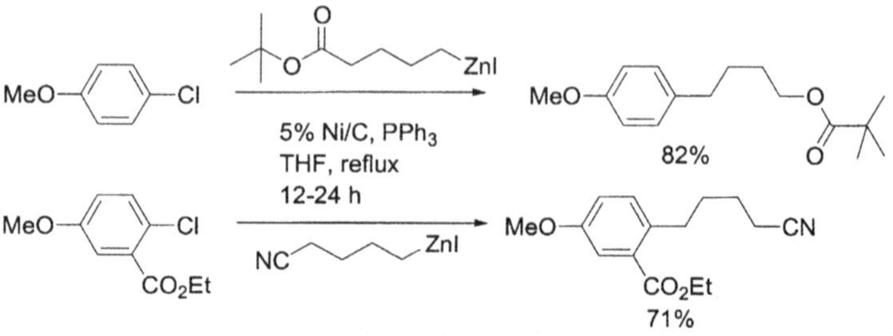

MeO–C$_6$H$_4$–Cl → 5% Ni/C, PPh$_3$, THF, reflux, 12-24 h → 82%

NC(CH$_2$)$_n$ZnI, CO$_2$Et → CN, CO$_2$Et, 71%

Scheme 49

9.3 Ligands for Coupling with Organic Chlorides

In order to maintain the activity of the Ni or Pd catalyst so as to attain 100% conversion with minimal amount of the catalyst, early developed reaction conditions called for 2 or more equivalents of triphenylphosphine to be added. Later, to suppress the problem of β-hydride elimination with alkylmetals, 1,1′-bis(diphenylphosphino)ferrocene was introduced by Kumada to overcome this problem [106]. Many other monodentate and bidentate phosphine and phosphite ligands have been introduced since for improving the coupling efficiency.

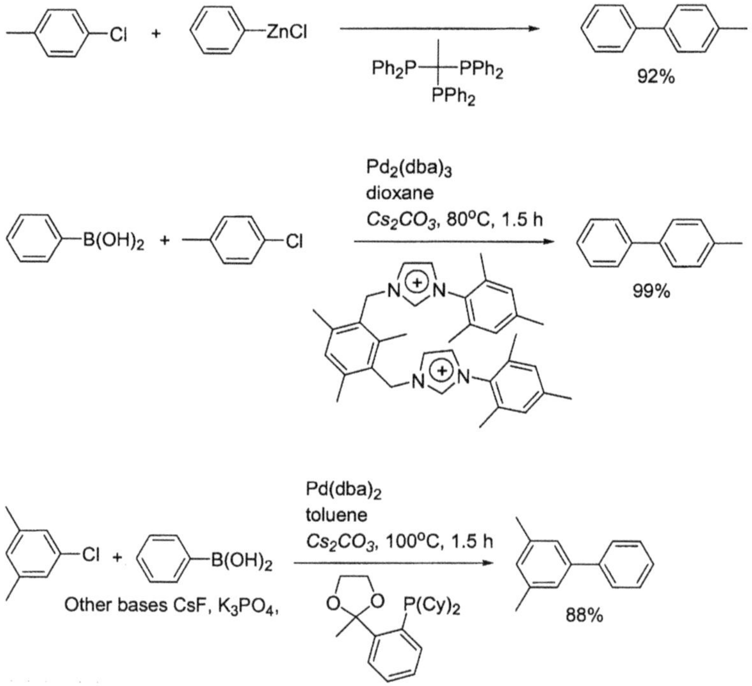

–C$_6$H$_4$–Cl + C$_6$H$_5$–ZnCl → (Ph$_2$P, PPh$_2$, PPh$_2$) → 92%

C$_6$H$_5$–B(OH)$_2$ + –C$_6$H$_4$–Cl → Pd$_2$(dba)$_3$, dioxane, Cs_2CO_3, 80°C, 1.5 h → 99%

Cl + B(OH)$_2$ → Pd(dba)$_2$, toluene, Cs_2CO_3, 100°C, 1.5 h, P(Cy)$_2$ → 88%

Other bases CsF, K$_3$PO$_4$,

Scheme 50

For a long time, one of the more serious deficiencies with the Ni- and Pd-catalyzed coupling reaction is the inability to utilize non-activated aryl chlorides satisfactorily. Recent introduction of numerous bulky mono-phosphines has essentially solved this problem (Scheme 50) [107] and certainly more will be introduced in the future.

10 Conclusion

Since the discovery of Kumada reaction in the 1970s, subsequent discoveries and use of other organometallics containing Al, Zr, Zn, Cd, Sn, B, and Si in cross-coupling reactions have greatly expanded the versatility of this reaction. All of these organometallics are much more chemoselective resulting in their popularity for complex organic synthesis. Out of this group of organometallic species, Sn, B, and Si compounds possess the highest chemoselectivity character. Due to the environmentally unfriendliness of Sn, it is mostly used in small-scale preparations. With B being quite innocuous to the environment, its use in large-scale preparation has been much more desirable. Although quite promising, the use of Si has not yet been as wide spread as the others in pharmaceuticals preparation. Besides the issue of chemoselectivity with Grignard reagents, other organometallic nucleophiles (RM), in some cases, provide better cross-coupling yields. For example, comparing the coupling of bromobenzothiazole with three different phenyl metals showed that phenyltrimethylstannane and phenylzinc chloride gave excellent yields but phenylmagnesium bromide gave substantially poorer yield than the other two [108]. However, when chemoselectivity is not an issue, the Kumada reaction is generally as effective as the other coupling protocols. Thus, if possible, screening of the different coupling reactions during the early phase of development is recommended.

With the introduction of a variety of new ligands for the coupling reactions, now the organic electrophiles are not limited to iodides, bromides, and triflates. Even chlorides can be used in the coupling reaction with excellent results. This will lower the expense of the coupling reaction and makes it even more desirable on a commercial scale. With absolute certainty these Ni- and Pd-catalyzed coupling methods will even be more widely used in the future.

References

1. (a) Kumada M (1980) Pure Appl Chem 52:669; (b) Corriu RJP, Masse JP (1972) JCS Chem Comm 144
2. General reviews: (a) Diederich F, Stang PJ (1998) Metal-catalyzed cross-coupling reactions. VCH, Weinheim; (b) Trost BM (1991) Comp Org Syn, vol 3. Pergamon Press, Oxford, chap 2; (c) Genet JP, Savignac M (1999) J Organomet Chem 576:305; (d) Boudier A, Bromm LO, Lotz M, Knochel P (2000) Angew Chem Int Ed 39:4414; (e) Tsuji J (2001) Gousei Kagaku Kyokaishi 59:607; (f) Hassan J, Sevignon M, Gozzi C, Schulz E, Lemaire M (2002) Chem Rev 102:1359
3. Naso F, Babudri F, Farinola GM (1999) Pure Appl Chem 71:1485
4. (a) Hermkens PHH, Ottenheijm HCJ, Rees D (1996) Tetrahedron 52:4527; (b) Franzén R (2000) Can J Chem 78:957

5. Piarraud A, Lasne MC, Barré L, Vaugeois JM, Lancelot JC (1993) J Labelled Compd Radiopharm 23:253
6. Köhler B, Langer M, Mosandl T (1997) DE 19 607 135
7. Miller JA (2001) Tetrahedron Lett 42:6991
8. (a) Ruyle WV, Sarett LH, Matzuk A (1969) GB 1 175 212; (b) Jones H, Houser RW (1980) US Patent 4 225 730
9. Giodano C, Coppi L, Minsci F (1994) US Patent 5 312 975
10. Marzoni G, Varney MD (1997) Org Process Res and Dev 1:81
11. (a) Negishi E-I, Valente LF, Kobayashi M (1980) J Am Chem Soc 102:3298; (b) Negishi E-I (1999) Palladium and nickel catalyzed reactions of organozinc compounds – organozinc reagents. Oxford University, Oxford, p 213
12. Knochel P, Singer RD (1993) Chem Rev 93:2117
13. Rieke RD, Hanson MV (1997) Tetrahedron 53:1925
14. Anderson BA, Becke LM, Booher RN, Flaugh ME, Harn NK, Kress TJ, Varie DL, Wepsiec JP (1997) J Org Chem 62:8634
15. Shiota T, Yamamori T (1999) J Org Chem 64:453
16. (a) Angiolelli ME, Casalnuovo AL, Selby TP (2000) Synlett 905; (b) Stevenson TM, Selby TP, Koether GM, Drumm JE, Meng XJ, Moon MP, Coats RA, Thieu TV, Casalnuovo AE, Shapiro R (2002) ACS Symposium Series 800:85
17. Fisher MH, Wyvratt MJ, Schoen WR, DeVita RJ (1992) International Pat Appl WO 92/16524
18. Kosugi M, Sasazawa K, Shimizu Y, Migita T (1977) Chem Lett 301
19. Stille JK, Lau KSY (1977) Acc Chem Res 10:434
20. (a) Stille JK (1986) Angew Chem Int Ed 25:508; (b) Farina V, Krishnamurthy V, Scott WJ (1998) The Stille reaction. Wiley, New York
21. Kosugi M, Ohya T, Migita T (1983) Bull Chem Soc Jap 56:3855
22. Hoshino M, Degenkolb P, Curran DP (1997) J Org Chem 62:8341
23. For example: Vedejs E, Haight AR, Moss WO (1992) J Am Chem Soc 114:6556
24. Yasuda N, Yang C, Wells KM, Jensen MS, Hughes DL (1999) Tetrahedron Lett 40:427
25. Humphrey GR, Miller RA, Pye PJ, Rossen K, Reamer RA, Maliakal A, Ceglia SS, Grabowski EJJ, Volante RP, Reider PJ (1999) J Am Chem Soc 121:11261
26. Jensen MS, Yang C, Hsiao Y, Rivera N, Wells KM, Chung JYL, Yasuda N, Hughes DL, Reider PJ (2000) Org Lett 2:1081
27. Naito T, Hoshi H, Aburaki S, Abe Y, Okumura J, Tomatsu K, Kawaguchi H (1987) J Antibiot 40:991
28. Farina V, Baker SR, Sapino C Jr (1988) Tetrahedron Lett 29:6043
29. (a) Baker SR, Roth GP, Sapino C (1990) Synth Commun 20:2185; (b) Roth GP, Sapino C (1991) Tetrahedron Lett 32:4073
30. Kant J, Farina V (1992) Tetrahedron Lett 33:3563
31. Nagano N, Itahana H, Hisamichi H, Sakamoto K, Hara R (1994) Tetrahedron Lett 35:4577
32. Armitage MA, Lathbury DC, Sweeney JB (1995) Tetrahedron Lett 36:775
33. Barrett D, Terasawa T, Okuda S, Kawabata K, Yasuda N, Kamimura T, Sakane K, Takaya T (1997) J Antibiot 50:100
34. Yamamoto H, Terasawa T, Ohki A, Shirai F, Kawabata K, Sakane K, Matsumoto S, Matsumoto Y, Tawara S (2000) Bioorg Med Chem Lett 8:43
35. Kohrt J, Filipski KJ, Rapundalo ST, Cody WL, Edmunds JJ (2000) Tetrahedron Lett 6041
36. Tarkiainen J, Vercouillie J, Emond P, Sandell J, Hiltunen J, Frangin Y, Guilloteau D, Halldin C (2001) J Labeled Compd Radiopharm 44:1013
37. (a) Miyaura N, Suzuki A (1995) Chem Rev 95:2457; (b) Suzuki A (1999) J Organomet Chem 576:147; (c) Review on Suzuki-Miyaura cross-coupling reaction with aryl chlorides; Gröger H (2000) J Prakt Chem 342:334; (d) Saito S, Ohtani S, Miyaura N (1997) J Org Chem 62:8024; (e) Review on B-alkyl Suzuki-Miyaura cross-coupling reaction;

Chemler SR, Trauner D, Danishefsky SJ (2001) Angew Chem Int Ed 40:4544; (f) Kotha S, Lahiri K, Kashinath D (2002) Tetrahedron 58:9633
38. Selection of base especially counter cation is critical. Systematic screen is required for optimum performance. For instance, TlOH was the choice for the key Suzuki-Miyaura reaction for the total synthesis of palytoxin by Kishi: Kishi Y (1989) Pure Appl Chem 61:313
39. (a) Palucki M, Hughes DL, Yang C, Yasuda N (2001) US Patent 6 262 268; (b) Sasaki M, Tachibana K, Fuwa H (2001) International Pat Appl WO01/98308; (c) Johnson CR, Braun MP (1994) US Patent 5 359 110
40. Ishiyama T, Murata M, Miyaura N (1995) J Org Chem 60:7508
41. Murata M, Watanabe S, Masuda Y (1997) J Org Chem 62:6458
42. Marcuccio SM, Rodopoulos M, Weigold H (2001) US Patent 6 288 259
43. Ishiyama T, Takagi J, Ishida K, Miyaura N (2002) J Am Chem Soc 124:390
44. Old DW, Wolfe JP, Buchwald SL (1998) J Am Chem Soc 120:9722
45. Littke AF, Fu GC (1998) Angew Chem Int Ed 37:3387
46. Netherton MR, Dai C, Neuschütz K, Fu GC (2001) J Am Chem Soc 123:10099
47. Schlama T, Colobert, F, Castanet A-S (2001) International Pat Appl WO 01/42197
48. Inbasekaran M, Wu W, Woo EP (1998) US Patent 5 777 070
49. Larsen RD, King AO, Chen CY, Corley EG, Foster RS, Roberts FE, Yang C, Lieberman DR, Reamer RA, Tschaen DM, Verhoeven TR, Reider PJ, Lo YS, Rossano LT, Brookes AS, Meloni D, Moore JR, Arnett JF (1994) J Org Chem 59:6391
50. Naka T, Nishikawa K, Kato T (1991) Eur Pat 459 136
51. Bernhart CA, Perreaut PM, Ferrari BP, Muneaux YA, Assens J-LA, Clément J, Haudricourt F, Muneaux CF, Taillades JE, Vignal M-A, Gougat J, Guiraudou FR, Lacour CA, Roccon A, Cazaubon CF, Brelière J-C, Fur GL, Nisato D (1993) J Med Chem 36:3371
52. Bühlmayer P, Furet P, Criscione L, de Gasparo M, Whitebread S, Schmidlin T, Lattmann R, Wood J (1994) Bioorg Med Chem Lett 4:29
53. (a) Heitsch H, Wagner A, Yadav-Bhatnagar N, Griffoul-Marteau C (1996) Synthesis 1325; (b) Deprez P, Guillaume J, Becker R, Corbier A, Didierlaurent S, Fortin M, Frechet D, Hamon G, Heckmann B, Heitsch H, Kleemann H-W, Vevert J-P, Vincent J-C, Wagner A, Zhang J (1995) J Med Chem 38:2357
54. Yasuda N, Huffman MA, Ho G-J, Xavier LC, Yang C, Emerson KM, Tsay F-R, Li Y, Kress MH, Rieger DL, Karady S, Sohar P, Abramson NL, DeCamp AE, Mathre DJ, Douglas AW, Dolling U-H, Grabowski EJJ, Reider PJ (1998) J Org Chem 63:5438
55. Narukawa Y, Nishi K, Onoue H (1997) Tetrahedron 53:539
56. Yu MS, Lopez de Leon L, McGuire MA, Botha G (1998) Tetrahedron Lett 39:9347
57. Miyachi N, Yanagawa Y, Iwasaki H, Ohara Y, Hiyama T (1993) Tetrahedron Lett 34:8267
58. Desmond R, Dolling U, Marcune B, Tillyer R, Tschaen D (1996) International Pat Appl WO 96/08482
59. Friesen RW, Brideau C, Chan CC, Charleson S, Deschênes D, Dubé D, Ethier D, Fortin R, Gauthier JY, Girard Y, Gordon R, Greig GM, Riendeau D, Savoie C, Wang Z, Wong E, Visco D, Xu LJ, Young RN (1998) Bioorg Med Chem Lett 8:2777
60. Davies IW, Marcoux J-F, Corley EG, Journet M, Cai D-W, Palucki M, Wu J, Larsen RD, Rossen K, Pye PJ, DiMichele L, Dormer P, Reider PJ (2000) J Org Chem 65:8415
61. Potter GA, Hardcastle IR, Jarman M (1997) Org Prep Proceed Int 29:123
62. (a) Castro CE, Stephens RD (1963) J Org Chem 28:2163; (b) Stephens RD, Castro CE (1963) J Org Chem 28:3313
63. Campbell IB (1994) In: Tayler RJK (ed) Organocopper reagents. IRL Press, Oxford, p 217. For additional review, refer to Sonogashira K, chap 5 in [2a] and chap 2.4 in [2b]
64. Publication numbers on Sonogashira reaction based on a SciFinder search are follows; 1993 (2); 1994 (1); 1995 (2); 1996 (6); 1997 (14); 1998 (19); 1999 (36); 2000 (91); 2001 (127)
65. Masuda T, Karim SMA, Nomura R (2000) J Mol Cat A Chem 160:125
66. Chandraratna RAS (1992) US Patent 5 089 509

67. Houpis IN, Choi WB, Reider PJ, Molina A, Churchill H, Lynch J, Volante RP (1994) Tetrahedron Lett 35:9355
68. Larock RC, Yum EK (1991) J Am Chem Soc 113:6689
69. Bleicher LS, Cosford NDP, Herbaut A, McCallum JS, McDonald IA (1998) J Org Chem 63:1109
70. (a) Stütz A (1987) Angew Chem 99:323; (b) Stütz A, Petranyi G (1984) J Med Chem 27:1539
71. (a) Beutler U, Mazacek J, Penn G, Schenkel B, Wasmuth D (1996) Chimia 50:154; (b) Nussbaumer P, Leitner I, Mraz K, Stütz A (1995) J Med Chem 38:1831
72. Dickman DA, Ku Y-Y, Morton HE, Chemburkar SR, Patel HH, Thomas A, Plata DJ, Sawick DP (1997) Tetrahedron Asymmetry 8:1791
73. Hay LA, Koenig TM, Ginah FO, Copp JD, Mitchell D (1998) J Org Chem 63:5050
74. Conde JJ, Mendelson W (2000) Tetrahedron Lett 41:811
75. Cooke JWB, Bright R, Coleman MJ, Jenkins KP (2001) Org Process Res Dev 5:383
76. Many of active pharmaceutical ingredients have been prepared by N-nucleophilic substitution on aryl halides in the presence or absence of copper catalyst. Here is a short list: Amodiaquine: US 2 474 819; 2 474 821 (Parke Davies). Flufenamic acid: Wilkinson F (1948) J Chem Soc 32. Lobenzarit: US 4 092 426 (Chugai); Mepacrine: US 2 113 357 (Winthrop Chemical); Minaprine: US 4 169 158 (Centre Etudes Exper.); Mirtazepine: US 4 062 848 (AKZO); Piribedil: US 3 299 067 (Science Union); Piromidic acid: US 3 673 184 (Dainippon); Prazosin: US 3 511 836 (Pfizer); Sulfachlorpyridazine: US 2 790 798 (American Cyamide); Thiethylperazine and Thioridazine (Sandoz): Bourquin J-P, Schwarb G, Gamboni G, Fischer R, Ruesch L, Guldimann S, Theus V, Schenker E, Renz J (1958) Helv Chim Acta 41:1072; Tolfenamic acid: US 3 313 848 (Parke Davis)
77. Kosugi M, Kameyama M, Migita T (1983) Chem Lett 927
78. Recent reviews on N-C and O-C couplings: (a) Wolfe JP, Wagaw S, Marcoux, J-F, Buchwald SL (1998) Acc Chem Res 31:805; (b) Hartwig JF (1998) Angew Chem Int Ed 37:2046; (c) Torraca KE, Huang X, Parrish CA, Buchwald (2001) J Am Chem Soc 123:10770; (d) Kuwabe S-I, Torraca KE, Buchwald SL (2001) J Am Chem Soc 123:12202
79. Raddy NP, Tanaka M (1997) Tetrahedron Lett 38:4807
80. Old DW, Wolfe JP, Buchwald SL (1998) J Am Chem Soc 120:9722
81. (a) Nishiyama M, Yamamoto T, Koie Y (1998) Tetrahedron Lett 39:617; (b) Yamamoto T, Nishiyama M, Koie Y (1998) Tetrahedron Lett 39:2367
82. Wolter M, Nordmann G, Job GE, Buchwald SL (2002) Org Lett 4:973
83. Hong Y, Tanoury GJ, Wilkinson HS, Bakale RP, Wald SA, Senanayake CH (1997) Tetrahedron Lett 32:5607
84. Tanoury GJ, Senanayake CH, Hett R, Kuhn AM, Kessler DW, Wald SA (1998) Tetrahedron Lett 39:6845
85. Wang C-LJ, Gregory WA, Wuonola MA (1989) Tetrahedron 45:1323
86. Madar DJ, Kopecka H, Pireh D, Pease J, Pliushchev M, Sciotti RJ, Wiedeman PE, Djuric SW (2001) Tetrahedron Lett 42:3681
87. Mase T, Houpis IN, Akao A, Dorziotis I, Emerson K, Hoang T, Iida T, Itoh T, Kamei K, Kato S, Kato Y, Kawasaki M, Lang F, Lee J, Lynch J, Maligres P, Molina A, Nemoto T, Okada S, Reamer R, Song JZ, Tschaen D, Wada T, Zewge D, Volante RP, Reider PJ, Tomimoto K (2001) J Org Chem 66:6775
88. Benzophenone imine as an amine surrogate: (a) Wolfe JP, Åhman J, Sadighi JP, Singer RA, Buchwald SL (1997) Tetrahedron Lett 38:6367; (b) Mann G, Hartwig JF, Driver MS, Fernández-Rivas C (1998) J Am Chem Soc 120:827
89. (a) Hiyama T, Hatanaka Y (1994) Pure Appl Chem 66:1471; (b) DeShong P, Handy CJ, Mowery ME (2000) Pure Appl Chem 72:1655; (c) Denmark SE, Wang Z (2001) Org Lett 3:1073
90. Negishi E-I, Liu F (1998) Pd- or Ni-catalyzed cross coupling with organometals containing Zn, Mg, Al, and Zr. In: Stang PJ, Diederich F (eds) Cross coupling reactions. VCH, Weinheim, p 1

91. Chen C-Y, Lieberman DR, Larsen RD, Reamer RA, Verhoeven TR, Reider PJ, Cottrell IF, Haughton PG (1994) Tetrahedron Lett 35:6981
92. (a) Labelle M, Belley M, Gareau Y, Gauthier JY, Guay D, Gordon R, Grossman SG, Jones TR, Leblanc Y, McAuliffe M, McFarlane C, Masson P, Metters KM, Ouimet N, Patrick DH, Piechuta H, Rochette C, Sawyer N, Xiang YB, Pickett CB, Ford-Hutchinson AW, Zamboni RJ, Young RN (1995) Bioorg Med Chem Lett 5:283; (b) King AO, Corley EG, Anderson RK, Larsen RD, Verhoeven TR, Reider PJ, Xiang YB, Belley M, Leblanc Y, Labelle M, Prasit P, Zamboni RJ (1993) J Org Chem 58:3731
93. (a) Daluge SM (1990) Eur Pat Appl EP 349242; (b) There are some asymmetric synthesis of Abacavir from academics: Crimmins MT, King BW (1996) J Org Chem 61:4192; (c) Olivo HF, Yu J (1998) J Chem Soc Perkin Trans 1:391
94. Freer R, Geen GR, Ramsay TW, Share AC, Slater GR, Smith NM (2000) Tetrahedron 56:4589
95. Gauthier DR, Sumigala RH Jr, Dormer PG, Armstrong JD III, Volante RP, Reider PJ (2002) Org Lett 4:375
96. Nobumitsu K (2001) JP 20001055360
97. Polniaszek RP, Wang X, DePue JS, Pandit CR, Gadamasetti KG, Pendri Y, Martinez EJ (1999) US Patent 5,856,507
98. Shinkai I, King AO, Larsen RD (1994) Pure Appl Chem 66:1551
99. (a) Gardner JP, Miller WD (2001) International Patent WO 01/90055; (b) Parlevliet FJ, De Vries JG, De Vries AHM (2002) International Patent WO 02/00340
100. (a) Haber S, Manero J (2001) Eur Pat Appl EP 679 619; (b) Hermann WA, Kohlpaintner CW (1993) Angew Chem, Int Ed 32:1524
101. Cortes J, Moreno-Manas M, Pleixats R (2000) Eur J Org Chem 239
102. (a) LeBlond CR, Andrews AT, Sun Y, Sowa JR Jr (2001) Org Lett 3:1555; (b) Fan Q-H, Li Y-M, Chan ASC (2002) Chem Rev 102:3385
103. Marck G, Villiger A, Buchecker R (1994) Tetrahedron Lett 35:3277
104. Eicken K, Gebhardt J, Rang H, Rack M, Schafer P (1997) International Patent WO 97/733846
105. Lipshutz BH, Blomgren PA (1999) J Am Chem Soc 121:5819
106. Tamao K, Sumitani K, Kiso Y, Zembayashi M, Fujioka A, Kodama S, Nakajima I, Minato A, Kumada M (1976) Bull Chem Soc Japan 49:1958
107. (a) Miller JA, Farrell RP (2001) US Patent 6 194 599; (b) Guram A, Bei X, Powers TS, Jandeleit B, Crevier T (2000) US Patent 6 124 476; (c) Review: Littke AF, Fu GC (2002) Angew Chem, Int Ed 41:4176
108. Hanaki N, Suzuki A (2001) JP2001 031 656

Topics Organomet Chem (2004) 6: 247–262
DOI 10.1007/978-3-540-36966-0

Stereospecific Introduction of Cephalosporin Side Chains Employing Transition Metal Complexes

Joydeep Kant

Department of Process Research and Development, Bristol-Myers Squibb Pharmaceutical Research Institute, One Squibb Drive, New Brunswick, New Jersey 08903, USA
E-mail: Joydeep.Kant@bms.com

Abstract The unique structural and chemotherapeutic properties of β-lactams continue to attract the attention of synthetic community since they present a variety of challenges. To synthesize cephalosporins with enhanced biological properties, efforts were focused on the modifications of the C(3) position of cephems with all carbon substituents. For example, cephalosporins with olefinic and allylic side chains were found compatible with good, broad-spectrum activity with excellent pharmacokinetic profiles. Efficient and versatile synthetic approaches to these compounds with an aim of commercialization and study of structure-activity relationships therefore became highly desirable. The development of various organometallic methodologies for the stereospecific introduction of cephalosporin side chains has advanced remarkably in the last few years by keeping pace with fundamental research in organotransition chemistry. This chapter surveys some of these advances in the use of organotransition metal complexes to gain easy access to these novel cephalosporins.

Keywords Stereoselective · Transition metal complex · C-C Bond formation · Cephalosporin · Allene

1 Introduction

Major developments in the area of β-lactam research have often originated from the discovery of new synthetic methodologies. Cephalosporins being the most important class of antibacterial compounds, owe much of their commercial success to the invention of new synthetic methods that allowed for their semisynthetic modifications [1]. For example, discovery of a method for cleavage of the natural α-aminodipic acid side chain from Cephalosporin C led to the cost efficient preparation of 7-aminocephalosporanic acid (7-ACA) [2] (Eq. 1):

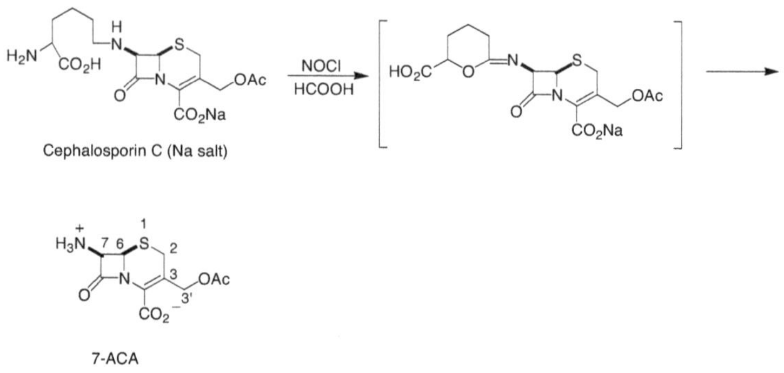

Cephalosporin C (Na salt) —NOCl / HCOOH→ [ ] → 7-ACA (1)

A second major breakthrough was the Morin's ring expansion of cheap penicillins to expensive cephalosporins. This discovery provided much of the driving force for the development of the first semisynthetic cephalosporins (Eq. 2) [3]:

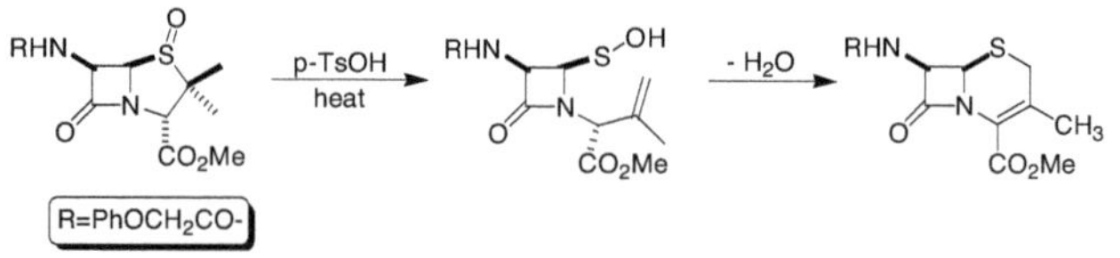

—p-TsOH / heat→ —- H_2O→ (2)

R=PhOCH$_2$CO-

Besides modifying the C-7 side chain, the most fruitful exercise has been introducing new C-3 side chains. Traditional synthetic activities centered around displacing the C-3′ acetoxy group in Cephalosporin-C, directly or in a stepwise fashion, with a variety of *N*- and *S*- based nucleophiles [1]. Many of the second and third generation cephalosporins were indeed synthesized in this fashion, and their commercial preparation was achieved in a few steps [4]. However, it was soon realized that a heteroatom at the C-3′ position was not an essential requirement for biological activity and studies were undertaken to explore novel C-3 side chains [1, 5]. Cephalosporins with all-carbon substituents at the C(3) position, especially olefinic side chains, have been shown to possess excellent biological and pharmacokinetic profiles; some representative examples of this class of antibiotics are Cefadroxil, Cephalexin, Cefixime, and Cefprozil (Scheme 1) [6, 7]. Efficient synthetic approaches to these compounds, with an aim of commercialization, therefore became highly desirable. Previous synthetic methodologies to form the carbon-carbon bond at the C(3) position relied on

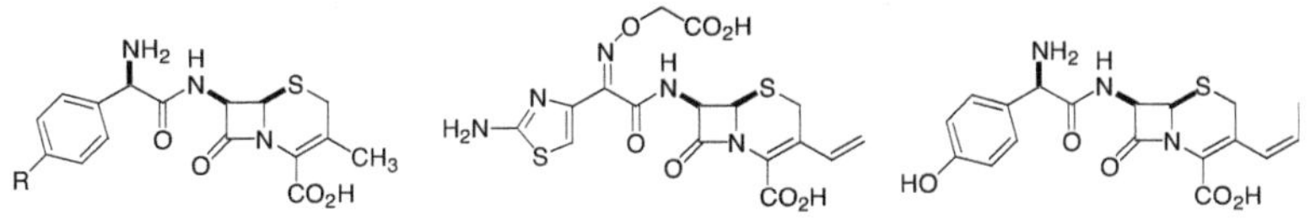

R = OH (Cefadroxil)
R = H (Cephalexin)

Cefixime

Cefprozil

Scheme 1

Friedel-Crafts reactions with 3-[(trifluoroacetoxy)methyl]ceph-2-em-4-carboxylic acids [8], reactions of 3-formylcephems with stabilized phosphoranes [9], Wittig reaction of 3-hydroxycephems with stabilized ylides [10], conjugate additions of organocuprates to 3-chloro- and 3-vinylcephems [11], and additions of Grignard reagents to 3-formylcephems [12]. Unfortunately, the scope of these procedures had been limited and often cephems were isolated as a mixture of Δ^2 and Δ^3 isomers. This chapter details some of the newer and efficient methods developed to date to synthesize these important class of cephalosporins employing organotransition metal complexes.

2 Cephalosporins via the Stille Chemistry

Palladium catalyzed coupling between vinyl triflates and organostannanes has been extensively investigated by the Stille group [13]. Farina and co-workers demonstrated the use of this methodology in the synthesis of a variety of 3-alkenyl-, 3-alkynyl-, and 3-arylcephems under exceptionally mild conditions using 3-trifloxycephems, available from 3-hydroxycephems [14], and vinyl stannanes (Eq. 3):

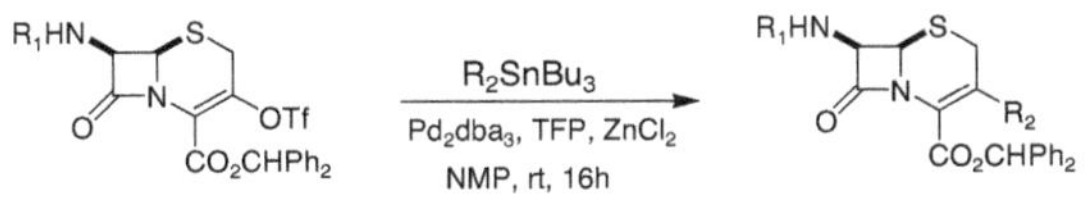

R_1 = *t*-Boc

R_2SnBu_3, Pd_2dba_3, TFP, $ZnCl_2$, NMP, rt, 16h

R_2	Yield
CH_2=CH	79%
(*Z*)-1-Propenyl	90%
$(CH_3)_2C$=CH	66%
CH_2=C(OEt)	52%
F_2C=CF	55%
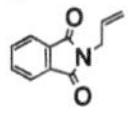	73%
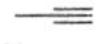 —≡	50%
H	65%
Me	85%
Bu	16%
p-$MeOC_6H_4$	57%
CH_2=CH-CH_2	48%
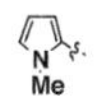 Me	89%

(3)

The strategy was further extended in the development of a stereocontrolled and cost-efficient synthesis of Cefprozil, a broad spectrum antibiotic (Eq. 4) [15]:

RHN, S, OTf, CO_2CHPh_2; R = *t*-Boc
$SnBu_3$; Pd_2dba_3, TFP, $ZnCl_2$; NMP, rt, 16h, 91%; >99% stereospecific
RHN, S, CO_2CHPh_2
NH_2, H, N, HO, O, S, CO_2H
Cefprozil
(4)

In a typical procedure, trifloxycephem (0.01 mole) was dissolved in dry NMP. The solution was degassed and $ZnCl_2$ (0.02 mole), tri(2-furyl)phosphine (0.4 mmol) and Pd_2dba_3 (0.2 mmol) was added. The solution was stirred for 10 min followed by the addition of stannane (0.01 mole). The reaction mixture was stirred at room temperature for 20 h, diluted with ethyl acetate, washed with water, brine, and dried over sodium sulfate. Filtration and concentration gave the crude product which was redissolved in acetonitrile and washed with pentane to remove the tin by-products. Distillation of solvent followed by crystallization or purification by chromatography provided isomerically desired cephem.

Under the reaction condition, transfer of the alkyl group was found to be difficult. While tetramethyltin readily coupled in excellent yield, tetrabutyltin only reacted under forcing conditions and in low yield; extensive decomposition of the triflate was observed. Allyltributyltin gave unsatisfactory results and mostly Δ^2 allylcephem was isolated.

While this approach was not satisfactory for the preparation of 3-allylcephems, coupling of readily available 3-(chloromethyl)cephems with stannanes provided a high yielding route to such allylcephems, 3-benzyl, and homoallylcephems via. the η^3-allylpalladium intermediate. (Eq. 5):

R_1HN, S, Cl, CO_2CHPh_2
R_2SnBu_3; Pd_2dba_3, TFP, THF, reflux
RHN, S, R_2, CO_2CHPh_2
(5)

R_1 = *t*-Boc
R_2= alkenyl, aryl, allyl, H (50-90% yields)

The reaction was found to be quite efficient and was carried out in refluxing THF. The corresponding bromomethyl cephems also underwent coupling reaction, albeit at a slower rate.

The choice of ligand was crucial in both cases. It was found that the triphenylphosphine based ligands were quite unsatisfactory. A much better ligand was tri(2-furyl)phosphine (TFP) which appreciably enhanced the rate of coupling of triflates with stannanes by making the Pd (II) intermediate more electrophilic and therefore more reactive. A ratio of 1:2 palladium:phosphine was employed which led to a very stable catalyst capable of at least 100 turnovers. Zinc chloride was used as an exogeneous halide source, but halide source was not absolute necessary in these reactions. The catalytic cycle proceeded via oxidative addition, transmetalation, followed by reductive elimination as outlined in

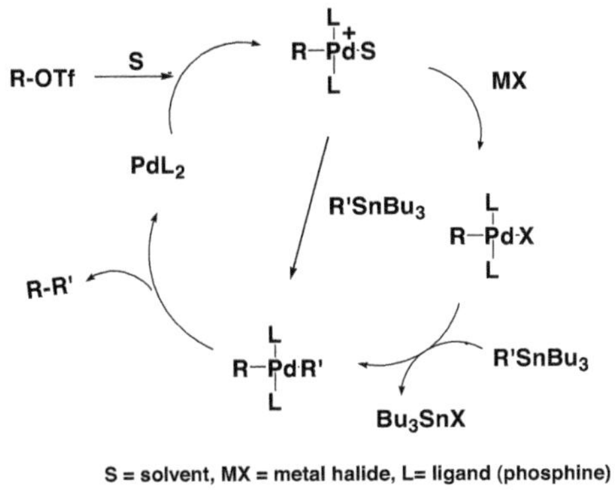

S = solvent, MX = metal halide, L= ligand (phosphine)

Scheme 2

Scheme 2. Overall, this method offered a practical approach to a variety of carbon based C-3 side chains from hydroxycephems.

A modification of Farina's approach employed simplified ligandless catalytic system for intermolecular coupling reaction between 3-trifloxycephem and vinylstannane. The reaction proceeded smoothly at room temperature using either d^{8} $Pd(OAc)_2$ or d^{10} $Pd_2(dba)_3$ and typically at a faster rate in the absence of phosphine ligands or added halide (Eq. 6) [16]:

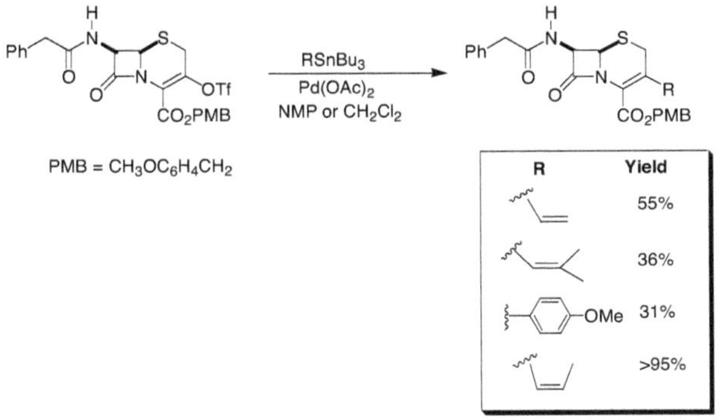

PMB = $CH_3OC_6H_4CH_2$

(6)

The reaction worked well in a variety of solvents; however, the preferred solvents were NMP or dichloromethane.

As expected, substituted alkenes were transferred with complete stereospecificity. Importantly, this catalyst reduced the reaction time from 16 h to 5 min and cooling had to be carried out on a multigram scale. The ligandless procedure, however, had limited scope as with most other stannanes catalyst decomposition ensued before completion of the reaction.

Roth demonstrated the use of fluorosulfonate cephems, cheap alternatives to trifloxycephems, as efficient electrophilic partners for palladium (0)-catalyzed cross coupling reactions [17]. Treatment of 3-hydroxycephem at -78 °C with a slight excess of diisopropylethylamine followed by fluorosulfonic anhydride furnished the desired crystalline cephem in 96% yield which subsequently underwent cross-coupling reaction with a variety vinylstannanes affording the desired 3-substituted cephems (Eq. 7):

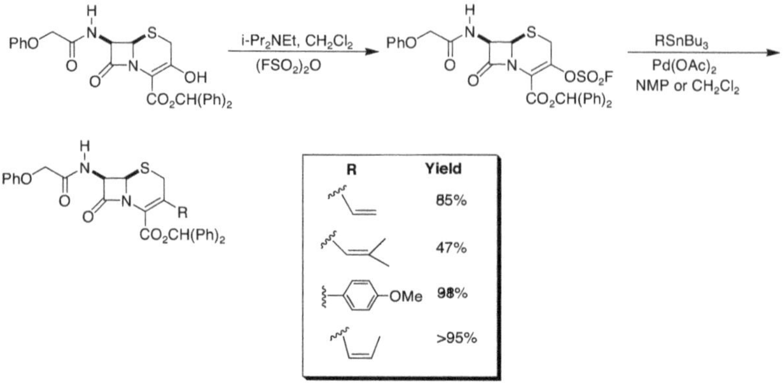

R	Yield
vinyl	85%
2-methylpropenyl	47%
4-methoxyphenyl	98%
(Z)-propenyl	>95%

(7)

Another modified version of Stille reaction in the introduction of the vinyl group at the C-3 position was reported by Umani-Ronchi. The synthesis employed cross-coupling reactions between 3-bromo or 3-mesyloxy cephem and vinyltributyltin in the presence of 10 mol% of $Pd(OAc)_2$ along with 2 equiv. of LiI. The reaction was performed under mild conditions at room temperature in anhydrous NMP. Addition of 2.0 equiv. of LiI was essential when 3-mesyloxy cephem was used. The reactive 3-iodo cephem formed via Finkelstein reaction underwent cross-coupling reaction with vinylstannane to give the desired product. In case of bromides, no appreciable rate enhancement was observed upon the addition of LiI (Eq. 8) [18]:

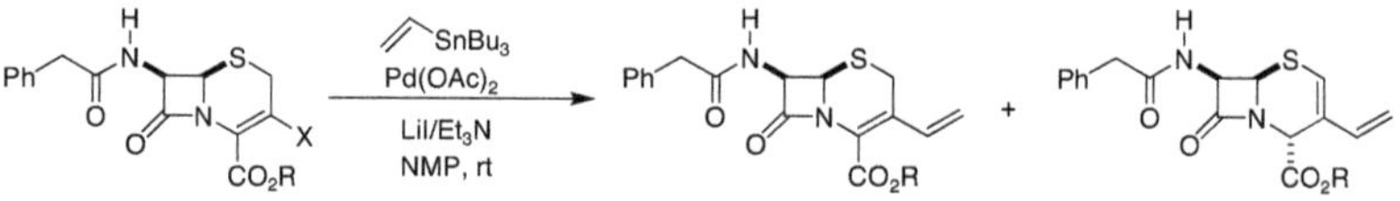

(8)

1. $X=OSO_2CH_3$, $R=CHPh_2$ (85% yield)
2. $X=OSO_2CH_3$, R=PNB (85% mostly Δ^2 isomer)
3. X=Br, $R=CHPh_2$ (85% along with Δ^2 isomer)
4. X=Br, R=PNB (85%)

Tanaka and his colleagues demonstrated yet another modified cross-coupling reaction to append the C-3 side chain using 3-trifloxy or chloro cephems with vinyl bromides in an Al/cat.$PbBr_2$/cat.$NiBr_2$(bpy) (bpy: 2,2′-bipyridine) redox system [19] (Eq. 9):

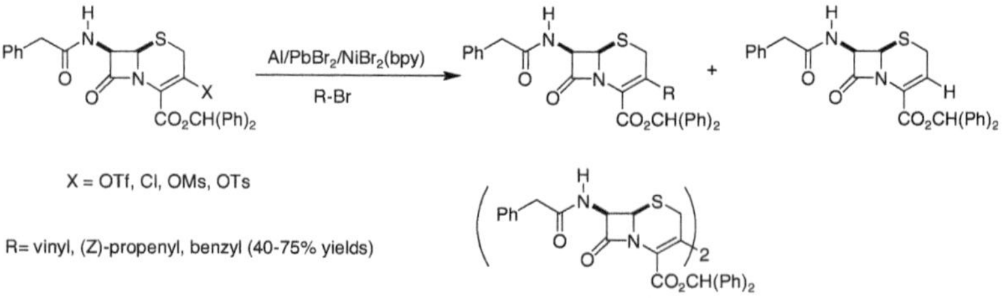

(9)

X = OTf, Cl, OMs, OTs

R= vinyl, (Z)-propenyl, benzyl (40-75% yields)

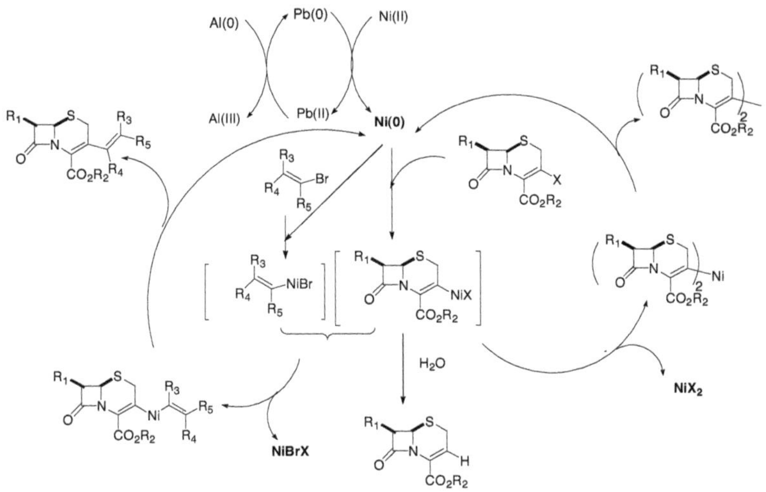

Scheme 3

A plausible mechanism proposed by Tanaka involved Ni(0)/Ni(II), Pb(0)/Pb(II), and Al(0)/Al(III) redox-promoted reactions (Scheme 3).

The Al/$PbBr_2$/$NiBr_2$(bpy) redox system generates Ni(0) which undergoes oxidative addition to 3-trifloxy- or chlorocephem and vinyl bromide, forming Ni(II) complexes of vinyl bromide and cephem, respectively. Subsequent transfer of the vinyl group from vinyl nickel complex produces the Ni(II) vinyl complex which via. reductive elimination furnishes the 3-substituted cephems. In addition to the desired product, <10% of the homo-coupled product and norcephalosporin are also produced (Scheme 3). The formation of norcephalosporin is reasonably understood by assuming hydrolysis of the intermediary Ni(II) cephem complex with moisture present in the reaction media. Indeed, when the reaction was carried out in wet DMF, norcephalosporin was obtained predominantly.

3 Cephalosporins via Addition-Elimination Reactions of Activated Triflates and Related Intermediates with Organocuprate and Organozinc Reagents

Reactions on substrates containing carbon bound leaving groups by organocuprates are conceptually among the most straightforward operations for the formation of carbon-carbon bonds [20]. The use of organocopper chemistry in cephems was first demonstrated by Spry et al. [11, 12]. C(3)-*n*-Butyl and *n*-hexyl cephems were prepared by the conjugate addition of 3-chloro or 3-thiophenyl substituted cephems with lower-order cuprates. Isomeric mixtures (Δ^3/Δ^2) of 3-alkyl cephems were isolated in low yields along with isomerized starting materials (Eq. 10):

Me_2CuLi, THF, -50 °C, 30 min (10)

1. X=Cl (50% yield)
2. X=SPh (28%)

Similarly, treatment with *n*-butylcuprate afforded a mixture of Δ^2 product and starting material (Eq. 11):

$(n\text{-}Bu)_2CuLi$, THF, -78 °C/ 40 min, 45-50% (11)

Cuprates, being basic in nature, have excellent potential for causing the ubiquitous (and undesired) Δ^3 to Δ^2 double bond isomerization.

Following the McMurray's chemistry on the stereospecific coupling between enol-trifluoromethanesulfonates (enol-triflates) and organocuprates [21], Kant successfully demonstrated the utility of lower and higher-order cuprates for substituting C-C for C-OTf bonds at the C3 sp^2 carbon in cephalosporins [22]. Treatment of 3-trifloxycephems with a variety of organocuprates in the presence of $BF_3{\cdot}OEt$ afforded 3-alkyl, aryl, and alkenylcephalosporins in moderate to high yields and in isomerically pure Δ^3 products (Eq. 12):

R_2CuLi, $BF_3.OEt$, THF, -78 °C, 2-5 h (12)

R	Yield
Me	75%
Et	75%
n-Bu	60%
t-Bu	70%
Ph	65%
CH_2=CH	35%
(Z)-Propenyl	29%

Addition of $BF_3{\cdot}OEt$ was necessary to impede the formation of the undesired Δ^2 isomer. Most likely, the free MeLi present in Gilman cuprate promotes isomerization by abstracting the C(2) proton from the cephem [23]. However, addition of $BF_3{\cdot}OEt$ in the cuprate generates a very reactive cuprate/Lewis acid combination by sequestering the free alkyllithium in solution thus preventing any isomerization [24]. The cuprate/$BF_3{\cdot}OEt$ combination worked well with alkyl and aryl cuprates, but lower yields were obtained with alkenylcuprates. Attempts to use higher-order cyanocuprates afforded a mixture of Δ^2 and Δ^3 isomers.

To prevent the formation of undesired isomers, Normant cuprates (dialkylmagnesiocuprates), evidently less basic in nature, were found to be the reagents of choice by Kant. A variety of structurally diverse alkyl, cycloalkyl, aryl, alkenyl, and allyl organocuprates generated from the corresponding Grignard reagents and copper (I) bromide-dimethyl sulfide complex at -78 °C afforded isomerically pure C-3 substituted cephems in good to excellent yields (Eq. 13):

$R_2CuMgBr$
-78 °C, 15 min-1h
THF

R	Yield
Me	95%
Et	89%
n-Pr	73%
t-Bu	78%
n-hexyl	81%
cyclohexyl	62%
cyclopentyl	82%
phenyl	85%
vinyl	85%
allyl	82%

(13)

It is important to generate these cuprates using stoichiometric amounts of copper salt, as in all instances a catalytic amount gave very low yield of the product. The transfer of (*Z*)-propenyl group was stereospecific (>99%) under reaction conditions which enabled the preparation of the key intermediate of Cefprozil in 80% yield (Eq. 14).

$(\text{propenyl})_2CuMgBr$
-78 °C, 1.5 h
THF
80% yield

(14)

Cefprozil

In a typical reaction, copper (I) bromide-dimethyl sulfide complex (1.0 mmol) in THF (2 mL) was placed in a two necked flask under an inert atmosphere. The flask was cooled to -78 °C. A solution of Grignard reagent (2.0 mmol) was added dropwise to the stirred suspension. The ice batch was removed, and the suspension was stirred until a dark colored homogeneous solution was observed (ca. 10–15 min.). The cuprate solution was re-cooled to –78 °C and a solution of trifloxycephem in THF (0.5 mmol) was added. The dark colored solution was stirred until completion of reaction and quenched into a solution of saturated NH_4Cl. The aqueous layer was extracted with ethyl acetate, washed with 10% $NaHCO_3$ solution and brine, dried, and evaporated to afford the cephem which was further purified by crystallization or chromatography.

Studies on the composition of Normant's cuprate prepared using 1.0 equiv. of copper bromide and 2.0 equiv. of methylmagnesium bromide indicated that the cuprate exists as a single dimeric species Cu_2MgMe_4 along with $MgBr_2$ [25]. The successful implementation of Normant's cuprate could be due to a reaction between trifloxycephem and Cu_2MgR_4 The reactivity of Cu_2MgR_4, a single entity, is quite different when compared to LO or HO cuprates, which usually exist as an equilibrium mixture of different entities, probably with differing degrees of reactivity and selectivity. Independently, Lipshutz also demonstrated the high yielding coupling of allylic cyanocuprates with triflates [26] (Eq. 15):

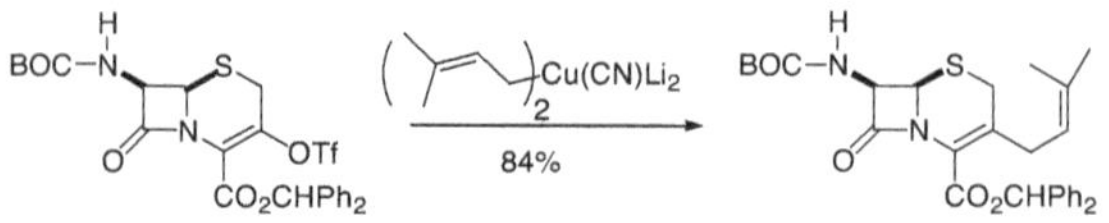

(15)

Stereochemical integrity about the olefinic center of the educt is completely maintained in these coupling reactions.

In order to make the process more economical, the reaction with the corresponding enol fluorosulfonates was demonstrated by Roth and Kant [27] (Eq. 16):

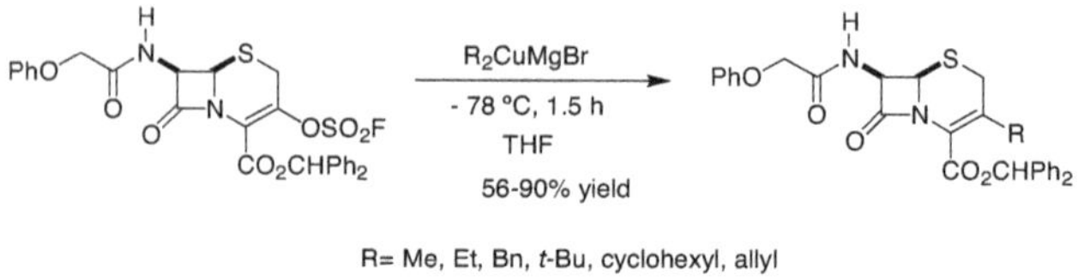

(16)

R= Me, Et, Bn, *t*-Bu, cyclohexyl, allyl

Torii and co-workers also reported the preparation of 3-alkenyl-Δ^2-cephems via. copper(I) chloride promoted alkenylation of 2-trifloxycephem with a variety of stannanes [28] (Eq. 17):

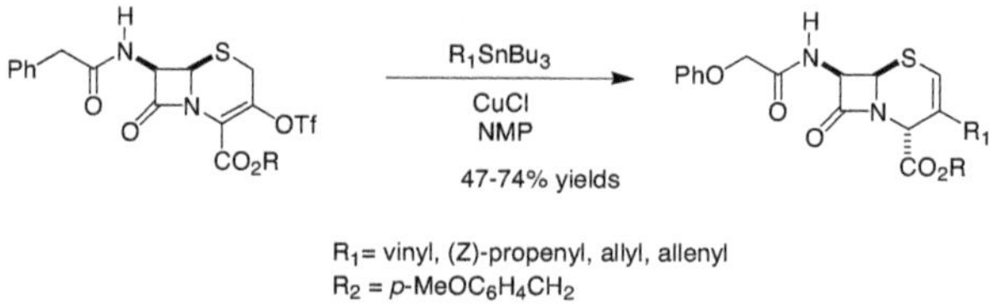

(17)

R_1= vinyl, (Z)-propenyl, allyl, allenyl
R_2 = *p*-$MeOC_6H_4CH_2$

The presence of copper(I) chloride was necessary as the lack of copper(I) chloride resulted in the recovery of the starting material.

A plausible mechanism involving a six-membered allenic intermediate formed by 1,2-elimination reaction of the triflate followed by addition of vinyl copper in a Michael fashion was proposed by Torri (Scheme 4).

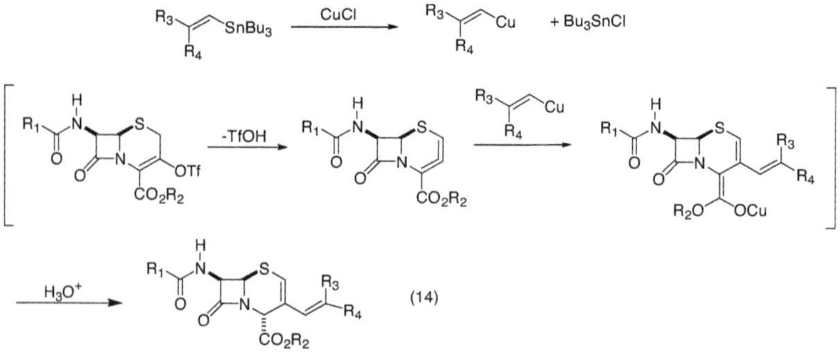

Scheme 4

The potential of other vinylic sulfonates to engage in coupling reactions was also investigated by Kant [22a]. It was discovered that vinyl tosylate and nosylate (*p*-nitrobenzenesulfonate) reacted equally well with the alkyl and vinyl cuprates to form carbon-carbon bonds (Eq. 18):

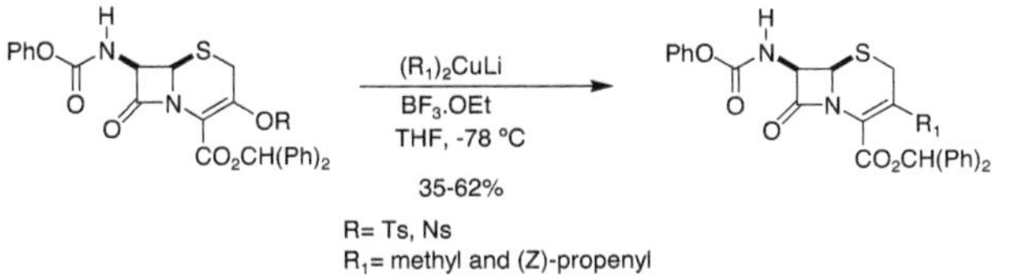

$(R_1)_2CuLi$, $BF_3.OEt$, THF, -78 °C, 35-62%

R= Ts, Ns
R_1= methyl and (Z)-propenyl

(18)

Treatment of dimethylcuprate with 3-mercapto tetrazole afforded the corresponding 3-methylcephem in high yield and isomeric purity (Eq. 19):

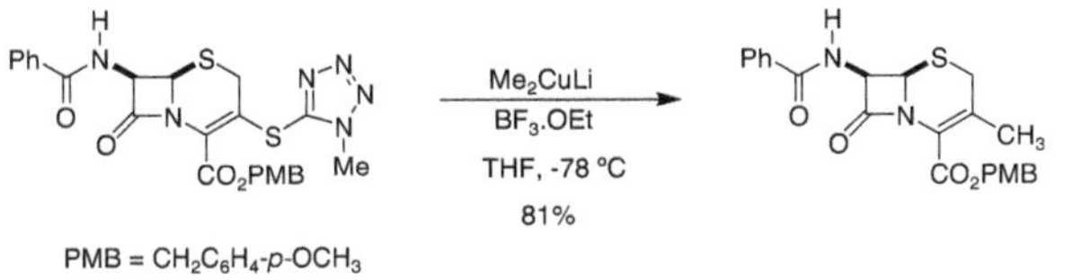

Me_2CuLi, $BF_3.OEt$, THF, -78 °C, 81%

PMB = $CH_2C_6H_4$-*p*-OCH_3

(19)

However, these substrate displayed limited scope in the reactions with vinyl or allylcuprates.

Lipshutz demonstrated Cu(I)-catalyzed substitution reactions of activated vinyl triflates with functionalized organozinc reagents mediated by halocyanocuprates [29]. The coupling required a modest 3 mol% Cu(I) in the form of a commercially inexpensive, stable copper(I) salt. No competing 1.2-addition was observed, and most couplings were complete in an hour or less at reasonable substrate concentrations (ca. 0.25 mol/l). The methyl moiety was found to be the desirable non-transferable (dummy) group on zinc. On the other hand, replacement of methyl with 2-thienyl group also afforded a coupling of roughly comparable efficiency and rate [30, 31] (Eq. 20):

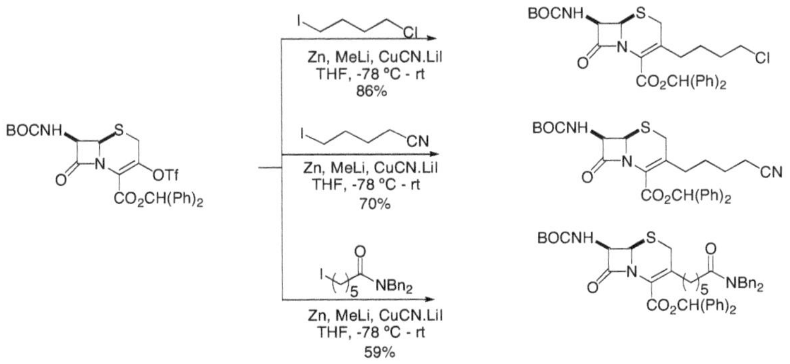

Zn, MeLi, CuCN.LiI, THF, -78 °C - rt: 86%; 70%; 59%

(20)

The likely sequence of events that enables copper to function in a catalytic mode is regeneration of solubilized copper cyanide, presumably kept in solution by LiCl present or the triflate salt formed as a by-product (Scheme 5). This methodology provides an excellent opportunity to synthesize C-3 substituted ce-

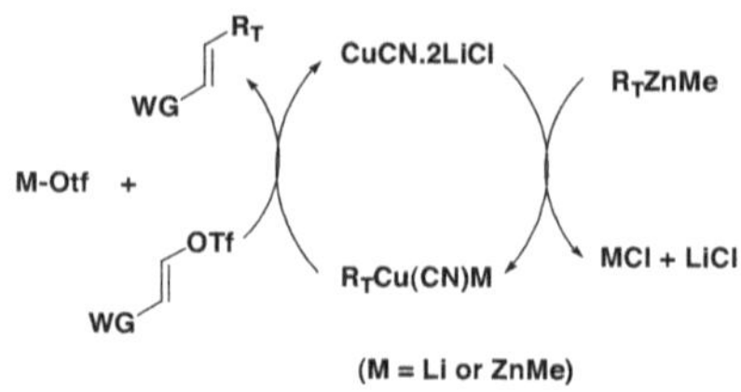

R$_T$ WG M-Otf + OTf WG CuCN.2LiCl R$_T$ZnMe R$_T$Cu(CN)M MCl + LiCl (M = Li or ZnMe)

Scheme 5

phems bearing ω-electrophilic functionality appendages which are otherwise difficult to obtain.

4 Cephalosporins from Allenylazetidinone. A Cyclization Strategy via Tandem Cuprate Addition-Sulfenylation Reaction

Kant and Farina reported yet another novel approach to the synthesis of 3-substituted cephems bearing carbon-based substituents of choice at the C(3) position from inexpensive penicillins using organocopper reagents [32–34]. This strategy involved synthesis of an allenylazetidinone from penicillin sulfoxide followed by the addition of an organocuprate at low temperature. Organocuprate underwent 1.4-conjugate addition at the central allenic carbon of the allenylazetidinone to form a carbon-carbon bond which was followed by ring closure via an intramolecular sulfenylation reaction. The chemistry was applied to the synthesis of a variety of 3-substituted cephems bearing substituents such as alkyl, cycloalkyl, aryl, alkenyl, and allyl. Precursors to the synthesis of important antibiotics, i.e., Cefadroxil, Cefixime, and Cefzil, are available from this novel approach. The methodology is not limited to carbon-based 3-substituted cephems, but provides access to some 3-norcephalosporins as well (Eq. 21):

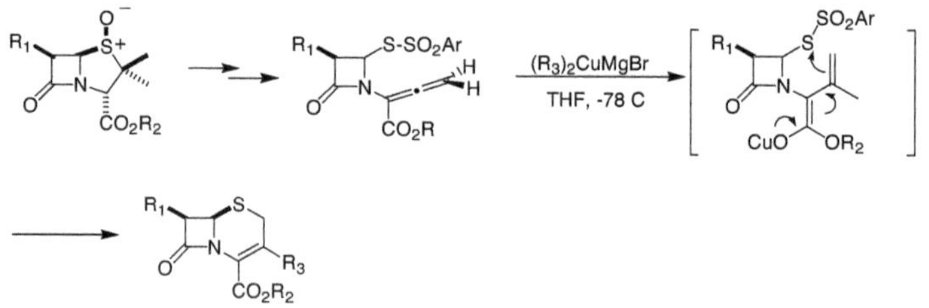

R$_1$ O S+ N O CO$_2$R$_2$ → R$_1$ S-SO$_2$Ar N O H H CO$_2$R $\xrightarrow{(R_3)_2CuMgBr,\ THF,\ -78\ C}$ [R$_1$ SO$_2$Ar S N O CuO OR$_2$] (21)

→ R$_1$ S N O R$_3$ CO$_2$R$_2$

Allenylazetidinone was synthesized in five steps from commercially available penicillin sulfoxide (Scheme 6).

Ring opening of penicillin sulfoxide with 2-mercaptobenzothiazole followed by the treatment with Na salt of *p*-toluenesulfinic acid afforded the key olefin [34]. Alternatively, the olefin can be prepared by the direct ring opening using Torii's condition [35]. Subsequent steps to convert the olefin to the allenylazetidinone employed oxidation of olefin using ozone or $OsO_4/NaIO_4$ [36], synthesis of enol triflate, and elimination. After purification, the allene was obtained as

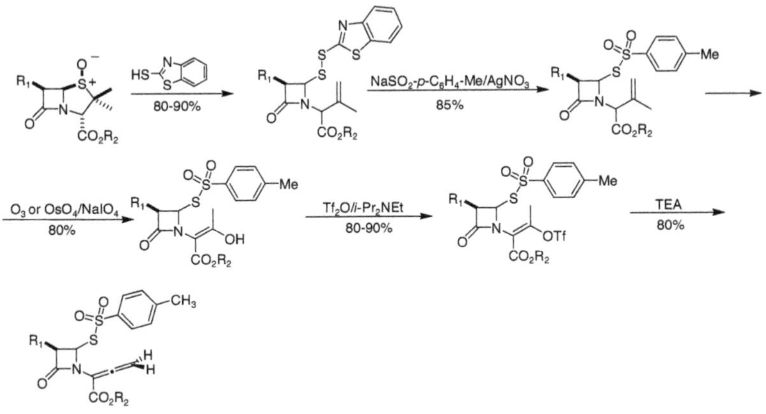

Scheme 6

an off-white amorphous compound which can be stored at 4 °C for months without any decomposition.

Treatment of allene with 1.2–1.5 equiv. of organocuprates derived from Grignards and copper bromide-dimethyl sulfide complex in THF at –78 °C afforded a variety of structurally diverse 3-substituted cephalosporins in high yield and purity (Eq. 22):

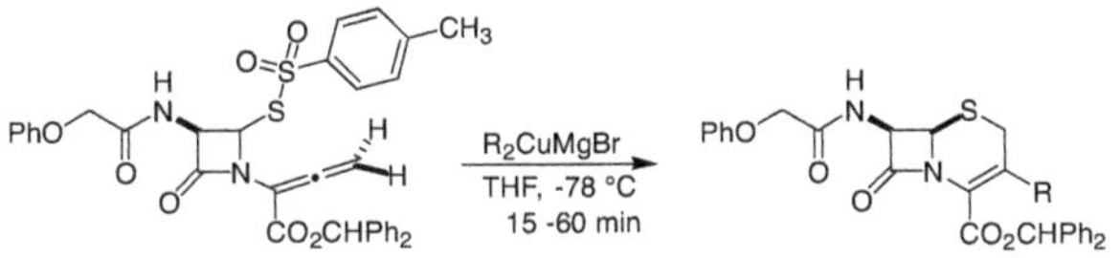

R	Yield
R = Me	95%
R = Et	85%
R = Ph	75%
R = $CH(CH_3)_2$	78%
R = $CH{=}CH_2$	78%
R = $CH_2\text{-}CH{=}CH_2$	82%
R = Cyclohexyl	75%
R = Cyclopentyl	68%
R = C_6H_{13}	78%
R = *t*-butyl	78%
R = (*Z*)-1-propenyl	75%

(22)

A complete stereoselective transfer of the (*Z*)-1-propenyl moiety was observed during the reaction with (*Z*)-1-(propenyl)$_2$CuMgBr and allenylazetidinone. The intermediate produced was converted to the antibiotic Cefprozil in a few steps. Normant's cuprate turned out to be the ideal reagents in order to introduce the desired substituents under cuprate addition-cyclization conditions. The most troublesome Δ^3/Δ^2 isomerization as seen with lower or higher-order cuprates was avoided with these reagents. It was noteworthy to observe a fine balance of reactivity and selectivity between allenyl azetidinones and organocuprates. Since reactive cuprates, such as higher-order or lower-order lithio cuprates are less selective, low yields of isolated cephems along with by-products were observed. The successful use of Normant's cuprates could be the consequence of a reaction between Cu_2MgR_4 (a single entity) and the allene, which is found to be more selective towards the allene, affording higher yields of cephem.

A similar strategy in the synthesis of alkenyl substituted cephems employing the use of alkenyltributyltins in conjunction with copper(I) chloride as opposed to preformed organocuprates was reported by Torri. The reaction, presumably, generates an organocopper species which eventually adds to allenylazetidinone affording the alkenyl cephems in fair to good yields. However, the chemistry is limited to the synthesis of C(3) alkenyl or phenyl substituted cephems (Eq. 23) [37].

R_1 = p-$MeOC_6H_4CH_2$; $RSnBu_3$, CuCl, NMP

R	Yield
R = CH=CH$_2$	70%
R = C(CH$_3$)=CH$_2$	86%
R = CH$_2$=C(CH$_3$)$_2$	56%
R = C=C=CH$_2$	40%
R = Ph	40%
R = (*Z*)-1-propenyl	66

(23)

In a different approach using allenes, Torii demonstrated the synthesis of 3-allyl- and benzyl- substituted cephems via. sequential reductive addition and cyclization of allene with allylic and benzylic halides under a three-metal redox system consisting of aluminum metal and catalytic amounts of $[NiCl_2(bipy)]$ and $PbBr_2$ in NMP. The chemistry, however, has limited scope and often 2-exo-methylenepenam was isolated along with the cephem (Eq. 24) [38]:

R_1 = p-$MeOC_6H_4CH_2$; RBr, $[NiCl_2(bipy)]/PbBr_2/Al$; 0-30 %

R	Yield
R = CH$_2$CH=CH$_2$	85-26%
R = CH$_2$C(CH$_3$)=CH$_2$	73%
R = CH$_2$CH=CH-CH$_3$	82%
R = CH$_2$CH=CH-Ph	60%
R = CH$_2$CH=C(CH$_3$)$_2$	35%
R = CH$_2$Ph	83%
R = CH$_2$C$_6$H$_4$Me-*p*	32%
R = CH$_2$C$_6$H$_4$OMe-*p*	20%

(24)

Interestingly, electrolysis of allenyl azetidinone in NMP containing $[NiCl_2(bipy)]$, $PbBr_2$, and allyl bromide afforded the 3-allyl-Δ^3-cephem (53%) together with the 2-exo-methlenepenem (11%).

5 Conclusion

This chapter summarizes some of the efficient and versatile approaches to the synthesis of carbon-based 3-substituted cephalosporins developed within the last ten years with an aim of commercialization or developing structure-activity relationships to identify new antibiotics. A variety of 3-substituted cephems is available from coupling of 3-trifloxycephems with organostannanes by a modified Stille coupling or via. complimentary organocuprate coupling. A radically different approach to cephalosporins has also been developed which involves the use of novel allenylazetidinones (a modification of the Morin rearrangement) which undergo chemoselective addition of cuprates to the central allenic carbon, followed by rapid cyclization by intramolecular sulfenylation. The starting Grignards are readily accessible, being either available commercially or easily prepared. The allenylazetidinone is readily synthesized from penicillin, yet another inexpensive substrate available from the natural chiral pool. The chemistry can be utilized to attach many desirable carbon tethers to the C(3) position of cephalosporins and therefore should prove valuable to medicinal chemists engaged in the field of cephalosporin chemistry. Furthermore this methodology nicely overcomes the troublesome problem of Δ^3/Δ^2 isomerization, frequently experienced in working with cephalosporins. The combination of these two methods fills a gap in the cephalosporin chemistry, by providing, for the first time, a practical and general approach to almost any conceivable C-3 side chains.

References

1. Durckheimer W, Blumbach J, Lattrell R, Scheunemann KH (1985) Angew Chem Int Ed Engl 24:180
2. Morin RB, Jackson BG, Flynn EH, Roeske RW (1962) J Am Chem Soc 84:3400
3. Morin RB, Jackson BG, Mueller RA, Lavagnino ER, Scanlon, WB, Andrews SL (1963) J Am Chem Soc 85:1896
4. (a) Nagata W, Narisada M, Yoshida T (1982) In: Morin RB, Gorman M (eds) Chemistry and biology of β-lactam antibiotics. Academic Press, New York, vol 2, p 1; (b) Walker DG, Brodfuehrer PR, Brundidge SP, Shih KM, Sapino C (1988) J Org Chem 53:983
5. (a) Kukolja S, Chauvette RR (1982) In: Morin RB, Gorman M (eds) Chemistry and biology of β-lactam antibiotics. Academic Press, New York, vol 1, p 93; (b) Abraham EP, Loder PB (1972) In: Flynn E (ed) Cephalosporins and penicillins, chemistry and biology. Academic Press, New York, p 1
6. Webber JA, Ott JL (1977) In: Pearlman D (ed) Structure activity relationships among the semi synthetic antibiotics. Academic Press, New York, p 161
7. (a) Newall CE (1988) In: Bently PH, Southgate R (eds) Recent advances in the chemistry of β-lactam antibiotics. The Royal Society of Chemistry, London, special publication no 70, p 365; (b) Tomatsu K, Shigeyuki A, Masuyoshi S, Kondo S, Hirano M, Miyaki T, Kawaguchi H (1987) J Antibiot 40:1175
8. Peter H, Rodriguez H, Muller B, Sbiral W, Bickel H (1974) Helv Chim Acta 57:2024
9. Webber JA, Ott JL, Vasileff RT (1975) J Med Chem 18:986
10. Scartazzini R (1977) Helv Chim Acta 60:1510
11. Spry DO, Bhala AR (1985) Heterocycles 23:1901
12. Spry DO, Bhala AR (1986) Heterocycles 24:1799
13. (a) Scott WJ, Crisp GT, Stille JK (1984) J Am Chem Soc 106:4630; (b) Scott WJ, Stille JK (1986) J Am Chem Soc 108:3033; (c) Pena MR, Stille JK (1989) J Am Chem Soc 111:5417
14. (a) Scartazzini R, Bickel H (1984) Helv Chim Acta 57:1919; (b) Kukolja S, Chauvette RR (1982) In: Morin R, Gorman M (eds) Chemistry and biology of β-lactam antibiotics. Academic Press, New York, vol I, p 93

15. (a) Farina V, Baker SR, Sapino C (1988) Tetrahedron Lett 29:6043; (b) Farina V, Baker SR, Benigni, DA, Hauck SI, Sapino C (1990) J Org Chem 55:5833
16. Baker SR, Roth GP, Sapino C (1990) Synth Commun 20:2185
17. (a) Roth GP, Sapino C (1991) Tetrahedron Lett 32:4073; (b) Roth GP, Fuller CE (1991) J Org Chem 56:3493
18. Contento M, Da Col M, Galletti P, Sandri S, Umani-Ronchi A (1998) Tetrahedron Lett 39:8743
19. Tanaka H, Zhao J, Kumase H (2001) J Org Chem 66:570
20. (a) Posner GH (1980) An introduction to synthesis using organocopper reagents. Wiley, New York; (b) Posner GH (1975) Org React 22:253; (c) Lipshutz BH (1987) Synthesis 325; (d) Lipshutz BH, Wilhelm RS, Kozlowski JA (1984) Tetrahedron 40:5005
21. (a) McMurry JE, Scott WJ (1980) Tetrahedron Lett 4313; (b) McMurry JE, Scott WJ (1983) Tetrahedron Lett 1979
22. (a) Kant J, Sapino C, Baker S (1990) Tetrahedron Lett 31:3389; (b) Kant J (1993) J Org Chem 58:2296
23. (a) Ashby EC, Watkins JJ (1977) J Am Chem Soc 99:5312; (b) Lipshutz BH, Kozlowski JA, Breneman CM (1985) J Am Chem Soc 107:3197
24. Lipshutz BH, Ellsworth EL, Siahaan T (1989) J Am Chem Soc 111:1351
25. Ashby EC, Goel AB (1983) J Org Chem 48:2125
26. Lipshutz BH, Elworthy TR (1990) J Org Chem 55:1695
27. Roth GP, Peterson SA, Kant J (1993) Tetrahedron Lett 34:7229
28. Tanaka H, Sumida S-I, Torri S (1996) Tetrahedron Lett 37:5967
29. Lipshutz BH, Vivian RW (1999) Tetrahedron Lett 40:2871
30. Lipshutz BH (1994) In: Schlosser M (ed) Organometallics in synthesis: a manual. Wiley, New York, p 283
31. Malmberg H, Nilsson M, Ullenius C (1982) Tetrahedon Lett 23:3823
32. Farina V, Kant J (1992) Tetrahedron Lett 33:3559
33. Kant J, Farina V (1992) Tetrahedron Lett 33:3563
34. Kant J, Roth JA, Fuller CE, Walker DG, Benigni DA, Farina V (1994) J Org Chem 59:4956
35. (a) Torii S, Tanaka H, Shiroi H, Sasaoka M, Saito N, Matsumura K (1986) US Patent 4 566 996; (b) Tanaka H, Kameyama Y, Yamauchi T, Torii S (1992) J Chem Soc Chem Commun 1793
36. Thomas J, Kant J (1993) Synthesis 293
37. Tanaka H, Kameyama Y, Sumida S, Torii S (1992) Synlett 33:7029
38. Tanaka H, Sumida S-I, Sorajo K, Kameyama Y, Torii S (1997) J Chem Soc Perkin Trans 637

Topics Organomet Chem (2004) 6: 263–283
DOI 10.1007/978-3-540-36966-0

Removal of Metals from Process Streams: Methodologies and Applications

Jeffrey T. Bien · Gregory C. Lane · Matthew R. Oberholzer

Bristol-Myers Squibb, Process Research and Development, P.O. Box 191, New Brunswick, NJ, 08903-0191, USA
E-mail: jeffrey.bien@bms.com, gregory.lane@bms.com

Abstract Metal-mediated processes are used within the pharmaceutical industry to prepare drug intermediates and drug substances. As such, the process research and development scientist is often faced with the challenge of removing the spent metal from the process stream. There exist many tactics that can be applied towards this goal. This review summarizes options for metal removal from process streams against a backdrop of factors to be considered for scaleup.

Keywords Metals removal · Process development · Pharmaceutical · Palladium · Catalyst

1 Introduction

Metal catalysts are of interest in academic and industrial settings because they effectively mediate a wide range of bond-making and bond-breaking reactions [1]. During the development of these transformations, special attention needs to be given to the workup protocol that will be used to remove the metal from the process stream, pharmaceutical intermediate, drug substance, or waste stream. Residual metal may have deleterious effects on downstream processing or metal remaining in the final product may raise quality and safety concerns, particularly for pharmaceutical compounds. The inability to adequately remove metal can have serious implications, and literature examples describe how changes in either the step chemistry or the overall synthetic route were required to avoid metal contamination [2]. Additionally, the process research and development scientist needs to be aware that metal contamination can also occur from unintended sources [3].

Understanding the cause of metal contamination is a good starting point for the selection of a metal removal strategy. The choice may be made after tracing the metal through the sequence of unit operations; that is, where the metal is introduced, what happens to it during the process, and where the metal finally resides. Even when the source of metal contamination is known, the development of a metal removal procedure can require a considerable investment in time. For example, unacceptably high levels of palladium found in a product after palladium on carbon (Pd/C) catalytic hydrogenation could result from (1) leaching of Pd from the carbon support into the product stream, (2) leaching of Pd coupled with binding of Pd by the product, or (3) passing of fine Pd/C particles during catalyst filtration. The first scenario may be remedied by a simple acid extraction if Pd (II) is the culprit. The second case is more complicated and could require extraction or precipitation with a specialized reagent. Crystallization could also be employed to reject preferentially the metal complex. The third scenario is different in that the contamination can be traced to physical causes such as particle attrition and/or poor filtration.

Many approaches for metal removal are available, but unfortunately no single removal procedure will solve every metal contamination problem. In fact, multiple procedures may be required after a single metal-catalyzed reaction. The selection of a metal removal process from among the variety of options may be complicated by the inherent sensitivity to process conditions. The purpose of this work is to outline the strategies for reducing the level of metals in process streams and to review the application of these approaches at the pilot plant scale. Because the impact of metal contamination is of critical concern for active pharmaceutical ingredients, regulatory requirements specific to these products are also reviewed.

2
Before the Metal is Added

Perhaps the best place to consider metals removal is during process development work in the laboratory. If product contamination by metal catalysts or reagents is expected and unavoidable, then optimizing the metal charge may minimize the effort needed to remove it.

For a metal-catalyzed reaction, it is worthwhile to study how all of the parameters affect the process. Statistical design of experiments (DOE) [4] can be used to gauge the sensitivity of the reaction to process parameters. In some cases the results may indicate that the process is less sensitive to catalyst loading and more sensitive to other process variables [5]. Perhaps those process variables could be adjusted to compensate for the effect of a reduction in the catalyst loading.

The DOE technique was utilized for the process development of a β-elimination reaction mediated by zinc (Scheme 1) [6].

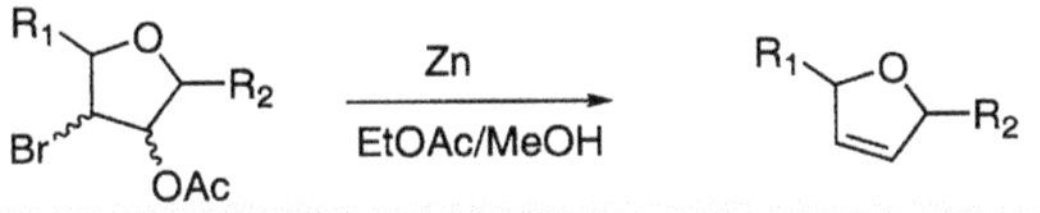

Scheme 1 β-Elimination step studied by DOE

The lab procedure specified the use of 100-micron zinc particles. The DOE study examined the effect of several process parameters on the reaction rate (Table 1); the results of a fractional factorial trial set indicated that temperature, metal equivalents and solvent had the greatest effect on the reaction rate. It was determined that the reaction could be run successfully with a lower metal loading, which in turn would lighten the burden of removal.

When the metal loading cannot be reduced, switching the form of the metal catalyst may be advantageous. Homogeneous catalyst systems, such as those prepared by dissolving a palladium-ligand complex or generating the Pd complex in situ, may be replaced with heterogeneous catalyst systems. Suzuki and Heck type coupling reactions have been mediated by Pd/C catalysts [7], resin and clay based catalysts [8], and supported liquid phase catalysts [9]. With heterogeneous catalysts, the bulk of the metal can be removed directly by filtration [10]. In some instances, metal leaching was reduced and the catalyst was successfully recycled [11].

Table 1 Parameters and ranges for the DOE study of a β-elimination reaction

Parameter	Range lower limit (-)	Range upper limit (+)
Temperature (°C)	25	30
Stirrer speed (RPM)	100	250
Acid equivalents	0.21	0.82
Solvent	THF	Methanol

3 Extraction

Metals can be removed from process streams using extraction techniques. These approaches are often a component of an isolation protocol that involves additional adsorptive or polishing techniques. The development philosophy can be viewed in two ways: (1) adjusting the process stream to the extraction reagent or (2) finding an extraction reagent that works directly on the process stream. The basis for these approaches suggests caution when selecting an extraction reagent with functionality similar to the product compound. In those cases, the reagent must simply have a higher affinity for the metal under the processing conditions.

In biphasic aqueous systems, two scenarios are possible: (1) The metal-rich organic solution can be treated with aqueous acid or aqueous base [12] or (2) the metal can be extracted as a complex from the aqueous rich phase into the organic rich phase or vice versa using either an organic or aqueous soluble extraction reagent; this process is accelerated by addition of a phase-transfer catalyst.

The reagents listed in Table 2 are capable of removing metals by extraction, and many are commercially available. The metal removed is typically a platinum group metal (PGM) in an oxidation state greater than zero. PGMs consist of platinum, palladium, rhodium, and ruthenium, but these extraction reagents can also be applied to the removal of metals such as nickel, copper, iron, and zinc. The ideal extraction agent should be non-toxic, selective for the metal in the process stream, and removable upon isolation of the product.

The metal form is likely to influence the selection of an extraction reagent. The metal may exist in the process stream as a neutral metal complex, such as palladium (0), in which case lab development should focus on those reagents that coordinate metals. The metal may exist as a negatively charged metal complex, such as $PdCl_4^{2-}$ formed by reaction of Pd (II) with hydrochloric acid [13]. Palladium (II) tetrafluoroborates and tetraphenylborates [14] exist in solution as positively charged metal complexes. Charged metals can be removed from process streams by extraction reagents that can form the appropriate counter ion under process conditions. In some cases, the metal may exist in the process stream as a mixture of different valences. The activity of some heterogeneous palladium catalysts had been attributed to the presence of Pd (0) in solution [15]. Lastly, colloidal catalysts may release some metal to the process stream, such as disaggregated metal colloidal particles or palladium black [16]. Complete removal of metal in these cases may require both coordination and counter-ion extraction reagents.

Many of the problems encountered in the scale up of batch extraction can be observed and addressed in the laboratory. Emulsification, precipitation, slow coalescence time, and rag layer formation are some of the typical issues that may be encountered on scale, and simple optimization of process conditions may mitigate them. For example, the temperature or volume of liquid phases can be increased to prevent precipitation of the product or the extracted complex, coalescence time may be reduced by increasing the ionic strength of the aqueous phase with sodium chloride, and an excessive rag layer may be removed by filtration prior to phase separation. It is essential to use actual plant grade materi-

Table 2 Metal extraction and precipitation reagents

Reagent	Mechanism
Acid or base [12]	Extraction, precipitate
Amines	
1. *N*-Octylaniline and mixed trialkylamines (Alamine 300 and 336) [17]	Extraction
2. Quaternary ammonium salts (Aliquat) [18]	Extraction
Calixarenes [22]	Extraction
Carboxylic acids	
1. Citric acid [19]	Extraction
2. Ethylenediamine tetraceticacid (EDTA – tetrasodium salt – Sequestrene 200) [20]	Extraction
3. Iminodiacetic acid [23]	Extraction
4. Lactic acid [21]	Extraction
5. Tartaric acid [19]	Extraction
Crowns and lariats [22]	Extraction
Hydroxyaromatics	
1. 7-[4-ethyl-1-methyloctyl]-8-quinolinol] Kelex-100 [23]	Extraction
2. 1-(2-Pyridylazo)-2-Naphthol (PAN) [24]	Extraction
Hydroxylamine [22]	Extraction
Oximes	
1. Aldoximes (Acorga CLX-50) [25]	Extraction
2. 5-Dodecyl salicylaldoxime (LIX-860)	Extraction
3. Hydroxy-5-nonylacetophenone oxime (LIX-84I)	Extraction
4. Hydroxy-5-nonylbenzophenone oxime (LIX-65N) [26]	Extraction
5. LIX 860+LIX 65N (LIX 865)	Extraction
6. Phenyloximes [27]	Extraction
Phosphorus containing [28]	
1. Phosphine oxides [29]	Extraction
2. Phosphonium salts [30]	Extraction
3. Tributylphosphine [31]	Extraction
4. Triphenylphosphine [32]	Extraction
5. Tris(hydroxymethyl)phosphine [33]	Extraction
Sulfur containing	
1. Dioctylsulfide[23]	Extraction
2. Dithiocarbamates (Aquamet) [34]	Precipitate
3. CYANEX 301 bis (2,4,4-trimethylpentyl) dithiophosphinic acid [35]	Extraction
4. CYANEX 471 triisobutylphosphine sulfide [38]	Extraction
5. Thiourea [36]	Extraction
6. Cysteine and *N*-acetyl cysteine [37]	Extraction
7. Thiosulfate [38]	Extraction
8. 2,4,6-Trimercapto-*s*-triazine (TMT) [39]	Precipitate
9. Sulfur containing monoamides [40]	Extraction
10. Xanthates [41]	Extraction, precipitate
Ureas [22]	Extraction

als in the lab at some point in development to determine if impurities exacerbate unclean phase boundaries. In cases where extraction efficiency depends on pH, the sensitivity to pH should be thoroughly examined in the laboratory.

Some aspects of extraction scale up require more careful consideration. Efficient extraction requires intimate contact between the aqueous and organic layer; consequently, the ratio of organic volume to aqueous volume becomes a critical parameter. Sufficient aqueous phase should be present to form a continuous phase around droplets of the organic phase, as suggested by Carpenter [42]. The interphase mass transfer will depend on how well the phases are mixed, and mixing in turn depends on the reactor and impeller design. Empirical correlations for dispersal based on stirring speed and on specific power input can aid in the estimation of scaled agitation rates [43].

In general, extraction processes are more efficient when carried out in a semi-continuous manner with countercurrent flow. Specialized equipment such as agitated towers, perforated plate columns and centrifugal extractors ensure efficient contact between the product stream and the aqueous stream [44].

Supercritical fluid extraction has also received attention for its ability to remove metals from process streams. It has been shown that supercritical carbon dioxide can extract both nonfluorous and fluorous ligand-metal complexes from aqueous environments. Fluorous ligands are linear or branched perfluoroalkyl chains with high carbon number [45]. The metal-fluorous ligand chelates perform effectively due to their higher solubility in the supercritical CO_2. Overall extraction efficiency was found to be effected by choice of ligand, chelate solubility in supercritical CO_2 and the extraction matrix. Fluorous hydroxamic acids and dithiocarbamates [46] have also been reported to extract metals into supercritical CO_2.

4 Crystallization

Crystallization can be used to reduce metal contamination when the metal species (i.e., free metal, metal-ligand complex, or metal-product complex) has a substantially larger solubility than the metal-free product. Crystallization will not be effective if (1) the metal species co-crystallizes with the desired product, (2) the metal species adsorbs on the product crystal surfaces, or (3) the metal becomes trapped inside the product crystals as they form. In some cases, the presence of the metal contaminant may inhibit crystallization [47], or lead to formation of the incorrect polymorph [48]. High metal loads may be successfully reduced through crystallization (e.g., from several thousand ppm to several hundred ppm); generally, crystallization is not very effective at low metal loads (e.g., <100 ppm). The benefits of this approach should be weighed against potential reductions in product yield.

Before scaling up a crystallization process, accurate solubility of the pure product over a range of process temperature and solution conditions must be measured [49]. Measurement of product concentration in solution during the crystallization helps to elucidate crystal growth kinetics [50]. These data can then be used to develop a crystallization control strategy: controlling crystal nucleation by seeding and controlling crystal growth by temperature cycling, adjusting rates of cooling and/or anti-solvent addition [51]. The chances of rejecting metal species from the crystal lattice are better if spontaneous nucleation and rapid agglomeration can be prevented. It may be possible to enhance the solubility of a metal by adding a soluble stabilizer. Merck researchers have shown that Pd (0) could be stabilized in solution with tributylphosphine and removed from the product slurry by filtration [31].

After the crystal cake is isolated from the metal-rich crystallization liquor, fresh solvent is used to displace the remaining liquor from the voids between crystal surfaces. The wash solvent should be similar in composition to the process stream to prevent the precipitation and/or crystallization of the metal species. The wash solvent volume depends on the filtration characteristics of the cake; typically, it will equal two to three times the cake volume.

5 Precipitation of Residual Metals

The extreme case of uncontrolled crystallization is precipitation – as is often seen in the formation of organic salts in solvents of low polarity. If only a small amount of an unwanted impurity is forced from solution, precipitation can prove quite useful. In fact, for trace metal removal at the plant scale, precipitation of the metal or metal complex away from the product may offer excellent efficiency compared to crystallization of the product away from the metal.

The ideal precipitation process involves charging the precipitating agent to the process solution containing the metal and mixing the batch for a short time at moderate temperature while large crystalline particles precipitate. The precipitate would be removed easily by simple filtration with minimal rinsing and no loss of product to the cake. Additionally, the precipitating agent would be non-toxic, would not affect any reactions downstream of the precipitation process, and ultimately would remain behind in solution when the desired product is isolated.

Of course, not every precipitation works this way. There are several things to consider prior to application on larger scale. A complex forms when the metal interacts with the precipitation agent. This is dependent upon concentration, mixing, the complex stability itself and complex stability under the process conditions. The complex may precipitate as fine particles or in an amorphous state, both of which can be difficult to filter. The removal of fines may require the use of several different filtration devices or filter aids (e.g., Celite) [52]. Product may be lost to the precipitate cake or the filtration train. Product recovery could require multiple cake and/or line rinses with concomitant increases in the batch volume, capital costs and reduction in process throughput. Dense or agglomerated precipitates can lead to rapid settling or flocculation, which could plug a reactor bottom outlet.

Successful precipitation procedures have been carried out on a pilot plant scale. Researchers at Bristol-Myers Squibb described the use of 2,4,6-trimercapto-*s*-triazine (TMT) to precipitate palladium from an aqueous acetonitrile product stream [39]. The optimized process involved an additional treatment with charcoal and diatomaceous earth and Pd levels were reduced from 300–600 ppm to <3 ppm. Residual TMT was absent from the product. Additionally, the authors note that TMT has favorable toxicology characteristics.

In another example, Bristol-Myers Squibb researchers were able to remove a zinc chelate from an ethyl acetate product solution by adding aqueous potassium carbonate; zinc carbonate precipitated from the mixture [6]. Precipitation alone was not sufficient; a filter aid was needed to form filterable granules and several washes of the precipitate cake were added to maximize product recovery. In addition, two aqueous EDTA extractions were required to reduce the zinc level below the 10 ppm target.

6 Adsorption

6.1 Batch Adsorption

One of the easiest metal removal treatments to conduct in the laboratory is adsorption, in which the metal partitions out of the solution phase and onto an insoluble (and otherwise inert) material added directly to the reaction vessel. After the system comes to equilibrium, the metal-enriched adsorbent is separated from the process stream by filtration. Typical adsorbents include activated carbon, functionalized polymer resins, silica (including functionalized silicas), alumina, zeolites and clays. Examples are shown below in Table 3.

The mechanism of adsorption consists of three events: (1) transport of the metal through the solution to the neighborhood of the solid particle, (2) diffusion of the metal through the fluid boundary layer to the surface of the particle, and (3) adsorption onto the surface. If the interior of a porous particle is available for absorption then the metal must diffuse into the pores and then adsorb there.

Of the three events, the third is the easiest to measure. The capacity of the adsorbent for the metal contaminant can be determined in the lab by varying the amount of adsorbent added to a fixed volume of solution of known product concentration. After appropriate mixing at either room or elevated temperature, the adsorbent is filtered off and the product is tested for residual metal content.

Several examples of adsorbent use and performance are noted here. SMOPEX isothiouronium fibers (1.5 g) were used to reduce homogeneous Pd catalyst levels from 395 ppm to 3 ppm in 130 g of an aqueous *N,N*-dimethylformamide reaction stream [72]. The SiliCycle thiol silica – 4 mmol relative to catalyst input – was used to reduce Pd(II) levels from 1000 ppm to <1 ppm in a THF-rich reaction stream [71]. SMOPEX sulfonic acid fibers (0.25 g) and pyridinium fibers (0.25 g) were used after filtration of a Pd/C catalyst to reduce the leached metal from 23 ppm to <1 ppm in 50 ml in a methanol-rich reaction stream [72]. Cationic rhodium could be reduced from 105 ppm to 2 ppm using SMOPEX sulfonic acid fibers (1.0 g) and thiourea (0.5 g) in 50 ml of an ethanol rich reaction stream [72]. The SiliCycle triamine silica was reported to reduce ruthenium levels from 1000 ppm to 86 ppm [71].

Table 3 Sorptive methodologies for metals removal

Methodology
1. Alumina – Activated [53]
2. Amberlite IRA, IRC-718, GT-73 and XAD resins [54]
3. Carbon/charcoal (e.g., DARCO) [55]
4. Celite [52]
5. Cellulose – functionalized [56]
6. Cellulose – unfunctionalized (e.g., Solka Floc) [57]
7. Chelex iminodiacetate resins [58]
8. Chitosan [59]
9. Clays [60]
10. Deloxan aminopropylated polysiloxane [61]
11. Dowex cation and anion exchange resins [62]
12. Keratin – reduced [63]
13. Membranes [64]
14. Polymer, dendritic supported phosphines, amines, alkoxides [65]
15. Polymer supported 2,4,6- trimercapto-*s*-triazine (TMT) [66]
16. Polymer supported 8-hydroxyquinoline [67]
17. Polymeric *N,N*-2-pyridylamides [68]
18. POLYORGS purazole, imidazole, pyrazole, amine fibers, granules, powders [69]
19. Polythioamides [63]
20. ScavNet [70]
21. SiliCycle urea, thiol, triamine, triaminetetraacetate Functionalized silica [71]
22. SMOPEX pyridinium, (di/tri)methylammonium, sulfonic acid, carboxylic acid, thiol, isothiouronium, polyethylene or cellulose based fibers [72]
23. Zeolites [73]

The first two events, transport of the metal through the solution to the solid adsorbent and diffusion of the metal through the fluid boundary layer onto the surface of the adsorbent, are more sensitive to scale because they are both mass transfer processes involving heterogeneous systems. The efficiency of metal transport in solution depends greatly upon thorough mixing of the solid particles within the entire reaction mixture. Different impellers have a different specific power input for a given stirring speed, and correlations for suspending solids are available [74].

The optimum impeller type may not be available in the pilot plant, so to assess the influence of mixing on the adsorption process one should consider the "scale down" approach. Here, batch adsorption is performed in a straight-walled lab reactor [75] with an impeller that mimics the geometry of that used in the plant. The impeller height above the reactor bottom should reflect the fraction of the batch depth above and below the impeller expected in the plant. Adsorption kinetics then can be measured as a function of stirring speed, taking care to explore agitation speeds at which uniform mixing does not occur.

Physical adsorption of metal contaminants can be influenced by temperature, pH, and solvent composition and it is worthwhile to research these parameters in the lab before scale up. In one example of palladium removal by activated carbon, batch adsorption treatment after a Suzuki coupling proceeded efficiently in heptane but performed poorly with 10% or more ethyl acetate in heptane (by volume) [76]. Ethyl acetate was the reaction solvent, so the switch to heptane had to be carefully monitored.

Batch adsorption processes are conducted on both laboratory and pilot plant scale, but there are some typical shortcomings of the method. The adsorbent may require some type of pretreatment to wash out reaction byproducts, to displace unwanted residual solvent (e.g., water), or even to activate it for adsorption. Valuable product may adsorb onto the solid particles in addition to the metal, and yield losses to the filter cake can be substantial. Adsorption of organic molecules on activated carbon is the basis for much of the world's wastewater treatment, so product loss to carbon should be expected.

Excessive power input during agitation may have the unpleasant consequence of particle attrition, especially if polymer resins or granular activated carbon are used. Powdered carbon is perhaps the most insidious adsorbent as it seems fine particles always find their way to all parts of the equipment. Removing powdered carbon from any reactor often requires manual scrubbing of the reactor walls. The effective cycle time for an adsorption process can be inflated substantially by cleaning time [77].

6.2 Column Adsorption

The essential difference between batch and column adsorption is how the product solution contacts the solid phase, and in this difference lies the advantage. In batch operation, because the solid adsorbent and process stream are mixed together at the start of the process, adsorption kinetics depend on how easily metal in solution finds free sites on the adsorbent surface. As adsorption proceeds, the concentration of the metal decreases and the effective concentration of available adsorption sites decreases as well. Consequently the driving force for adsorption decreases. In column operation, the concentration of metal decreases as the process stream travels through a column. However, the stream continually encounters fresh adsorbent, so a higher driving force for adsorption is sustained as long as fresh adsorbent is available (i.e., if the column is long enough).

Even though in a packed column the process stream is forced through the tortuous network of interstitial void space, resistance to mass transfer remains in the form of diffusion. The linear velocity, vo, (i.e., volumetric flow rate divided by the empty column cross sectional area) of the process stream is a key parameter in tuning the efficiency of column adsorption. If vo is too high, then the metal has insufficient time to diffuse and adsorb, and therefore the packing is not fully utilized. In this scenario, a longer column and more adsorbent will be needed to complete the removal. If *vo* is too low, then axial dispersion may lower the effective concentration of product, thereby slowing the adsorption kinetics and diluting the product exiting the column.

Perhaps the most important design consideration for column adsorption is whether the metal or metal-product complex adsorbs reversibly or irreversibly onto the solid phase. Irreversible adsorption means that the column can be washed with fresh solvent to recover co-adsorbing product without fear of metal desorption and recontamination. Once the solid phase reaches its capacity for the metal, it must be removed and replaced. Reversible adsorption behaves very much like chromatography in that both the clean product and the metal eventually elute from the solid phase and the solid phase may (perhaps) be reused. The challenge is to design *vo* and column length *L* within reasonable limits so that the two do not co-elute.

Before column design and operation can be studied, batch adsorption data (adsorption isotherms) need to be obtained in the laboratory. A variety of adsorbents compatible with the process stream should be tested for efficacy through a batch adsorption screening protocol [78], taking care to minimize attrition of particles. It is worthwhile to test both the dry isolated product and dry adsorbent after treatment for metal content and complete a mass balance to verify adsorbent performance. Candidate adsorbents should be considered not only for capacity (kg metal/kg adsorbent) but also for how much they contribute to the overall cost of the product. Successful candidates can then be tested in small scale columns using a preparative HPLC apparatus to determine the optimum linear velocity, *vo*, for purity and recovery.

The flow rate, column length and particle size cannot be chosen arbitrarily because all of these parameters influence the amount of pressure needed to force the process stream through the bed [79]. Higher pressures are needed to pump liquid through a bed of smaller particles and/or a bed of lower volume fraction of void space. Therefore, fine powders and highly compressible solids should be excluded from consideration.

Scale up of column adsorption requires the consideration of several factors. Scaling on linear velocity will work only if the same particles are used in the lab as in the plant, and the column efficiency is the same. Large scale columns may not run as efficiently as smaller columns because of uneven packing of the adsorbent and/or unequal liquid distribution entering the column. These could lead to channeling, where the process stream does not flow uniformly through the bed and consequently some fraction of the adsorbent is not fully utilized. These factors can be checked and should not prevent consideration of this as a technique to metals removal on scale.

Other equipment is available for fixed bed adsorption in the plant, including cartridge filters and pad filters. These devices are designed to separate process streams from catalysts and typically incorporate a fine mesh filter or membrane coupled with an activated carbon core to remove residual metals from solution. Examples include Koch Membrane Systems, Inc. CARBO-COR fixed columns and Cuno, Inc. ZETACARBON cartridges and housings [80]. Prior to use, one should determine their capacity for the metal and the contact time of the process fluid with the adsorbent required for metal removal. In this way, recirculation may be avoided. These units offer lower pressure drop, easy installation and clean up. In fact, many of these companies have systems that are designed to capture catalysts through to manufacturing scale.

7
Decantation

Decantation, also known as siphoning, can be used in place of filtration to separate the process stream from solid metal particles. Decanting is useful for gross separations, as in the case of removing water from Raney nickel, but it can be impractical to perform on scale. Decanting requires time to allow metal to settle below the suction (siphon) inlet. Fine metal particles can be difficult to remove and when present in large amounts, can plug filtration equipment. In addition, the remaining metal particles must be thoroughly cleaned from the reactor [81]. Decantation also leaves behind some of the product rich process stream; additional solvent, followed by decantation, would be required to improve product recovery thus increasing cycle time and batch volume.

8
Reality in the Pilot Plant: Unintended Sources of Metal

Perhaps the most frustrating complications to scale up are those that result from unexpected yet avoidable conditions. Contamination from corroded process equipment ranks highly on the list of common but unanticipated scale-up problems. Regular inspection and testing of plant equipment is an integral part of preventative maintenance and should be able to identify corrosion problems, but only after the damage has started.

High levels of iron in a batch may suggest corrosion of steel. This type of corrosion can be quite severe (Fig. 1). The three-blade retreat-curve impeller shows

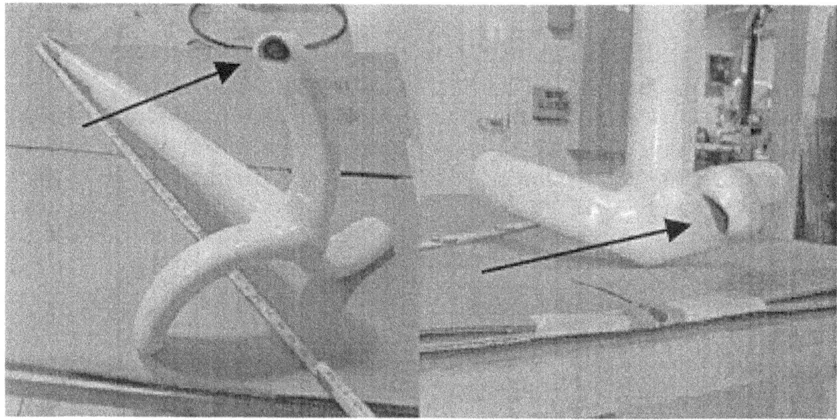

Fig. 1 Corroded impeller as source of metal contamination

extensive damage. The agitator was originally coated with glass, but the coating either degraded with time or had chipped during use. The steel became exposed to process liquids, some of which were very corrosive; a recent bromination reaction was suspect. The damage shown (Fig. 1) went unnoticed until a compound prepared in the glass-lined vessel failed metals analysis due to high levels of iron [82].

Energy-Dispersive X-Ray Analysis (EDAX) [83] of product samples indicated the presence of iron oxide, but no corrosion was observed in the portable equipment or the transfer lines. Because of its location, the damage to the impeller was not visible until the entire reactor system was inspected by video boroscope.

Reactor glass lining can degrade through repeated scouring by solid reagents or products or by ultrasound vibration. Carbon steel and stainless steel piping are susceptible to corrosion but are less expensive to install than more resistant alloys such as tantalum and Hastelloy-C [84]. This piping is also much less fragile than glass. Process lines, distillation overhead piping, fittings, filter housings, and reactor internals should be examined on a regular basis. In addition to the agitator damage mentioned previously, corrosion was found in three other places in the same reactor set: (1) the stainless steel vent pipe, (2) a stainless steel pressure gauge which had separated from its protective Teflon [84] backing, and (3) a stainless steel pipe used to transfer distillate into a receiver [82]. Fortunately, these areas were not in contact with the process stream. Material compatibility testing – performing the lab scale reaction and workup in the presence of a small piece of Hastelloy-C [84] or stainless steel or borosilicate glass – may reveal corrosion problems well in advance of the plant campaign. Extensive corrosion tables have been published for most materials of construction in contact with many reagents and solvents [85].

Metals can enter the process stream in other unexpected ways. During the pilot plant processing of a candidate protease inhibitor, aqueous potassium carbonate was charged from a polyethylene-lined drum to a 300 gallon (1200 l) glass-lined reactor containing a solution of the product in isopropyl acetate [86]. The batch was mixed and allowed to settle, during which time both yellow liquid and a brown gelatinous phases appeared. Laboratory testing showed that a 2% solution of citric acid added to the batch dissolved the gel and yielded a clear "water white" upper phase and a cloudy orange colored lower layer. Iron contamination was suspected, but there was no obvious corrosion from transfer piping, connections or ancillary equipment. The source of contamination was determined to be the drum containing the aqueous potassium carbonate solution – the thin polyethylene liner had ruptured during overnight storage, and the potassium carbonate solution had contacted the inner surface of the carbon steel drum. Iron was extracted into the potassium carbonate solution. In this case, a small amount of metal complicated a routine extraction. To avoid reoccurrence, more rugged plastic drums are now used in place of the lined steel drums for corrosive aqueous mixtures.

9
Analytical

Several analytical techniques are useful for metals analysis. Those most commonly utilized for the quantitation of residual metals are inductively coupled plasma atomic emission spectroscopy (ICP-AES), inductively coupled plasma mass spectroscopy (ICP-MS), atomic absorption spectroscopy (AAS), and graphite furnace atomic absorption spectroscopy (GFAAS) [87] using a suitable metal standard reference material (SRM) [88]. The SRM matrix should be similar to that in the samples. The background in ICP-MS and GFAAS is very low, and can potentially provide for greater sensitivity. The following sensitivities are possible: ICP-MS (parts per quadrillion/trillion), ICP-AES (parts per billion/million), AAS (parts per billion/million), GFAAS (parts per billion/million). ICP and AAS are ideal for ultra-low sample sizes. Sample preparation is of vital importance to measurement reproducibility and accuracy; it is the most time consuming step in the analytical process. Samples could be prepared by an acid mediated decomposition/digestion – which must be complete. Typical analysis times are in the range of 2–6 min. Along with EDAX [83], X-ray fluorescence (XRF) spectroscopy should also be mentioned. XRF requires abundant material; the technique is nondestructive and can analyze solids, liquids, and gases. Analysis time is from 10 to 30 min, and is sensitive to heavier elements (sodium to uranium) at the part per million level. Classical methods for analysis, such as spectrophotometric detection of colored metal complexes, have also been reported for palladium [89].

10
Regulatory Issues for Pharmaceuticals

During the development of a pharmaceutical product, specifications must be developed for safety, quality, and efficacy that fulfill the regulatory requirements of the intended market. Process research and development scientists are charged with the responsibility of delivering drug substances that meet these predefined specifications. One of the areas in process development that causes considerable concern is that of impurities, both organic (isomers, starting materials, process-related impurities, residual solvents) and inorganic (heavy metals or residual metals, catalysts, inorganic salts, filter aids) [90]. The International Conference on Harmonization of Technical Requirements for Registration of Pharmaceuticals for Human Use (ICH) publishes scientific guidelines for pharmaceuticals with the intent of providing global specifications that will be recognized by the regulatory agencies of the United States (Food and Drug Administration, FDA), European Union (Committee for Proprietary Medicinal Products, CPMP) and Japan (Ministry of Health, Labour, and Welfare, MHLW) [91]. As far as inorganic impurities are concerned, the ICH guideline on new drug substances states:

> "Inorganic impurities are normally detected and quantified using pharmacopoeial or other appropriate procedures. Carry-over of catalysts to the new drug substance should be evaluated during development. The need for inclusion or exclusion of inorganic im-

purities in the new drug substance specification should be discussed. Acceptance criteria should be based on pharmacopoeial standards or known safety data [92]."

Thus, the use of a catalyst or reagent containing a metal in a synthetic scheme will require that its level be monitored as the synthetic scheme advances or at a minimum, at the final stage of drug substance testing.

The most common test for heavy metals is described in the United States Pharmacopoeia (USP), with similar methods reported in the European Pharmacopoeia (EP) and the Japan Pharmacopoeia (JP) [93]. The test consists of a visual comparison between a lead (Pb) standard that has been treated to generate a colored sulfide and the test substance treated under similar conditions. According to the USP, heavy metals that will usually give a positive result include lead (Pb), mercury (Hg), bismuth (Bi), arsenic (As), antimony (Sb), tin (Sn), cadmium (Cd), silver (Ag), copper (Cu), and molybdenum (Mo). There can be considerable variation in the levels of heavy metals that are acceptable in a pharmaceutical drug substance. Factors such as the dosage strength, mode of administration (oral vs injectable), treatment population and duration (chronic vs short term), known toxicity of the metal in question and the ability of manufacturing process to control the metal levels will all play a role in setting a residual metal specification. Ultimately, patient safety is the primary consideration. Since the USP test uses a Pb standard of 10 μg/ml (10 ppm), the heavy metals specification for drug substances is often set at 10 ppm (for solids: 10 μg in 1 g). However, the presence of metals such as As, Cd, Cr, and Hg in a synthetic sequence will immediately raise concerns due to their known toxicities [94]. As such, limits for these metals in particular are often set well below 10 ppm. There are separate USP methods for As and Hg which can detect levels of 3 ppm and 1 ppm, respectively [95].

It should be noted that the USP heavy metals test and associated compendia tests have been criticized for their lack of specificity [96]. Newer analytical techniques, such as inductively coupled plasma-mass spectroscopy (ICP-MS), can detect and quantitate levels of nearly every inorganic element of interest to the synthetic chemist, often with sub-ppm accuracy [97]. For cases where compendia methods do not exist, such as for palladium, ICP-MS can be used to detect Pd in the 0.1 ppm range [98]. An overview of analytical techniques available to track inorganic impurities was provided earlier in this review.

The European Agency for the Evaluation of Medicinal Products (EMEA) has published a guidance on specification limits for residual metal catalysts in active substances [99]. To determine an acceptable limit, permissible daily exposure (PDE) levels have been suggested based upon dietary sources of metal(s) and toxicology literature [100]. The EMEA guidance proposes two options for setting limits of residual metals in active ingredients. With the first option, an assumed dose of 10 g per day is used with the average body weight of 60 kg. Under these circumstances, the limits for platinoids (which includes the elements Pt, Pd, Ir, Rh, Ru, Os) are 5 ppm for an orally administered drug. For the metals Mo, V, Ni, and Cr, the limits are 10 ppm. Higher limits are allowed for Cu, Mn (15 ppm) and Zn, Fe (20 ppm). For parenteral drugs, the concentrations limits are typically ten times lower (e.g., 0.5 ppm for the platinoids to 2 ppm for Zn,

Fe). The guidance also makes clear that if more than one element is present (e.g., Fe and Cu), and then the total limit should not exceed 20 ppm for orally administered drugs. Platinoids, however, have a group limit such that the total limit remains 5 ppm when there are two or more such metals present. The EMEA guidance contains a second provision for cases where it is not possible to achieve the concentration limits described above. In these instances, limits can be based upon actual daily doses of the active ingredient. For the purpose of setting specifications, a dosage of 1 g per day is recommended, thus allowing a tenfold increase of levels of residual metal(s). The guidance points out that these values represent the upper limits of what is allowable and that specifications should be based upon the lower limits of what is readily achievable in production. More importantly, the guidance document suggests that the higher levels of metal residues described above may be invoked only after manufacturing data and process optimization to remove the metal(s) in question demonstrate that it is not possible to achieve the lower limits specified in the guidance (e.g., 5 ppm for platinoids to 20 ppm for Zn, Fe). Thus, the burden to control the levels of residual metals in an active ingredient rests with the process scientist.

With the production of a drug substance intended for use in humans, current Good Manufacturing Practices (cGMP) are applicable. ICH guidelines have been published for GMPs (Q7A) for active pharmaceutical ingredients [101]. While most GMP guidelines tend to focus more heavily on drug product development, they are equally applicable to the production of drug substances. Process scientists must have procedures in place that can control the levels of impurities and thus ensure that batches of drug substances are produced which consistently meet their predetermined specifications [102]. In a GMP environment, it is not acceptable to have one batch with a heavy metal content of 2 ppm (within specifications) and then have a batch run under similar circumstances yield heavy metal impurities of 200 ppm (well over specifications). As the process matures during research and development and eventually to manufacturing, maintaining consistent control over the levels of residual metals throughout the process will demonstrate that the process itself is well understood and under control [103].

11 Conclusion

Metal-mediated chemistries are of significant synthetic utility; metals removal need not be an impediment to their application on the larger scale. Options were presented for the removal of metals from process steams – from the perspective of the process research and development scientist. The options include extraction, crystallization, precipitation, adsorption, and decantation. They should be screened on representative process streams, selected for their ability to remove the metal as gauged by appropriate analytical methods, then reexamined under conditions that approximate those that may be encountered on scale. The isolated product should then be examined for its ability to work in the next step. Additionally, the process research and development scientist should perform a thorough inspection of the equipment train and remain vigilant to possible sources of adventitious metal contamination. The metal content can be determined through the use of very sensitive analytical methods such as ICP or AAS. The target level would be based on a range that allows successful use of the material in the next step and, in the case of a drug substance prepared in a final step, regulatory guidelines. Taken together, the metal mediated process is well positioned for success.

Acknowledgments The authors would like to acknowledge Steven Beck, Pat Confalone, Joydeep Kant, Melanie Miller, Richard Mueller, William Nugent, and John Scott for reviewing the manuscript. MRO wishes to thank Gregory Harris and Mark Angelino for several helpful discussions.

References

1. Larsen RD (1999) Curr Opin Drug Discovery Dev 2:651; (b) Totleben MJ, Prasad JS, Simpson JH, Chan SH, Vanyo DJ, Kuehner DE, Deshpande R, Kodersha GA (2001) J Org Chem 66:1057; (c) Crettaz R, Waser J, Bessard Y (2001) Org Proc Res Dev 5:572; (d) Miyagi M, Takehara J, Collet S, Okano K (2000) Org Proc Res Dev 4:346; (e) Smith GB, Dezeny GC, Hughes DL, King AO, Verhoeven TR (1994) J Org Chem 59:8151; (f) Blaser HU, Indolese A, Schnyder A, Steiner H, Studer M (2001) J Mol Catal A:Chemical 173:3; (g) Chen CK, Singh AK (2001) Org Proc Res Dev 5:508
2. Maryanoff CA, Mills JE, Stanzione RC, Hortenstein JT (1988) Catalysis from the perspective of an organic chemist: common problems and possible solutions. In: Chemistry & industry 33 (catalysis of organic reactions). Marcel Dekker, New York Basel, p 359, chap 18
3. Giannousis P, Carlson J, Leimer M (1999) Process research and development of CGS 19755, an NMDA antagonist. In: Gadamasetti KG (ed) Process chemistry in the pharmaceutical industry. Marcel Dekker, New York Basel, p 173
4. Carlson R (1992) Design and optimization in organic synthesis. Elsevier Science BV, Amsterdam, The Netherlands
5. Köhler K, Heidenreich RG, Krauter JGE, Pietsch J (2002) Chem Eur J 8:622
6. Matthew R. Oberholzer (personal communication)
7. (a) Ennis DS, McManus J, Wood-Kaczmar W, Richardson J, Smith GE, Carstairs A (1999) Org Proc Res Dev 3:248; (b) Gala D, Stamford A, Jenkins J, Kugelman M (1997) Org Proc Res Dev 1:163
8. (a) Varma SV, Naicker KP, Liesen PJ (1999) Tetrahedron Lett 40:2075; (b) Zhang TY, Allen MJ (1999) Tetrahedron Lett 40:5813
9. (a) Leese MP, Williams JMJ (1999) Syn Lett 10:1645; (b) Mizra AR, Anson MS, Hellgardt K, Leese MP, Thompson DF, Tonks L, Williams JMJ (1998) Org Proc Res Dev 2:325
10. Zhao F, Bhanage BM, Shirai M, Arai M (2000) Chem Eur J 6:843
11. (a) Drewry DH, Coe DM, Poon S (1999) Med Res Rev 19:97; (b) Akiyama R, Kobayashi S (2001) Angew Chem Int Ed 40:3469
12. (a) Chen-Chou C (1997) European Patent EP 0 791 393 A1; (b) Bunel EE (1994) United States Patent US 5,324,851
13. Li Y, Hong XM, Collard DM, El-Sayed MA (2000) Org Lett 2:2385
14. (a) Meneghetti SP, Lutz PJ, Kress J (2001) Organometallics 20:5050; (b) Reddy KR, Surekha K, Lee GH, Peng SM, Liu ST (2001) Organometallics 20:5557; (c) Tsuji S, Swenson DC, Jordan RF (1999) Organometallics 18:4758; (d) Done MC, Rüther T, Cavell KJ, Kilner M, Peacock EJ, Braussaud N, Skelton BW, White A (2000) J Organomet Chem 607:78; (e) Burk MJ, Feaster JE, Nugent WA, Harlow RL (1993) J Am Chem Soc 115:10125; (f) Gao JX, Yi XD, Xu PP, Tang CL, Zhang H, Wan HL, Ikariya T (2000) J Mol Catal A Chem 159:3
15. (a) Davies IW, Matty L, Hughes DL, Reider PJ (2001) J Am Chem Soc 123:10139; (b) Toebes ML, van Dillen JA, de Jong KP (2001) J Mol Catal A 173:75

16. (a) Biffis A, Zecca M, Basato M (2001) J Mol Catal A Chem 173:249; (b) Mayer ABR, Mark JE, Hausner SH (1998) J Appl Polym Sci 70:1209
17. (a) Alamine is a trademark of Cognis Corporation; (b) Lokhande TN, Anuse MA, Chavan MB (1998) Talanta 46:163
18. (a) Aliquat is a trademark of Cognis Corporation; (b) Foulon C, Pareau D, Durand G (1999) Hydrometallurgy 51:139
19. (a) Rose JB (1964) United States Patent US 3,125,560; (b) Wasay SA, Barrington SF, Tokunaga S (1998) Environ Technol 19:369
20. Sequestrene is a trademark of BASF Corporation
21. Chen CY, Dagneau P, Grabowski EJJ, Oballa R, O'Shea P, Prasit P, Robichaud J, Tillyer R, Wang X (2003) J Org Chem 68:2633
22. (a) Yordanov AT, Roundhill DM (1998) Coord Chem Rev 170:93; (b) Shukla JP, Sawant SR, Kumar A, Singh RK, Varadarajan N (1996) Nucl Sci J 33:39
23. (a) Kelex is a trademark of the Sherex Chemical Company Inc; (b) Al-Bazi SJ, Freiser H (1989) Inorg Chem 28:417
24. Al-Bazi SJ, Freiser H (1987) Solv Extr Ion Exch 5:997
25. (a) Acorga is a trademark of Avecia Limited; (b) Wisniewski M (2000) J Radioanal Nucl Chem 246:693
26. (a) LIX is a trademark of the Cognis Corporation; (b) Rong Q, Freiser H (1987) Solv Extr Ion Exch 5:923
27. (a) Van Der Puy M, Soriano DS, Dimmit JH (1986) European Patent EP 0 166 151 A1; (b) Fabiano MD, Kordosky GA, Mattison PL, Virnig MJ (1992) International Patent WO 92/08813
28. Juang RS (1999) Proc Nat Sci Council ROC(A) 23:353
29. Ahn YM, Yang K, Georg GI (2001) Org Lett 3:1411
30. Yamashoji Y, Matsushita T, Kawaguchi T, Tanaka M, Shono T, Wada M (1990) Anal Chim Acta 231:107
31. Larsen RD, King AO, Chen CY, Corely EG, Foster BS, Roberts FE, Yang C, Lieberman DR, Reamer RA, Tschaen DM, Verhoeven TR, Reider PJ, Lo YS, Rossano LT, Brookes AS, Meloni D, Moore JR, Arnett JF (1994) J Org Chem 59:6391
32. Seeverens PJH, Klaassen EJM, Maessen FJMJ (1983) Spectrochim Acta 38B:727
33. (a) Maynard HD, Grubbs RH (1999) Tetrahedron Lett 40:4137; (b) Paquette LA, Schloss JD, Efremov I, Fabris F, Gallou F, Méndez-Andino J, Yang J (2000) Org Lett 2:1259
34. Aquamet is a trademark of National Starch and Chemical Company
35. CYANEX is a trademark of Cytec Industries Inc; (b) Sarkar SG, Dhadke PM (2000) Ind J Chem Tech 7:109
36. Gefvert DL (1994) United States Patent US 5,284,633
37. (a) Köngisberger K, Chen GP, Wu RR, Girgis MJ, Prasad K, Repič O, Blacklock TJ (2003) Org Proc Res Dev 7:733; (b) Villa M, Cannata V, Rosi A, Allegrini P (1998) International Patent WO 98/51646
38. Nowottny C, Halwachs W, Schügerl K (1997) Sep Purif Tech 12:135
39. (a) Rosso VW, Lust DA, Bernot PJ, Grosso JA, Modi SP, Rusowicz A, Sedergran TC, Simpson JS, Srivastava SK, Humora MJ, Anderson NG (1997) Org Proc Res Dev 1:311; (b) Winkle DD, Schaab KM (2001) Org Proc Res Dev 5:450; (c) Bailey JR, Hatfield MJ, Henke KR, Krepps MK, Morris JL, Otieno T, Simonetti KD, Wall EA, Atwood DA (2001) J Organomet Chem 623:185; (d) Nakamura Y, Morioka I, Umehara A, Yamada I (1973) German Patent DE 2240733
40. Hagemann JP, Kaye PT (1999) Tetrahedron Lett 55:869
41. Chowdhury DA, Kamata S (1994) Chem Lett 3:589
42. Carpenter KJ (1997) Agitation. In: Sharratt PN (ed) Handbook of batch process design. Blackie Academic and Professional, London, chap 4
43. Lines PC, Carpenter KJ (1990) IchemE Symp Ser 121:167
44. McCabe WL, Smith JC, Harriott P (2001) Unit operations of chemical engineering, 6th edn. McGraw-Hill, NewYork, p 748

45. (a) Ashraf-Khorassani M, Combs MT, Taylor LT (1997) Talanta 44:755; (b) Gladysz JA (1994) Science 266:55; (c) Horváth IT, Rábai J (1994) Science 266:72
46. Glennon JD, Hutchinson S, Walker A, Harris SJ, McSweeney CC (1997) J Chrom A 770:85
47. Kubota N (2001) Cryst Res Technol 36:749
48. Olives AI, Martin MA, del Castillo B, Barba C (1996) J Pharm Biomed Anal 14:1069
49. (a) Mullin JW (2001) Crystallization, 4th edn. Butterworth-Heinemann, Oxford, chap 3; (b) Lewiner F, Klein JP, Puel F, Févotte G (2001) Chem Eng Sci 56:2069
50. (a) Kim S, Myerson AS (1996) Thermodynamic properties of supersaturated solution and their use in determination of crystal growth kinetic parameters. In: Myerson AS, Green DA, Meenan P (eds) Crystal growth of organic materials. American Chemical Society, Washington DC, p 157; (b) Mullin JW (2001) Crystallization, 4th edn. Butterworth-Heinemann, Oxford, chap 6; (c) Kirwan DJ, Feins IB, Mahajan AJ (1996) Crystal growth kinetics of complex organic compounds. In: Myerson AS, Green DA, Meenan P (eds) Crystal growth of organic materials. American Chemical Society, Washington DC, p 116
51. Beckmann W, Nickisch K, Budde U (1998) Org Proc Res Dev 2:298
52. (a) Celite is a trademark of the Celite Corporation; (b) Vo L, Fan L, Bhattacharyya S, Deegan T, Wright P, Labadie J (1999) Suzuki coupling on the Trident system. Parallel solid-phase synthesis of biaryl carboxamides. Application Note APN #020 Argonaut Technologies, Inc. San Carlos, CA
53. Sood A, Fleming HL, Novak JW Jr (1989) United States Patent US 4,824,576
54. (a) Amberlite is a trademark of the Rohm and Haas Company; (b) Rovira M, Hurtado L, Cortina JL, Arnaldos J, Sastre AM (1998) React Funct Polymers 38:279
55. Darco is one of the most commonly used activated carbons. Darco is a trademark of NORIT Americas Inc., Suite 250, Building C, 5775 Peachtree Dunwoody Rd. Atlanta, GA 30342
56. Mentasti E, Sarzanini C, Gennaro MC, Porta V (1987) Polyhedron 6:1197
57. (a) Solka Floc is a trademark of International Fiber Corporation, 50 Bridge Street, North Tonawanda, New York 14120; (b) For examples of its use on multikilogram scales, see: Kowalczyk BA, Dyson NH (1997) Org Proc Res Dev 1:117; (c) Yasuda N, Huffman MA, Ho G-J, Xavier LC, Yang C, Emerson KM, Tsay F-R, Li Y, Kress MH, Rieger DL, Karady S, Sohar P, Abramson NL, DeCamp AE, Mathre DJ, Douglas AW, Dolling U-H, Grabowski EJJ, Reider PJ (1998) J Org Chem 63:5438
58. (a) Chelex is a trademark of Bio-Rad Laboratories; (b) Nakamura Y, Hirai H (1976) Chem Lett 2:165
59. (a) Guibal E, Von Offenberg Sweeney N, Vincent T, Tobin JM (2002) React Funct Poly 50:149; (b) Ruiz M, Sastre AM, Zikan MC, Guibal E (2001) J Appl Poly Sci 81:153
60. Suraj G, Iyer CSP, Rugmini S, Lalithambika M (1997) Appl Clay Sci 12:111
61. Deloxan is a trademark of Degussa-Huls Aktiengesellschaft
62. (a) Gronbeck DA, O'Connell KM, Burke WA, Gaudet MN, Caporale SJ (1997) United States Patent US 5,702,611; (b) Dowex is a trademark of the Dow Chemical Company
63. Ottenheym JH, Dassen BHN (1976) United States Patent US 3 931 002
64. (a) Hestekin JA, Bachas LG, Bhattacharyya D (2001) Ind Eng Chem Res 40:2668; (b) Hayashita T, Hamada M, Takagi M (1985) Chem Lett 6:829
65. (a) Urawa Y, Miyazawa M, Ozeki N, Ogura K (2003) Org Proc Res Dev (Technical Note) ASAP article; (b) Boussie T, Hall K, LaPointe AM, Murphy V, Powers T, Van Beek JAM (1998) International Patent WO 98/56796
66. Ishihara K, Nakayama M, Kurihara H, Itoh A, Haraguchi H (2000) Chem Lett 1218
67. Dierssen H, Balzer W, Landing WM (2001) Marine Chem 73:173
68. Sinner F, Buchmeiser MR, Tessadri R, Mupa M, Wurst K, Bonn GK (1998) J Am Chem Soc 120:2790
69. (a) POLYORGS Certificate 68715.27.04.81 (RV) have also been described as Russian Chelating Sorbents; (b) Vlašánková R, Sommer L (1999) Chem Papers 53:200

70. ScavNet Homogeneous catalyst removal agent, is a trademark of Engelhard Corporation
71. (a) SiliCycle is a trademark of the SiliCycle, Inc; (b) Beatty ST, Fischer RJ, Hagers DL, Rosenberg E (1999) Ind Eng Chem Res 38:4402; (c) Tikhomirova TI, Fadeeva VI, Kudryavtsev GV, Nesterenko PN, Ivanov VM, Savitchev AT, Smirnova NS (1991) Talanta 38:267
72. (a) SMOPEX is a trademark of the Johnson Matthey PLC. Ekman K, Peltonen R, Sundell M, Teichman RA III (2002) International Patent WO 02/33135 A1; (b) Jaskari T, Vuorio M, Kontturi K, Manzanares JA, Hirvonen J (2001) J Cont Rel 70:219; (c) Karinen RS, Krause AOI (2001) Ind Eng Chem Res 40:6073; (d) Mäki-Arvela P, Salmi T, Sundell M, Ekman K, Peltonen R, Lehtonen J (1999) Appl Catal A 184:25
73. Grant DC, Skriba MC (1992) European Patent EP 0 491 533 A2
74. McCabe WL, Smith JC, Harriott P (2001) Unit Operations of Chemical Engineering, 6th edn. McGraw-Hill, NewYork, chap 9
75. See for example Chemglass part no. CG1930-02. Chemglass, Inc., 3861 North Mill Rd., Vineland, NJ 08360
76. Oberholzer M (1999) (unpublished data)
77. Anonymous (1989) Standard practice for determination of adsorptive capacity of activated carbon by aqueous phase isotherm technique. ASTM Standard D 3869-89a. American Society for Testing and Materials, Philadelphia, PA
78. (a) Anonymous (1992) Ambersorb carbonaceous adsorbents. Rohm and Haas Company, Philadelphia, PA; (b) Anonymous (1989) Standard practice for determination of adsorptive capacity of activated carbon by aqueous phase isotherm technique. ASTM Standard D 3869-89a. American Society for Testing and Materials, Philadelphia, PA; (c) Anonymous (1991) Measuring the adsorption capacity of powdered activated carbon. Technical Report AN58-3. American Norit Company, Inc, Atlanta, GA
79. Bird RB, Stewart WE, Lightfoot EN (1960) Transport phenomena. Wiley, New York, p 200
80. (a) CARBO-COR is a trademark of Koch Membrane Systems, Inc., 805 Main Street, Wilmington, MA 01887–3388; (b) ZETACARBON is a trademark of CUNO Inc., 400 Research Parkway, Meriden, CT 06450; (c) Kaster JA, Bickler R (2002) Optimization of catalyst removal on flash activated carbon – Biotage, Inc. A Dyax Corp. Company, Abstracts of Papers, 223rd ACS National Meeting, Orlando, FL, American Chemical Society, Washington DC, ORGN-297
81. Raggon JW, Snyder WM (2002) Removal of palladium black on the inside of reactor walls. Org Proc Res Dev 6:67
82. Yule R (2001) Personal communication to M Oberholzer
83. EDAX is a nondestructive elemental analysis technique that offers the ability to map the composition of solid samples within regions of micron dimensions. As such, it is useful for analysis of mixed solids. It is typically coupled to a scanning electron microscope (SEM)
84. (a) Teflon is a trademark of E.I. DuPont De Nemours and Company; (b) Hastelloy-C is a trademark of Haynes International, Inc
85. (a) Pruett KM (1995) Chemical resistance guide for metals and alloys. Compass, La Mesa, California; (b) Schweitzer PA (1995) Corrosion resistance tables. Marcel Dekker, New York
86. Oberholzer M (personal communication)
87. Jia X, Wang T, Wu J (2001) Talanta 54:741
88. Lewen N, Schenkenberger M, Raglione T, Mathew S (1997) Spectroscopy 12:14
89. (a) Kavlentis E (1988) Anal Lett 21:1681; (b) Lee JS, Uesugi K, Choi WH (1995) Anal Proc 32:279; (c) Sarkar A, Thokdar TK, Paria PK, Majumdar SK (1988) Indian J Chem Sect A 27A:650
90. Anderson NG (2000) Practical process research and development. Academic Press, London, UK

91. International Conference on Harmonization (1999) Q6A, Specifications: test procedures and acceptance criteria for new drug substances and new drug products: chemical substances
92. International Conference on Harmonization (2002) Q3A(R), Impurities in new drug substances
93. United States Pharmacopoeia (USP) heavy metals test 231, USP25-NF20, 2002. European Pharmacopoeia (EP) heavy metals test 2.4.8, 2002. Japan Pharmacopoeia (JP) heavy metals test 21. The USP and JP have three alternate test methods, and the EP has five. The choice of which test to run is based upon the handling characteristics of the test sample
94. (a) The FDA's Center for Food Safety & Applied Nutrition has published a series of guidances on certain metals (As, Cd, Cr, Ni, Pb) in shellfish. The documents, published in 1993, are titled "Guidance Document for Arsenic (*) in Shellfish," (*)-Substitute Cadmium, Chromium, Nickel or Lead for the titles of the other guides. Each guidance contains extensive background information and references on the extent of the metals found in shellfish, consumption limits for the various metals and analytical methods for tracking the metals. Guidances are available from: Office of Seafood, Food and Drug Administration (HFS-400), 5100 Paint Branch Parkway, College Park, MD 20740-3835, Telephone: (301) 436-2303. The documents are available online at: http://www.cfsan.fda.gov/~frf/guid-sf.html; (b) Extensive literature exists for mercury, particularly organomercury (e.g., methylmercury) compounds: Committee on the Toxicological Effects of Methylmercury, Board on Environmental Studies and Toxicology, Commission on Life Sciences, National Research Council (NCR), "Toxicological effects of methylmercury", National Academy Press, Washington DC, 2000. ISBN: 0-309-07140-2. This book is available to be read online: http://www.nap.edu/books/0309071402/html/
95. (a) USP arsenic test 211, USP25-NF20, 2002; (b) USP mercury test 261, USP25-NF20, 2002
96. Ciciarelli R, Jäkel D, König E, Müller-Käfer R, Pavel J, Röck M, Thevenin M, Ludwig H (1995) Determination of metal traces – a critical review of the pharmacopoeial heavy metal test. USP Pharm Forum 21:1638
97. Wang T, Wu J, Hartman R, Jia X, Egan RS (2000) J Pharm Biomed Anal 23:867
98. Lewen N, Shenkenberger M, Larkin T, Conder S, Brittain HG (1995) J Pharm Biomed Anal 13:879
99. The European Agency for the Evaluation of Medicinal Products (2002) Note for guidance on specification limits for residues of metal catalysts. December. The EMEA provides recommendations to the Committee for Proprietary Medicinal Products (CPMP, the regulatory body for the European Union)
100. In addition to the references cited in the EMEA draft guidance, a division of the US Health and Human Services, the Agency for Toxic Substances and Disease Registry (ATSDR), has extensive toxicology data for many of the metals (as well as selected organic compounds) listed in this review. For more information, contact: ATSDR, Division of Toxicology, 1600 Clifton Rd, Mail Stop E29, Atlanta, GA. The agency's web site provides extensive toxicology and regulatory data on several metal species, including the following elements: Al, Ag, As, Ba, Be, B, Cd, Co, Cr, Cu, F, Hg, Mn, Ni, Pb, Sb, Se, Sn, Tl, V, and Zn. See: http://www.atsdr.cdc.gov/
101. International Conference on Harmonization, (2000) Q7A, Good manufacturing practice guide for active pharmaceutical ingredients
102. Willig SH (2001) Bulk pharmaceutical chemicals. In: Good manufacturing practices for pharmaceuticals. Marcel Dekker, Inc., chap 14, p 265
103. For a review of process validation see: Berry IR, Harpaz D (1997) Validation of bulk pharmaceutical chemicals. Interpharm Press, Inc

Author Index Volumes 1–6

Subject Index

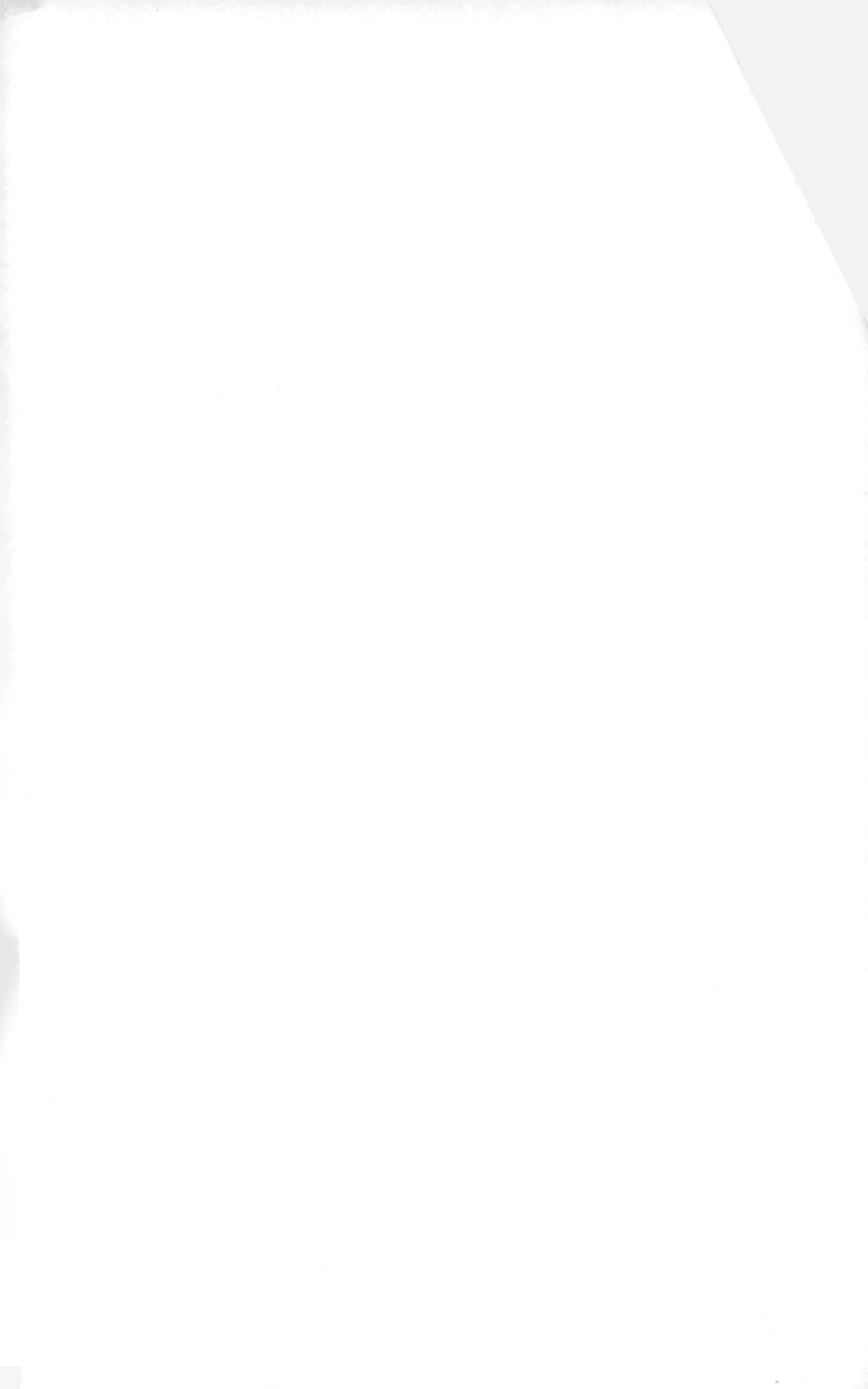

GPSR Compliance
The European Union's (EU) General Product Safety Regulation (GPSR) is a set of rules that requires consumer products to be safe and our obligations to ensure this.

If you have any concerns about our products, you can contact us on

ProductSafety@springernature.com

In case Publisher is established outside the EU, the EU authorized representative is:

Springer Nature Customer Service Center GmbH
Europaplatz 3
69115 Heidelberg, Germany

www.ingramcontent.com/pod-product-compliance
Ingram Content Group UK Ltd.
Pitfield, Milton Keynes, MK11 3LW, UK
UKHW021833190726
13853UKWH00003B/1290
* 9 7 8 3 6 6 2 1 4 4 2 3 7 *